Security and Access Control Using Biometric Technologies

Robert Newman
Georgia Southern University

COURSE TECHNOLOGY
CENGAGE Learning™

Australia • Brazil • Japan • Korea • Mexico • Singapore • Spain • United Kingdom • United States

COURSE TECHNOLOGY
CENGAGE Learning™

Security and Access Control Using Biometric Technologies

Robert Newman

Vice President, Career and Professional Editorial: Dave Garza

Executive Editor: Stephen Helba

Managing Editor: Marah Bellegarde

Senior Product Manager: Michelle Ruelos Cannistraci

Editorial Assistant: Meaghan Orvis

Vice President, Career and Professional Marketing: Jennifer McAvey

Marketing Director: Deborah S. Yarnell

Senior Marketing Manager: Erin Coffin

Marketing Coordinator: Shanna Gibbs

Production Director: Carolyn Miller

Production Manager: Andrew Crouth

Content Project Manager: Andrea Majot

Art Director: Jack Pendleton

Cover illustration: Image copyright 2009. Used under license from Shutterstock.com

Production Technology Analyst: Tom Stover

Library of Congress Control Number: 2009929994

ISBN-13: 978-1-4354-4105-7

ISBN-10: 1-4354-4105-2

Course Technology
20 Channel Center Street
Boston, MA 02210
USA

Cengage Learning is a leading provider of customized learning solutions with office locations around the globe, including Singapore, the United Kingdom, Australia, Mexico, Brazil, and Japan. Locate your local office at: **international.cengage.com/region**

Cengage Learning products are represented in Canada by Nelson Education, Ltd.

For your lifelong learning solutions, visit **course.cengage.com**

Visit our corporate website at **cengage.com**

Printed in the United States of America
1 2 3 4 5 6 7 12 11 10 09

Brief Table of Contents

Table of Contents

CHAPTER 2

PART 2 Technology

CHAPTER 3

CHAPTER 9

System Integrity and Accessibility

CHAPTER 10

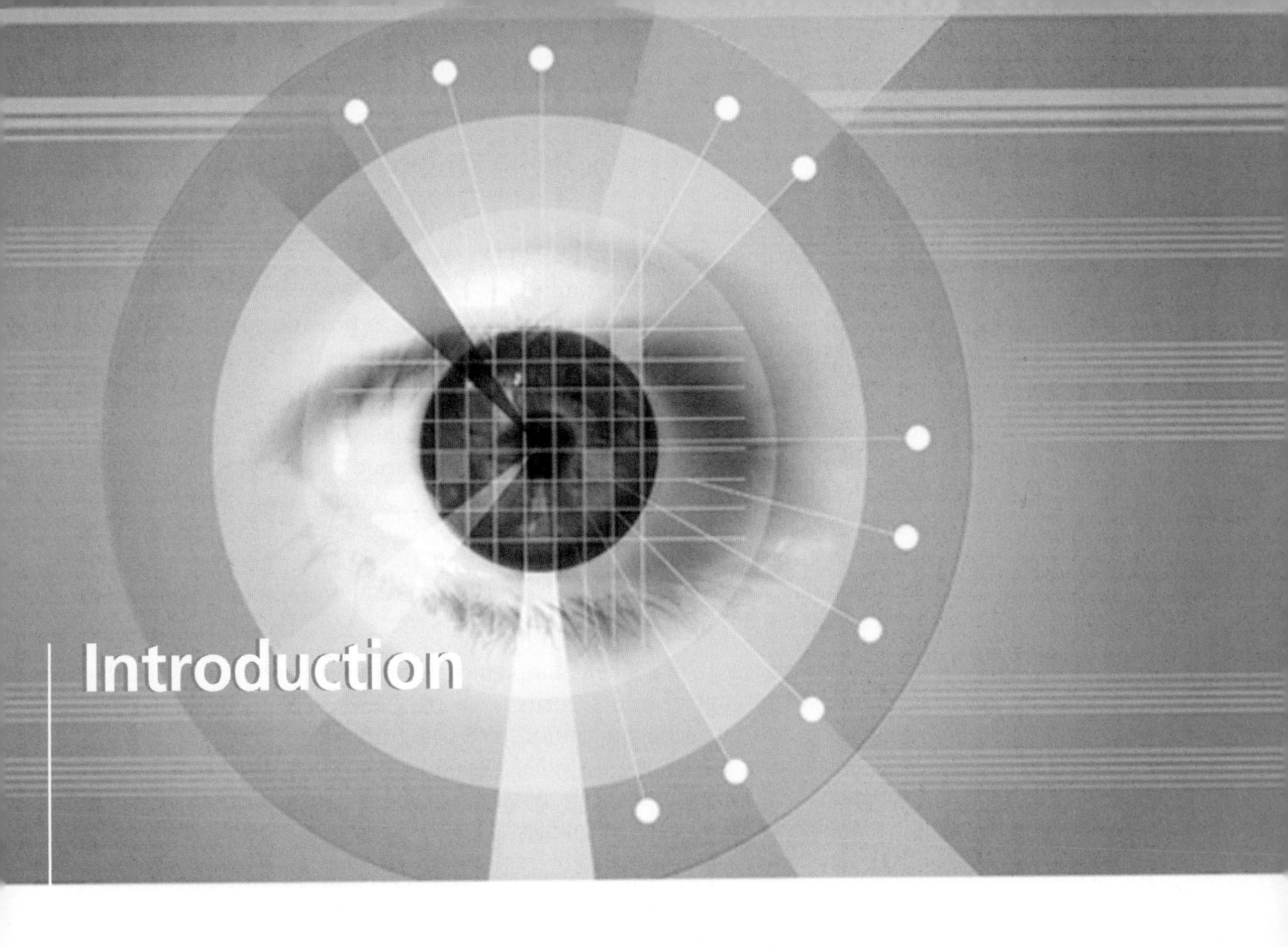

Introduction

The modern rapid advancements in networking, communication, and mobility have increased the need for reliable ways to verify the identity of any person. Biometrics is the study of methods for uniquely recognizing humans based on one or more intrinsic physical or behavioral traits. In information technology, biometric authentication refers to the technology that measures and analyzes human physical and behavioral characteristics for authentication purposes in computer and network-based systems and physical access control. Authorization is the assignment of a privilege or privileges (i.e., access to a building or network) verifying that a known person or entity has the authority to perform a specific operation. Authorization is provided after authentication. Finally, access control is the process of granting access to information system resources only to authorized users, programs, processes, or other systems.

A new revelation about some Internet or electronic commerce (e-commerce) crime is splashed across the news media every day: someone has compromised a corporation's database, which threatens the security of the population's financial resources; someone has hacked into a computer system and stolen information that can be used in identity theft schemes; terrorists are using identity theft techniques to support terrorism. Millions of dollars are at stake—citizen's, corporate, and government.

On a larger scale, numerous scam and fraud schemes are being directed toward Internet users. A serious issue involves identify theft and identify fraud, where Internet criminals and terrorists use personal information to steal personal assets from savings and checking accounts. Numerous property crimes are committed using the Internet, computers, and electronic devices. Assets are compromised because of illegal or improper access to organizational resources—both physical and electronic.

The current computer and wireless networking market has been growing at an unbelievable rate, and security preparedness has not kept pace. In large part, this is due to the expansion of Internet access to almost all sectors of society. Everyone with a computer now has access to the Web. Numerous electronic devices, such as PDAs, cell phones, and portable computers are saturating the network. The opportunity for expansion of electronic services is astronomical. This book is part of a program that is designed to provide a broad working knowledge of biometrics that can be applied to the security issues that permeate today's computer and network systems.

This book is divided into three parts, namely Fundamentals, Technology, and Administration. We introduce biometric and access control technologies and methods along with security and biometric terms. Both hardware and software system solutions and applications that would be used in a biometric security environment are introduced. There are a number of security and privacy issues that arise from the use and implementation of a biometrics system. These include both industry standards and legal issues. Also included are opportunities and solutions that address physical access situations.

Last but not least are the project management, implementation, and operation issues that must be addressed with access control and security systems. New access control systems, such as intrusion detection and intrusion prevention, are being introduced into the marketplace. These legacy-type systems are being integrated into state-of-the-art biometric technologies. These multibiometric, multimodal security systems will be used to control access into the nation's digital and physical assets and resources.

This book provides a wide range of information relating to access control for computer and network resources and assets where nefarious activities could exploit computer and electronic devices. Emphasis is placed on access control techniques that can lessen vulnerabilities and threats that are inherent in the Internet and networking environment. These are the areas that can benefit from the use of biometric security methods and solutions. Efforts are made to present techniques and suggestions for corporate security personnel and administrators to effectively provide protection to an organization's assets and resources, particularly the database. In summary, this book is oriented toward the reader who will apply the information provided for addressing the numerous issues relating to organizational asset and resource access, control, and security.

Intended Audience

The material included in this book is geared toward the information system and information technology student and also to network management and security administration personnel in today's computer and network organizations. The content and topics presented are not highly technical or theoretical, and are aimed at a wide population of students, managers, administrators, and practitioners. Readers of this book—individuals, businesspeople, government employees, or students—will learn the basics of biometrics and biometric systems for access control, security, and management.

Chapter Descriptions

This book is divided into three major parts: Fundamentals, Technology, and Administration. Each part should be addressed in this sequence as it depends on the understanding of information previously presented.

Part One: Fundamentals

Part One, "Fundamentals," presents biometric basics and includes an overview of biometric methods and technologies. This part sets the stage for the remainder of the book and introduces numerous terms and definitions associated with biometrics and security.

Chapter 1, "Access Control Using Biometrics," introduces basic concepts that make up the field of biometrics. Biometric and security terms and definitions are introduced and an overview of various biometric methods is provided. This chapter discusses various security issues that can be addressed using biometric solutions.

Chapter 2, "Biometric Traits and Modalities," describes the technical details of biometric technologies so readers will understand the features, functions, and relevance of each biometric method. Technologies such as scanning, imaging, recognition, and verification are introduced.

Part Two: Technology

Part Two, "Technology," explores the working components of biometric methods and processes with discussions oriented toward both hardware and software elements and features. Biometric applications and solutions are covered.

Chapter 3, "Biometric Applications and Solutions," describes numerous solutions that are available to address access and security requirements. A search on the Web reveals a tremendous listing of products that can solve a problem. This chapter guides readers to determine which product or service addresses a particular security issue. Applications are generally classified according to citizen, employee, or customer facing.

Chapter 4, "Repositories for Database and Template Storage," discusses access and security to database systems and repositories. Since security systems must be employed that provide various levels of access and rights control, these topics hold a high priority in today's e-commerce environment. Topics such as the size of the database asset, response time, and the disk latency required when accessing and processing security requirements are discussed.

Chapter 5, "Legacy and Biometric Systems," provides examples of hardware and software systems that might be considered in a search for a biometric security system. It compares features and functions available on security surveillance and management systems for access control. This chapter presents options for hardware and software system solutions that address network and computer security issues.

Chapter 6, "Biometric Multi-Factor System Design" looks at how real systems are protected using multiple types of biometric solutions to provide a high level of security. This chapter presents various configurations for performing access and physical security functions.

Part Three: Administration

Part Three, "Administration," builds on Part One and Part Two and provides the reader with the tools to manage security and access control issues by applying biometric applications and solutions. Management faces numerous issues in the day-to-day management and control of organizational assets and resources. Confidentiality, integrity, and accessibility (CIA) issues are addressed.

Chapter 7, "Policy and Program Management," establishes the framework for understanding the environmental and management concerns relating to legacy and biometric applications and systems. Various policies are introduced along with the programs that implement them. The material presented is especially relevant to the potential CIOs, managers, and administrators who will be responsible for ensuring a secure operating environment.

Chapter 8, "Security and Access Technologies," explores the hardware and software products and services that are used to provide security of an organization's assets and resources. These products are implemented in physical settings, computer installations, and networking systems. Discussions are directed to various software, hardware, and networking security configurations.

Chapter 9, "System Integrity and Accessibility," describes the issues of confidentiality, integrity, and accessibility (CIA). This chapter explains the basic concepts of performance, false positives, and false negatives. Other topics such as false match rate, false accept rates, false reject rates, and biometric identification performance matrices of reliability, selectivity, recall, and precision are discussed.

Chapter 10, "Security and Privacy Issues," covers the numerous security and privacy issues that must be addressed when employing any security or access system or product. Identification and verification issues are addressed.

Chapter 11, "Implementation and Operation Issues," explores how a biometric solution has the opportunity to succeed or fail with any system implementation. There are many "gotchas" in implementing a security or access system that relies on biometric products and services. Security must be part of the original system design and not an afterthought. This chapter discusses issues that can affect a biometric solution, such as throughput of the primary system for access requirements. Additional issues that are discussed include demographics, user location, weather, HVAC, and enrollment in the system.

Chapter 12, "Standards and Legal Environment," explores in detail a number of standards that exist for security products and services, as well as the numerous legal issues that can arise from the use of biometric devices.

Features

This book includes many features designed to enhance your learning experience and help you fully understand computer and network security.

Chapter Objectives. Each chapter begins with a detailed list of the concepts to be mastered within that chapter. This list serves as both a quick reference to the chapter's contents and a useful study aid.

Chapter Summaries. Each chapter's text is followed by a summary of the concepts introduced in that chapter. These summaries provide a helpful way to review the ideas covered in each chapter.

Key Terms. The terms in bold text in each chapter are gathered in a Key Terms list with definitions at the end of the chapter, providing additional review and highlighting key concepts.

Review Questions. The end-of-chapter assessment begins with a set of review questions that reinforces the ideas introduced in each chapter. These questions help you evaluate and apply the material you have learned.

Discussion Exercises. Several essay-oriented questions are provided as additional review questions.

Hands-On Projects. Each chapter provides several Hands-On Projects aimed at providing you with practical experience through research activities and projects. A progressive hands-on project places readers in the role of a security consultant in a fictitious company called OnLine Access LLC. Users should practice safe Internet and computer use when working through the cases and exercises. Keep backup copies of everything and keep your virus protection updated!

Text and Graphic Conventions

Wherever appropriate, additional information and exercises have been added to this book to help you better understand the topic at hand. Icons throughout the text alert you to additional materials. The icons used in this textbook are described below.

Each Hands-On Project in this book is preceded by the Hands-On icon and a description of the exercise.

Instructor's Materials

The following additional materials are available when this book is used in a classroom setting. All of the supplements available with this book are provided to the instructor on a single CD-ROM (ISBN: 1435441060). You can also retrieve these supplemental materials from the Course Technology Web site, *www.course.com*, by going to the page for this book, under "Download Instructor Files & Teaching Tools."

Electronic Instructor's Manual. The Instructor's Manual that accompanies this textbook includes additional material to assist in class preparation, with suggestions for lecture topics, tips on setting up a lab for the Hands-On Projects, and solutions to all of the end-of-chapter materials.

ExamView Test Bank. This Windows-based testing software helps instructors design and administer tests and pretests. In addition to generating tests that can be printed and administered, this full-featured program has an online testing component that allows students to take tests at the computer and have their exams automatically graded.

PowerPoint Presentations. This book comes with a set of Microsoft PowerPoint slides for each chapter. These slides can be used as a teaching aid for classroom presentations, made available to students on the network for chapter review, or printed for classroom distribution. Instructors are also at liberty to add their own slides for other topics that they introduce.

Figure Files. All of the figures and tables in the book are reproduced on the Instructor Resources CD. Similar to PowerPoint presentations, these are included as a teaching aid for classroom presentation, to make available to students for review, or to be printed for classroom distribution.

Acknowledgments

The author and publisher would like to thank the following reviewers:

Robert Guess, Tidewater Community College
Herb Mattord, Kennesaw State University
Steve Strom, Butler Community College
Michael Whitman, Kennesaw State University

About the Author

Robert Newman is currently a lecturer of information systems in the College of Information Technology at Georgia Southern University. Previously, Robert was associate professor of telecommunications management at Decatur, GA campus of DeVry University. He is a long-time student and practitioner of data processing, data communications, and networking technologies. He started his career at the University of South Carolina in the Computer Science Department, where he held a number of positions in a large computer operation, including operations manager. He has taught in the computer information technology departments in a number of colleges and universities. Robert's formal education includes degrees from the University of South Carolina, Columbia, and Georgia State University in Atlanta, Ga. He has completed advanced degree work in computer science at the University of Alabama at Birmingham. He is currently active in computer forensics and enterprise information systems security education and is an active speaker on these topics.

Robert's professional experience includes many years in the telephone industry, in numerous positions at BellSouth and AT&T. He gathered hands-on networking knowledge in software development, broadband operations, and network management and surveillance at BellSouth. Early in his career, he developed a solid background in IBM mainframe hardware and software and business applications.

Before his life in the computer industry and interspersed over the years, he has been a member of federal, state, and county law enforcement agencies in Georgia, Alabama, and South Carolina. He is a graduate of the Northeast Georgia Police Academy and Georgia Post Certified (GaPOST) and a Certified Information Systems Security Professional (CISSP). He is an active member of the FBI's Coastal Empire Infragard Organization. He has accumulated a wealth of knowledge concerning the security and protection of data and computer resources and network administration. Security lectures are part of the network administration, management information systems, computer forensics, and data communications courses Robert currently teaches.

Comments and suggestions concerning this book can be sent to the author at newmanrc@georgiasouthern.edu.

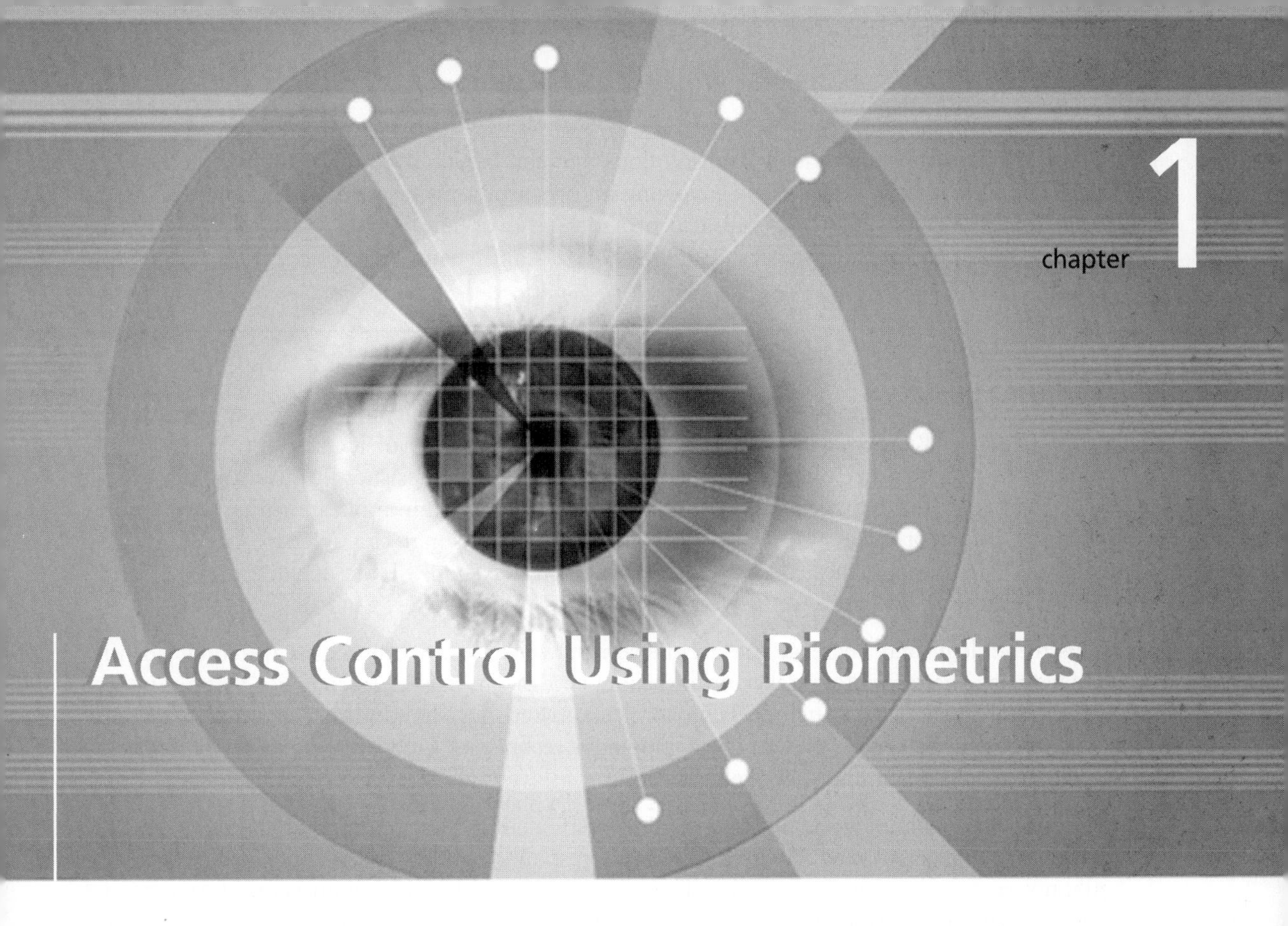

chapter 1 Access Control Using Biometrics

After completing this chapter, you should be able to do the following:

- Become familiar with biometric and security terms
- Look at various security concepts and processes
- Identify environments where biometric systems are viable
- Identify various usage, verification, identification, and screening techniques
- Describe access control security issues
- Review various cost parameters for asset security
- Establish a basic understanding of the biometrics environment

The vulnerabilities inherent in current computer and network systems should not come as a surprise to security professionals. Because most financial transactions are conducted over e-commerce systems, the potential for system compromises is significant. Every day, articles are presented in the news media where an organization has been compromised because of some access control situation. Because of the emergence of numerous identity theft scams, identification has become a major priority for mechanized computer systems. Valid users must provide some evidence that they are "who they say they are." Authentication and authorization techniques are required to validate access to a user's account or resource.

There are numerous approaches and solutions that can be deployed to ameliorate the negative consequences of lax access control involving an organization's resources. This chapter sets the stage for addressing this topic and also provides an overview into the very relevant security issue of access control for physical and logical assets and resources.

Introduction

Biometrics offers a secure means of limiting access to sensitive facilities and computer and networking assets. Biometrics refers to the automatic identification or identity verification of living persons using their enduring physical or behavioral characteristics. It includes the study of biological data and biometric-based authentication, which is an application that uses specific personal traits for access control. Many body parts, personal characteristics, and imaging methods have been suggested and used for biometric systems. These include fingers, hands, feet, faces, eyes, ears, teeth, veins, voices, signatures, typing styles, gaits, and odors.

This technology may increase the accuracy with which security systems can readily identify individuals. The technology relies on measurements of physical characteristics that are unique to an individual. These include behavioral characteristics, handwritten characteristics, and voice recognition. There are practical biometric techniques currently available that may be acceptable to users and also cost effective.

The biometric technology is not in wide use today, mainly due to a lack of standards and the expense of implementation. There is also a mistrust and usefulness recognition, who are generally not educated on the technology. These concerns include possible invasion of privacy and the association of fingerprinting techniques with criminals. This chapter presents an overview of the biometric environment and the various technologies that can be deployed to provide access control and other security implementations.

Why Biometrics?

Biometrics is a general term used to describe a characteristic or a process. As a characteristic, it is a measurable biological (anatomical and physiological) and behavioral trait that can be used for automated recognition. As a process, it is an automated method of recognizing an individual based on measurable biological (anatomical and physiological) and behavioral characteristics. It includes the study of biological data and biometric-based authentication, which is an application that uses specific personal traits for **access control**. This biometric method of identification is preferred over traditional methods involving passwords and PIN numbers for two primary reasons:

- The person to be identified is required to be physically present at the point of identification.
- Identification based on biometric techniques obviates the need to remember a password or carry a token.

Behavior is described as one of a recurring or characteristic pattern of observable actions or behavioral responses. Behavioral science examines human activities in an attempt to discover these recurrent patterns and to formulate rules. These rules can be used in the area of

biometrics. The other part of the biometric story is the biological component. **Physiology** is the branch of biology dealing with the functions and vital processes of an individual's parts and organs. Anatomy concerns the structure and detailed analysis of an organism or body. This knowledge is used in the development of the various biometric modalities.

Everyday life is getting more complicated, especially with the need to remember an ever-growing number of different user IDs, logons, passwords, PIN numbers, and lock combinations. There are many risks involved when using passwords or any other ID numbers:

- They can be forgotten.
- They can be copied or stolen.
- They need to be changed frequently.
- They are not accurate all the time.

Using a unique, physical attribute—such as a fingerprint or iris—to effortlessly identify and verify that "you are who you say you are," is the best and easiest solution in the market. That is the simple truth and power of biometrics technology today. Although biometric technology has been around a while, modern advances in miniaturization, coupled with big reductions in cost, have made biometrics readily available and affordable to the small business owner, larger corporations, and public sector agencies alike.

There are many new, innovative techniques using enterprise biometrics. Many proponents believe biometrics may soon see broad enterprise and government adoption, not only because of many promising new technologies, but also because organizations seek innovative identity management and access control methods. Technology can strengthen multifactor authentication systems in these entities.

There are many flavors of biometrics available today, from traditional fingerprint scanners and facial recognition systems to next-generation technologies that may change the product landscape. However, issues include:

- User and customer acceptance of biometrics
- Implementation issues and technology drawbacks
- Hardware and software requirements
- Integration with infrastructure systems
- Encrypted storage and transmission of biometric data
- Impact of biometrics on multifactor authentication strategies

These issues and many others will be addressed throughout this book. A brief overview of various biometric methods is also presented in this chapter.

Biometrics Fundamentals

The term biometrics has two distinct meanings: *bio* meaning living creature, and *metrics* meaning the ability to measure an object quantitatively. The use of biometrics has been traced back as far as the Egyptians, who measured people to identify them. The first modern biometric device was introduced on a commercial basis over 25 years ago when a machine that measured finger length was installed for a time-keeping application at a Wall Street establishment. In the ensuing years, hundreds of these hand geometry devices were installed at high-security facilities operated by both commercial and government installations [1].

Biometrics falls under the umbrella of an **automated identification and data capture** (AIDC) system, the term used to describe data collection by means other than manual notation or keyboard input. The significance of automatically captured data includes:

- A more efficiently run organization
- Improved and more timely decision making
- The efficient use of time, people, and materials

The family of AIDC technologies can be broken down into six categories:

- Biometrics
- Electromagnetics
- Magnetics
- Optical
- Smart cards
- Touch

Biometric practices in wide use today fall into one of two groups: identification or security. The underlying advantages to biometric identification include elimination of common problems such as illicitly copied keys, lost or broken mechanical locks, and forged or stolen personal identification numbers (PINs). These situations could lead to automatic teller machine (ATM) and checking account fraud. Additionally, biometric systems can be used for identification purposes involving security access systems in:

- Management information departments
- Government agencies
- ATMs or banks
- Law enforcement
- Prisons
- International border control
- Military agencies

Collectively, biometric technologies are defined as automated methods of verifying or recognizing the identity of a living person based on a physiological or behavioral characteristic. In analyzing the definition of biometrics, several distinct terms must be elaborated upon to completely understand the framework of biometric technology. The phrase "automated methods" refers to three basic methods connected with biometric devices:

- Provide a mechanism to scan and capture a digital or analog image of a living personal characteristic.
- Utilize compression, processing, and comparison of the image to a database of stored images.
- Interface with applications systems.

These methods can be configured in a number of different topographies depending upon the biometric device and application. For example, a common issue is whether the stored images or templates reside on a card, in the device, at a host, or in a database.

The term "living person" may seem obvious, but it is an important component in defining biometrics. A latex finger, digital audio tape, plaster hand, or prosthetic eye might be substituted for the real item; however, biometric devices can incorporate specific algorithms that determine whether there is a live characteristic being presented. The term "living" also separates the biometric industry from the forensic identification field, although basic principles transcend both areas [2]. An **algorithm** is a limited sequence of instructions or steps that tells a computer system how to solve a particular problem. A biometric system will have multiple algorithms; for example, image processing, template generation, comparisons, etc.

Biometric Perceptions

Widespread usage, implementation, and public acceptance of biometric technology have not established a foothold in corporate America. Trying to impose biometrics in a workplace might be seen as a "Big Brother" initiative. Biometrics introduces not just financial concerns but psychological issues as well. In a day and age where most people feel that they are monitored for one reason or another, biometrics can be seen as an invasion of privacy by some. Providing the neighborhood bank with a Social Security number and a password is accepted by nearly all of us, but providing a retinal pattern or computerized fingerprint to an international credit card company might seem too Orwellian.

The general population's perceptions of biometrics have been altered by false claims and data. It was reported that military pilots refused to use a retina scan system, believing it might impair their visual acuity. Although there was no evidence that the system affected eyesight, the system was removed. In other reported cases, retina scan users with watery eyes sometimes left data collection sensors moist, leading to concerns about eye diseases, transfer of body fluids, and AIDS. Although there is no known—or even alleged—case of injury or disease resulting from such a system, user concerns became so great that the system was withdrawn from the market. Using a hand for identification might be far more socially acceptable than an iris because people don't like the idea of repeatedly exposing an eye close to a camera. Likewise, many people could also be adverse to the criminal connotations of having their fingerprints taken.

However, some studies have reported positive findings about the acceptance of biometrics. Banks and financial institutions have been pioneers in implementing biometric applications for their customers. Subsequently, banks and other financial institutions that have tested biometric-based security on their clientele say consumers overwhelmingly have a pragmatic response to the technology. Anything that saves the information-overloaded citizen from having to remember another password or personal identification number comes as a welcome respite. Adding a statistical footing to this anecdotal evidence, a nationwide survey reported that a high percent of people approve the use of finger imaging, and don't feel it treats people as criminals. Many adults have already been fingerprinted for a job, government license, or other identification purpose.

The fundamental fear behind biometrics is the fear of the unknown. Much of the wariness of biometrics may come from the strangeness of a new technology. A poll asked approximately 100 people how they would react to a finger scan at a bank. Sixty percent of the people who only heard a description of the procedure reacted positively toward the idea, but once they tried it, favorability shot up to 90 percent. Legal, privacy, and user societal issues will be critical to selecting and implementing a successful biometric-based security and identification management solution.

Using Biometrics

The modern rapid advancements in networking, communication, and mobility have increased the need for reliable ways to verify the identity of any person. Nowadays, identity verification is mainly performed in two ways:

- Possession-based: Security is based on a "token" the user has, such as a credit card or a document. If it is lost, somebody else can use it to falsify their identity.
- Knowledge-based: Security is based on a password. Even if one uses the best encrypting algorithm, the entire security system is based on a key. If the password is too short, it is simple to guess it or crack it after several attempts. But if it is too complicated, it can't be remembered. The common user will have to write it down somewhere, and it can be lost or stolen.

These weaknesses of standard validation systems can be avoided if some body feature becomes the key. Particular characteristics of the body or habits are much more complicated to forge than a string, even if it is very long. It is evident that using biometrics adds a complexity to identification systems that would be hard to reach with a standard password-based approach. The main advantages of biometrics compared to a standard system are:

- Biometric traits can't be lost or forgotten.
- Biometric traits are difficult to copy, share, and distribute. The person who is being authenticated must be present at the time and point of authentication.
- Biometric systems can be used in conjunction with passwords or tokens, thus improving the security of existing systems without replacing them.

Biometric Methods

This section will include a brief overview of finger scanning, finger and hand geometry, iris and palm imaging, face, retina, and voice recognition, and signature verification biometric techniques. Table 1-1 provides a matrix of techniques along with their comparative accuracy. Additional methods and technical details will be presented in Chapter 2.

Finger Scanning

The user's finger is placed on a reader where a picture is taken of the fingerprint. The system then converts this picture into a map of minutiae points, which is then input into an algorithm for creating a binary template. This binary template is stored and compared during the authentication and verification process. Common fingerprint patterns are divided into three main groups, which consist of arches, loops, and whorls. Approximately 5 percent of the patterns are arches, 30 percent are whorls, and 65 percent are loops.

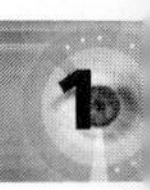

Technique	Comparative Accuracy
Finger scanning	1:500
Hand geometry	1:500
Iris imaging	1:131,000
Retina recognition	1:10,000,000
Signature verification	1:50
Speech recognition	1:50

Table 1-1 Techniques and Comparative Accuracy

Finger scanning imaging techniques include optical, thermal, tactile, capacitance, and ultrasound. Optical images can be captured from images made by a finger on a glass plate. Tactile and thermal images can be captured from the pressure of a temperature sensor. Capacitance images are generated from capacitance silicon sensors. Sound waves can generate ultrasound images from finger patterns. There are a number of automated fingerprint identification systems currently being used by a number of organizations. Finger scanning provides high accuracy, and fraudulent deception of the system is difficult. The equipment is easy to use and is readily accepted by users.

Finger Geometry

Some biometric vendors use **finger geometry** or finger shape to determine identity. Unique finger characteristics, such as finger length, width, thickness, and knuckle size are measured on one or two fingers. Two techniques are used to capture the images: The user inserts the index and middle fingers into a reader and a camera takes a three-dimensional image, or the user inserts a finger into a tunnel, where sensors take three-dimensional measurements.

Finger geometry systems are simple to use, very accurate, and are impervious to deception. Public acceptance, however, is somewhat lower than it is for finger scanning, because users must insert their fingers into a reader.

Hand Geometry

Some commercial biometric applications use the two-dimensional shape of the hand for access control. Users place their hand on a reader, aligning their fingers with specially positioned guides, and a camera captures an image. Measurements center on finger length and the shape of the fingers and knuckles. Commercial systems that perform three-dimensional shape analysis of the hand are under development.

Hand geometry systems are simple to use, accurate, and impervious to deception, and are readily accepted by users. Hand geometry, however, is less distinctive than finger scanning techniques, since people can have similar hand geometry.

Palm Imaging

Measurement of palms is performed by techniques similar to those for finger scanning. A scanner shaped to accommodate the palm scans the ridges, valleys, and minutiae found on the palm. Alternatively, latent or ink images of the palm can be scanned, and the minutiae is extracted, processed, and stored in the system. Palm images are useful in crime detection, and some vendors are developing commercial applications.

Like finger scanning, **palm scanning** is simple to use, very accurate, and impervious to deception. User acceptance, however, is not very high.

Iris Imaging

The human iris is a complex structure well suited for unique identification. Each iris contains a complex pattern of specific characteristics such as a corona, crypts, filaments, freckles, pits, and striations. A black and white video camera can be used to capture an image of the iris. Unique features of the iris are extracted from the captured image by the recognition system. These features are converted into a unique iris code, which is compared to previously stored iris codes for user recognition.

Artificial duplication of the pattern of an individual iris is virtually impossible and **iris imaging** provides high accuracy. The technique is relatively easy to use, and there appears little resistance from users. Fraudulent deception of the system is very unlikely.

Retina Recognition

The retina of each individual forms a unique pattern. The user views a green dot for a few seconds until the eye is sufficiently focused for a scanner to capture the blood vessel pattern. The retina pattern is captured by the scanner and then compared to previously stored patterns for identification.

Retina scanning provides very high accuracy, provided the user's eye is properly focused. Reflection from glasses can create interference. Users are resistant to this technique due to the infrared light scanner. Deception of the system is very unlikely.

Face Recognition

No direct physical contact is required with this system. The camera captures a face image and a number of points on the face are mapped by the system. From these measurements, a unique representation of the individual's face is created. A complete map of the entire face can be created. There is a downside to this system, since people change over time. Some systems compensate for this by combining recently stored information with previously stored images.

Face recognition offers reasonable accuracy but can be affected by poor lighting, glasses, facial hair, and aging. The system can be retrained and updated to recognize changes in users. The equipment is easy to use and readily accepted by users. Current systems can recognize and identify users from 2 to 32 meters.

Voice Recognition

Speaker recognition or **voice recognition** is a biometric modality that uses an individual's speech—a feature influenced by both the physical structure of an individual's vocal tract and the behavioral characteristics of the individual—for recognition purposes.

Voice recognition uses the acoustic features of speech that have been found to differ between individuals.

Signature Verification

Signature scanning techniques examine the way users sign their names. The system examines the dynamics of the signing process, rather than the signature. Extracted characteristics may include the angle at which the pen is held, the time taken to sign, the velocity and acceleration of the signing process, and the number of times the pen is lifted from the document during the signing process.

Ordinary forgery techniques do not work because behavioral characteristics are used rather than a signature. Signature data can be captured with a special pen that contains sensors or with a tablet that senses the motion of a stylus. Acoustic emission measurements of the pen can be captured. A number of signatures can be recorded to build a user's profile.

Signatures are one of the most common methods of establishing identity, and are therefore readily accepted by financial institutions and others that require conventional signatures. Recognition accuracy is high, and the method is considered impervious to deception.

Biometric Security Environment

Biometric technology can be used successfully to identify users; however, it is necessary for the application to be specifically designed, taking into consideration the ease of use, deployment, accuracy, and cost. Finger scanning appears to be the technology of choice. These systems are moderately accurate and easy to use. Acceptance by users is good and the technique is relatively inexpensive. The biometric methods depicted in Figure 1-1 may be acceptable alternatives for access control for some sensitive facilities.

Trusted Systems

Trust is the composite of availability, performance, and security, which includes the ability to execute processes with integrity, secrecy, and privacy. Technical solutions that employ

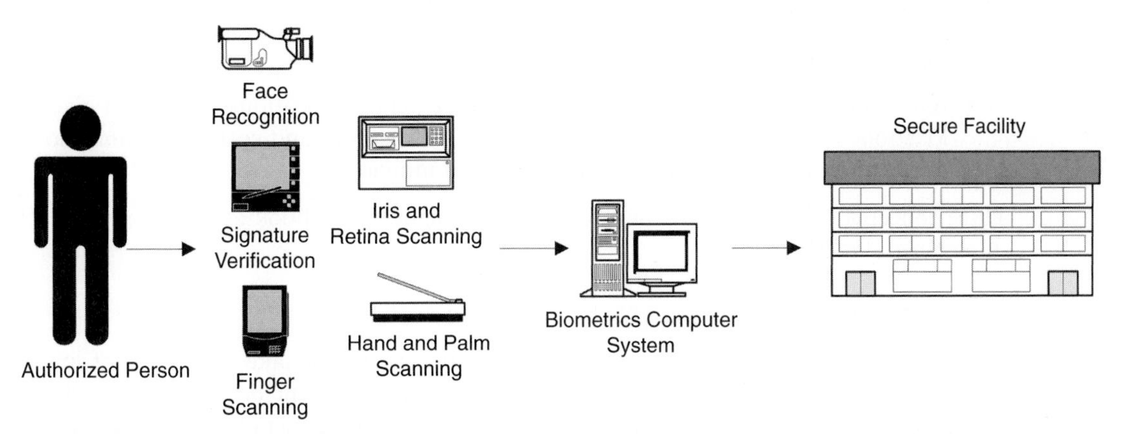

Figure 1-1 Biometrics Access

Course Technology/Cengage Learning

biometric applications to provide support to commercial trust systems will be presented in Chapter 3. These solutions include biometric systems that are:

- Citizen facing
- Employee facing
- Customer facing

Each of these systems provides a part of the total solution for the protection of the organization's computer and network assets. The task of the systems administrator and management is to develop and implement security systems that will protect the organization from the untrusted networks, such as the Internet. These systems can include legacy security techniques, biometric solutions, and a combination of both. These system solutions will be presented in Chapter 5.

Access Control

Numerous legacy applications and systems have been implemented to provide security in the form of access control. Many of these solutions involve network devices, such as routers and firewalls, that screen or filter out attacks and threats. The sophistication and volume of these efforts require increased efforts to control access into an organization's assets and resources. The major topic in this book is access control—using a combination of legacy and biometric solutions. These integrated solutions, presented in Chapter 6, are called multibiometric and multimodal systems.

Confidentiality, Integrity, and Accessibility

It is essential that readers understand the concepts of **confidentiality**, **integrity**, and **accessibility** (CIA). These concepts are especially relevant in the e-commerce, Web, and networking environments. The definitions are as follows:

- **Confidentiality** has been defined by the International Organization for Standardization (ISO) as "ensuring that information is accessible only to those authorized to have access," and is one of the cornerstones of information security.
- **Integrity** implies protection against unauthorized modification or destruction of information. This defines a state in which information has remained unaltered from the point it was produced by a source, during transmission, in storage, and eventual receipt by the destination.
- **Accessibility** is an aggregate measure of how reachable locations are from a given location. Common measures of accessibility are distance and cost. There are also new laws that define accessibility for handicapped individuals.

These concepts will be explored in more detail throughout this book as they apply to each of the biometric methods.

Using Biometric Systems

There are many individuals in today's society whose goal is to take advantage of innocent, commercial organizations and the government. Law enforcement organizations and corporate security departments are heavily involved in countering identity theft and fraudulent activities being committed by criminals and terrorists. In today's computer and networking environment, these unscrupulous individuals can appear anywhere in the world at any time, and cause distress to corporate, government, educational, health care, and public resources. This means legitimate users and organizations must initiate activities and actions that can minimize or eliminate these threats. Security boundaries using biometric solutions can be set up around groups and data centers within an organization. Such boundaries can make it difficult for outside attackers to penetrate an organization's network; however, these systems can also make it difficult for internal personnel to access restricted systems.

In the near past, most transactions were one-on-one and up-front and personal. Either cash or personal check was the common medium of exchange. Now, most citizens and organizations use some form of credit or debit card to conduct financial transactions. These cards are not only easier to use, but they provide an electronic audit trail of the transactions. Electronic funds transfers account for most transfers of money between banks and other financial institutions. Many organizations pay employees by direct deposit to the employees' bank account. Credit unions, utilities, insurance companies, e-commerce companies, mortgage companies, and other organizations can automatically process deductions against their client's bank accounts. It is now possible to make banking transactions and stock market transactions from portable and remotely deployed computer and networking devices. There are many opportunities inherent in this environment for fraud and financial loss.

Employees in commercial and government organizations and other authorized internal users are the source of most breaches of computer and network security. Measures that minimize these breaches often include passwords and other means to identify authorized users. By using biometric technologies to supplement these legacy security techniques, security and access control can be enhanced. Networks can be monitored for abnormal patterns, and employees can be educated in security awareness and practices.

One major challenge to the organization is to balance the cost of security against the risks being addressed. The cost of providing solutions and controls, both biometric and legacy, should be in proportion to the value of the assets being protected. It makes no sense to spend more than an asset is worth to protect it. An analysis of the situation must prove the feasibility and viability of the proposed security controls. This analysis needs to look at how critical the systems are to the business. Attention must be paid to the exposure of the systems to fraud and illegal disclosure of information, and the risk to assets from both internal and external sources must be explored. One of the first steps toward securing the organization's assets is to develop a site security policy and create a baseline.

Biometrics are of interest in any area where measurable physiological and/or behavioral characteristics can be used to verify the identity of an individual. They are of interest in any area where it is important to verify the true identity of an individual. Initially, these techniques were employed primarily in specialist high security applications; however, managers are now seeing their use and proposed use in a much broader range of public-facing situations.

Personal identification numbers (PINs) were one of the first identifiers to offer automated recognition. However, it should be understood that this means recognition of the PIN, and not necessarily recognition of the person who has provided it. The same applies to cards and other tokens. The token can be easily recognized, but it could be presented by anybody. Using the two together provides a slightly higher confidence level, but this is still easily compromised by determined individuals.

A biometric cannot be easily transferred between individuals and represents a possible unique identifier. If the verification procedure can be automated in a user-friendly manner, there is considerable scope for integrating biometrics into a variety of processes. It means that verifying an individual's identity can become both more streamlined by the user interacting with the biometric reader and considerably more accurate as biometric devices are not easily fooled.

In the context of air travel and tourism, for example, one immediately thinks of boarding gate identity verification, immigration control, and other security-related functions. However, there may be other potential applications in areas such as marketing, premium passenger services, online booking, and alliance programs, where a biometric may be usefully integrated into a given process at some stage. In addition, there are organization-related applications, such as workstation/LAN access, physical access control, and other potential applications. Biometric applications and solutions will be presented in Chapter 3.

Biometric Identification Advantages

There are a number of advantages to biometric identification technology:

- Biometric identification can provide extremely accurate, secured access to information; fingerprints, and retinal and iris scans produce absolutely unique data sets when done properly.
- Current methods, such as password verification, have many problems (people write them down, they forget them, they make up easy-to-hack passwords).
- Automated biometric identification can be done very rapidly and uniformly, with a minimum of training.
- Identities can be verified without resorting to documents that may be stolen, lost, or altered.

Access Control for Physical Facilities and Resources

Biometric identification devices use individual physical attributes such as fingerprints, palm prints, voice attributes, and retina patterns to identify authorized users. Biometrics is currently used as a network security technique at only a few U.S. companies and government installations. Fingerprint identification is the most popular biometrics technique and retina scanning is the most accurate.

Biometric identification has been expensive to implement; however, costs are decreasing. There is considerable end-user resistance to fingerprinting, which is associated with criminal activity. While biometrics promises more secure identification of authorized users than that of passwords, it also can be circumvented. Access control and security will be presented in Chapter 4.

Access Security

All efforts to secure an organization's resources will be in vain if the facilities that house them are not protected. Access control into computer and network facilities, sensitive areas, and research facilities has become a priority for many organizations. This is particularly true if the asset is expensive and replacing or recreating it would be difficult and cost-prohibitive. This section will look at the issues relating to egress and ingress to buildings that house computer and network resources.

It is essential that only authorized personnel be allowed unattended access into a facility. Not only is ingress an issue, egress can be just as important, as hardware, software, company secrets, and documentation tend to disappear if management controls are not present. Unauthorized individuals can cause several problems, including theft of equipment or data, destruction of equipment or data, and viewing of sensitive data and information. An access control program must be in place that includes internal personnel, contractors, partners, maintenance personnel, sales personnel, and visitors.

Access Control for Internal Personnel

The size of an organization and the personnel complement can have an impact on the level of security access to computer, network, and physical resources. In small offices, with limited computer and networking resources, everyone may have access to everything. If this is a stand-alone entity, security implications and implementations might be limited; however, if this small office is part of a corporate network, the situation changes. A short list of questions that should be answered includes the following:

- Where is the office entrance? Is it in a larger building with multiple tenants?
- Is the facility located on the ground floor?
- Do employees access the facility after hours and on holidays?
- Do employees perform minor maintenance?
- Is the computer system standalone or networked?
- What is the method of network communication and where are the communication devices located?
- Is the system in a secure room? Who has the keys?
- Is other equipment or supplies located along with the computer or networking components?
- Who should have access to the computer processor and console?
- Are data files stored off-campus? If so, who has access?

Access Control for Contractors, Maintenance, Vendors, and Suppliers

There are a number of support organizations, vendors, and suppliers that might need access to the physical facility. Maintenance personnel will probably require frequent access to the computer and network components. Software engineers and other specialists may also require access during all hours of the day. Communication suppliers and vendors often have a requirement to access the demarcation closet. This is a critical location in the corporate network and access must be adequately controlled. Different groups may require a different level of access control. Very few of these people should be allowed unescorted access into sensitive areas. Someone must be responsible for providing authorization for these personnel to access computer equipment and networking facilities.

Activities that external organizations might perform include the following:

- Office rearrangement
- Furniture additions

- Construction
- Electrical and communication circuit wiring
- Installation and maintenance of network services
- Installation of network and computer devices and components
- Maintenance of computer and database systems
- Software upgrades
- Inspections and audits
- Hands-on training
- Facility maintenance

Access Control for Visitors

Visitors can fall into a number of different categories, including relatives of employees, tour groups, competitors, sales representatives, trainers, and others. None of these visitors should be allowed to roam freely throughout the organization. Trainers and education providers belong to a different category than the others, as they may require access to both computer and networking components as part of a training session. The following suggestions may provide some guidance concerning access for the various classes of visitors.

- A clear and concise policy must identify security access requirements.
- An employee must accompany all nonemployees.
- Tour groups must be controlled and monitored by a tour guide.
- A supervisor must accompany sales representatives and competitors.
- Trainers must have clearance to view and access corporate systems.

Techniques for Preventing Theft and Destruction

The most successful method of preventing theft of computer, networking, and confidential proprietary resources is to control access. Access control devices can prevent access by unauthorized individuals and record access by those authorized. Mainframes, client/servers, and other large devices are difficult to get past the front door, however there are a number of devices and software that are easy to conceal. There are several approaches that can be taken to reduce the incidence of theft of these resources. These include preventing access, restricting portability, and detecting exit. Additional details concerning biometric methods and capabilities will be presented in several subsequent chapters. Information concerning system integrity and accessibility will be presented in Chapter 9.

Methods for protecting organizational assets include intrusion detection and intrusion prevention. **Intrusion detection** is the art of detection and response to situations involving computer, database, and network misuse. Intrusion detection systems (IDS) are network based, host based, or hybrid based. An intrusion system is a conglomeration of capabilities to detect and respond to threats. Intrusion detection systems are available that provide the following functions:

- Event log analysis for insider threat detection
- File integrity checking

- Network traffic analysis for perimeter threat detection
- Security configuration management

A newer security solution for protecting an organization's resources and assets is called an intrusion prevention system. **Intrusion prevention** is a preemptive approach to network security used to identify potential threats and respond to them swiftly. Like an intrusion detection system, an intrusion prevention system (IPS) monitors network traffic. As with IDS, IPS configurations include both host-based IPS and network-based IPS implementations. IPS systems have some advantages over intrusion detection systems, as they are designed to sit inline with network traffic flows and prevent attacks in real time.

Additional details for both system types will be provided in Chapter 6, which addresses multibiometric systems. Multimodal biometric systems integrate face recognition, fingerprint verification, and speaker verification in making a personal identification. They can be used to overcome some of the limitations of a single biometric method.

Computer Authentication and Authorization

Authenticate means to verify or guarantee the identity of a person or entity and to ensure that the individual or organization is really who he/it says he/it is. **Authentication factors** include pieces of information used to verify a person's identity for security purposes.

Authentication is the process of validating the claimed identity of an end computer or network user or a device such as a host computer, router, or some other physical element. It consists of a security measure designed to establish the validity of a transmission, message, or originator, or a means of verifying an individual's authorization to receive specific categories of information.

Authentication methods identify users. Once users are identified and authenticated, they may access resources based upon an authorization. An authentication system may be the most important part of a network or computer system. It requires a user supply some identification, such as a username and password. Simple password systems may be sufficient for small systems; however, if a higher level is required, then more advanced security may be necessary. These systems may provide various security capabilities such as biometric or token authentication devices.

Authorization is the act of granting access rights to a user, groups of users, system, or program. This process allows some user or element to act against some database or allows access into some area. Authorization is provided after authentication.

Authorization refers to securing the network by specifying which area of the network—whether it is an application, a device, or a system—a user is allowed to access. The authorization level is based on the "need to know," and thus varies from user to user. For example, payroll personnel have access to the payroll systems; engineers have access to design systems. This also means that previously authenticated access to system resources and services to users or processes is selectively granted. It is possible to explicitly grant or deny authorization. By explicitly denying access, only access to confidential information is restricted. This compares with explicitly granting access, which does not allow access to any information without specifically granting access. Authorization can be defined on a system-wide basis or limited to specific data elements. Authorization can be based on the type of access, which includes read, write, delete, or execute. There are different authorization processes for routers, gateways,

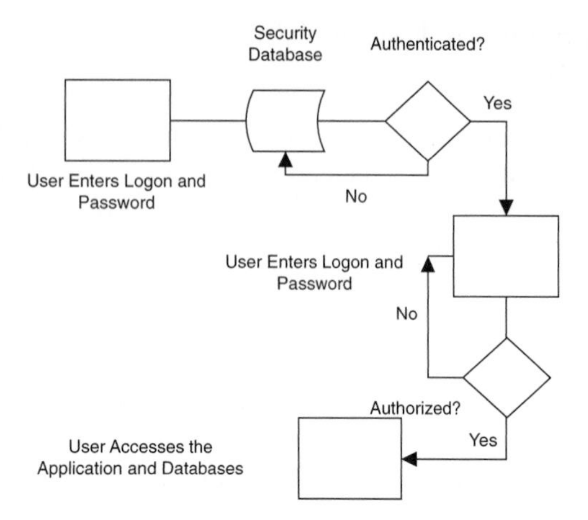

Figure 1-2 Authorization and Authentication Flow

Course Technology/Cengage Learning

servers, and switches. Figure 1-2 shows the authentication and authorization flow of a login session.

Note that the process in Figure 1-2 requires a separate login and a different password for the application or database access. Routers, gateways, and switches and other network devices require authorization so unauthorized personnel do not change configurations or compromise the system. Access to a server is limited in order to preserve directory integrity. A user is often allowed a maximum of three attempts to enter the correct authorization before the access attempt is terminated. Biometric security and access technologies will be presented in Chapter 8.

A basic requirement of a security system is to identify the user who is trying to gain access to the network and computer system. The owner of the network must determine whether this user is a friend or foe. It is essential that networks, computers, and information be protected from unauthorized disclosure. This concept is called confidentiality. In a computer system or network, there must be a method of identifying anyone who is allowed into the environment. This takes the form of identification, authentication, authorization, and accounting, collectively called **AAA** or **Triple-A.**

This section will examine techniques for identifying those active entities that are responsible for initiating specific actions on a computer system or network. This class of techniques is called identification. Computer-access identification consists of procedures and mechanisms which allow an entity that is external to a system resource to notify that system of its identity. The requirement to perform an identification technique occurs when it is necessary to associate an action with some user or entity. Computer systems and network devices can determine who invoked an operation by examining the reported identity of the entity that initiated the session in which the operation was invoked. This identity is usually established via a login sequence.

A system administrator may have assigned this login. The login is the simplest type of mechanism comprising identification. Once users are logged in, they are allowed to access various resources based on the rights and privileges assigned to their user accounts or the objects they possess. Identification is usually combined with another procedure, called authentication, which allows a system to determine if the identification sequence was correct.

Factor	Solution
Something you are	Biometric techniques
Something you have	RFD card, smart card, chip
Something you know	Password, logon

Table 1-2 Three-Factor Authentication Examples

Three-factor authentication consists of some method of determining "something you are," "something you have," and "something you know." Table 1-2 provides examples of three-factor authentication. Biometrics will add the additional security element of "something you are."

Biometric Terms and Concepts

Readers must understand a number of biometric terms and technical concepts before continuing on to the next chapters. This section will identify and explain the differences between verification, identification, screening, and continuity of identity system; introduce and define additional terms and concepts such as pattern recognition and matching and computer security exact password matching; and describe the differences between a token and a template.

Continuity of identity is concerned with human physical continuity. As the name implies, it asserts that our identity is supplied by the continuity of our physical form. An end user is the individual who will interact with the system to enroll, verify, or identify. Other user categories include cooperative user, indifferent user, noncooperative user, and uncooperative user. A cooperative user is an individual who willingly provides his or her biometric attribute to the biometric system for capture. An example of this is a worker who submits a biometric attribute to clock in and out of work.

Identification is the one-to-many process of comparing a submitted biometric sample against all biometric reference templates on file to determine whether it matches any of the templates and, if so, the identity of the enrollee. The biometric system that uses the one-to-many approach seeks to find an identity among a database rather than authenticate a claimed identity. An **identity** is the commonsense notion of personal identity. This includes a person's name, personality, physical body, and history, including such attributes as nationality, educational achievements, employer, security clearances, financial and credit history, etc. In a biometric system, identity is typically established when the person is registered in the system through the use of so-called "breeder documents," such as a birth certificate or passport. It is a subset of physical and/or behavioral characteristics by which an individual is uniquely recognizable. Identity is information concerning the person, not the actual person.

Biometric verification is the process of establishing the validity of a claimed identity by comparing a verification template to an enrollment template. Verification requires that an identity be claimed, after which the individual's enrollment template is located and compared with the verification template. Verification answers the question, "Am I who I claim to be?" Some verification systems perform very limited searches against multiple enrollee records. Note that verification is one-to-one process and identification is one-to-many process.

A **template** consists of data that represents the biometric measurement of an enrollee, which a biometric system compares against subsequently submitted biometric samples. The template becomes biometric data after it has been processed from its original representation (using a biometric feature extraction algorithm) into a form that can be used for automated matching purposes (using a biometric matching algorithm). Biometric data stored in a template format cannot be reconstructed into the original output image.

A **token** is a physical device that carries an individual's credentials. The device is typically small, for easy transport, and usually employs a variety of physical and/or logical mechanisms to protect against modifying legitimate credentials or producing fraudulent credentials. Examples of tokens include picture ID cards (e.g., state driver's licenses), smart cards, and USB devices.

Matching is the process of comparing a biometric sample against a previously stored template and scoring the level of similarity.

Pattern recognition is a subtopic of machine learning. It can be defined as the act of inputting raw data and taking an action based on the category of the data. A complete pattern recognition system consists of a sensor that gathers the observations to be classified or described; a feature extraction mechanism that computes numeric or symbolic information from the observations; and a classification or description scheme that does the actual job of classifying or describing observations, relying on the extracted features. **Pattern matching** is the act of checking for the presence of the constituents of a given pattern. In contrast to pattern recognition, the pattern is rigidly specified. An example is exact password matching, where the user provides the specific password to be compared.

Screening involves a reasonable examination of persons, cargo, vehicles, or baggage for the protection of the vessel, its passengers, and crew. **Covert screening** occurs in situations where biometric samples are being collected at a location that is not known to bystanders. An example of a covert environment might involve an airport checkpoint where face images of passengers are captured and compared to a watch-list without passengers' knowledge.

Enrollment is the process of collecting a biometric sample from an end user, converting it into a biometric reference, and storing it in the biometric system's database for later comparison.

Extraction is the process by which the biometric sample captured is transformed into an electronic representation. During enrollment, this electronic representation is known as the biometric template. During the authentication process, it is known as the live sample.

Password-based verification systems have several limitations and their information content (i.e., cryptographic strength) can be arbitrarily increased. This action is inconvenient for the user, who must remember a long, cryptic string. On the other hand, the cryptographic strength (accuracy) offered by biometric systems is limited by the inherent information content in the biometric characteristic, and the accuracy of the feature extraction and classification methods used by the system. In spite of three decades of research, designing highly accurate biometric systems remains a very challenging pattern-recognition research problem. The accuracy of a classification method depends on the closeness of a reading or indication of a measurement device to the actual value of the quantity being measured.

Major Issues and Concerns

Biometric technology is inherently individualizing and interfaces readily to database technology, making privacy violations easier and more damaging. Privacy must be designed into systems from the beginning, as it is difficult to retrofit complex systems for privacy functions.

Biometrics is no substitute for quality data about potential risks. No matter how accurately a person is identified, identification alone reveals nothing about whether a person is a terrorist, for example. Such information is completely external to any biometric ID system.

Biometric systems are only as good as the initial identification, which in any foreseeable system will be based on exactly the document-oriented methods of identification. The quality of the initial "enrollment" or "registration" is crucial.

Biometric identification is often more than is needed for the task at hand. It is not necessary to identify a person, and to create a record of his or her presence at a certain place and time, if all that needs to be known is whether that person is entitled to do something or be somewhere. Customers in a bar use IDs to prove they're old enough to drink, not to prove who they are, or to create a record of their presence.

Some biometric technologies are discriminatory. A nontrivial percentage of the population cannot present suitable features to participate in certain biometric systems. Many people have fingers that simply do not "print well." Even if people with "bad prints" represent 1 percent of the population, this would mean massive inconvenience and suspicion for that minority.

The accuracy of biometric systems is impossible to assess before deployment. Accuracy and error rates published by biometric technology vendors are not trustworthy, as biometric error rates are intrinsically manipulative. Biometric systems fail in two ways: through a false match, which incorrectly matches a subject with someone else's reference sample; and a false non-match, which fails to match a subject with his or her own reference sample. There is a trade-off between these two types of error, and biometric systems may be "tuned" to favor one over another. When subjected to real-world testing in the proposed operating environment, biometric systems frequently fall short of the performance promised by vendors.

The cost of failure is high. If an individual loses a credit card, it can be cancelled and a new one issued. If someone loses a biometric, it is lost forever. Any biometric system must be built to the highest levels of data security, including transmission that prevents interception, storage that prevents theft, and system-wide architecture to prevent both intrusion and compromise by corrupt or deceitful agents within the organization.

Despite these concerns, political pressure for increasing the use of biometrics appears to be informed and driven more by marketing from the biometrics industry than by scientists. Much federal attention is devoted to deploying biometrics for border security. This is an easy sell, because immigrants and foreigners are, politically speaking, easy targets. But once a system is created, new uses are usually found for it, and those uses will not likely stop at the border. With biometric ID systems, as with national ID systems, administrators and politicians must be wary of getting the worst of both worlds: a system that enables greater social surveillance of the population in general, but does not provide increased protection against terrorists. [3]

Security Cost Justification

Because network security measures and initiatives can require a substantial investment, it is imperative that they be undertaken with some confidence that the investments are warranted and effective. One national study found that large companies and government agencies were able to quantify an average loss of $1 million from security breaches. Security programs and measures, however, can also be very expensive, and still not provide sufficient security of assets and resources.

One method of determining the degree of security needed is to assess the value of the database element that may be placed at risk. This value would include the cost of collecting the data and re-creating the database. The original data may no longer be available, which means that the database could not be reconstructed. This scenario assumes that the database was not backed up, which is also a possibility. The organization's return on investment in network security would depend on the monetary value assigned to the data and database. This exercise is not a trivial pursuit.

Organizations spend large sums annually on security hardware, software, and services without any real objectives or plans. Protecting facilities and a computer network against damaging attacks and other risks is a complex task involving many aspects of the asset structure and operation. Specialists, who are well versed in computer and network security issues and solutions, may possess the qualifications to undertake this task.

Independent and vendor-supplied professional security consulting services are available. Consulting services can help network managers assess network vulnerability. They can also suggest security products and services and help develop effective security management practices. These firms can also provide managed-security solutions, which consist of access, monitoring, and problem resolutions. Personnel from these firms have continual exposure to security issues, challenges, and attack methods, and their experience can supplement a limited inhouse security staff.

Many organizations believe that they can solve security issues by buying a solution. This might be a false assumption, because security should be embedded in the core operations of the organization and not applied to the surface. Corporate systems design should incorporate security as it is being developed, not include it as an afterthought.

The proper location for implementing asset security is the organization itself, starting with the employees who need access to computer and networking resources in order to perform their functions. The most powerful security measure is to obtain and maintain the loyalty of these employees. Security measures should be designed and implemented to protect an organization's assets without unduly interfering with legitimate uses of the network. Chapter 5 will provide insight into cost issues for biometric systems.

Summary

- A hostile environment exists in and around various network and computer configurations. It is necessary to establish a robust resource-access program in order to protect commercial and government resources from a multitude of security challenges.
- Identification, authentication, and authorization methods and techniques can address primary access issues. These measures may take the form of biometric solutions that enhance legacy security systems.
- Vulnerabilities introduced by insecure communications can be counteracted by mechanisms and services, which are part of communications security. These services

include a means to verify the source and integrity of a message. Biometric techniques can help mitigate security issues.

- Biometric methods can include finger scanning; finger and hand geometry; iris and palm imaging; face, retina, and voice recognition; and signature verification techniques. Each of these methods has advantages and disadvantages, and must be matched to the access control and security program being implemented.
- Access to computer and network facilities, buildings, and other assets must be protected from unauthorized personnel and nonemployees. Biometric and legacy access control procedures are available for individuals who represent internal and external organizations that might require access to sensitive areas. It is also essential that deterrents be in place to keep hardware, software, documentation, and other resources from disappearing out the door.

Key Terms

Access control The process of granting access to information system resources only to authorized users, programs, processes, or other systems.

Accessibility A biometric feature easily presented to an imaging sensor.

Algorithm A sequence of instructions that tell a biometric system how to solve a particular problem. An algorithm will have a finite number of steps and is typically used by the biometric engine to compute whether a biometric sample and template are a match. Cryptographic algorithms are used to encrypt sensitive data files, to encrypt and decrypt messages, and to digitally sign documents.

Authenticate To verify (guarantee) the identity of a person or entity. To ensure that the individual or organization is really who he/she/it says it is.

Authentication A security measure designed to establish the validity of a transmission, message, or originator, or a means of verifying an individual's authorization to receive specific categories of information.

Authentication factors Includes pieces of information used to verify a person's identity for security purposes.

Authorization The act of granting access rights to a user, groups of users, system, or program.

Automatic identification and data capture (AIDC) The term used to describe data collection by means other than manual notation or keyboard input.

Behavior One of a recurring or characteristic pattern of observable actions or behavioral responses.

Biometric (adjective) A measurable physical characteristic or personal behavioral trait used to recognize the identity, or verify the claimed identity, of an individual. Facial images, fingerprints, and iris scan samples are all examples of biometrics. **(noun)** One of various technologies that utilize behavioral or physiological characteristics to determine or verify identity. A "finger scan is a commonly used biometric." The plural form is also acceptable: "Retina scans and iris scans are eye-based biometrics."

Biometric verification The process of establishing the validity of a claimed identity by comparing a verification template to an enrollment template.

CIA Acronym for confidentiality, integrity, and accessibility.

Confidentiality Defined by the International Organization for Standardization (ISO) as "ensuring that information is accessible only to those authorized to have access" and is one of the cornerstones of information security.

Continuity of identity Concerned with human physical continuity. As the name implies, it asserts that our identity is supplied by the continuity of our physical form.

Covert screening Occurs in situations where biometric samples are being collected at a location that is not known to bystanders.

Enrollment The process of collecting biometric samples from a user and the subsequent preparation, encryption, and storage of biometric reference templates representing that person's identity.

Extraction The process by which the biometric sample captured in the previous block is transformed into an electronic representation.

Face recognition Offers reasonable accuracy but can be affected by poor lighting, glasses, facial hair, and aging.

Finger geometry Finger shape to determine identity. Unique finger characteristics, such as finger length, width, thickness, and knuckle size are measured on one or two fingers.

Fingerprint Scanning Acquisition and recognition of a person's fingerprint characteristics for identifying purposes. This process allows the recognition of a person through quantifiable physiological characteristics that detail the unique identity of an individual.

Hand geometry Measurements center on finger length and the shape of the fingers and knuckles.

Identification The one-to-many process of comparing a submitted biometric sample against all biometric reference templates on file to determine whether it matches any of the templates and, if so, the identity of the enrollee whose template was matched. The biometric system using the one-to-many approach seeks to find an identity in a database rather than authenticate a claimed identity.

Identity The commonsense notion of personal identity. A person's name, personality, physical body, and history, including such attributes as nationality, educational achievements, employer, security clearances, financial and credit history, etc. In a biometric system, identity is typically established when the person is registered in the system through the use of so-called "breeder documents."

Integrity Implies protection against unauthorized modification or destruction of information. This defines a state in which information has remained unaltered from the point it was produced by a source, during transmission, storage, and eventual receipt by the destination.

Intrusion detection The art of detection and response to situations involving computer, database, and network misuse.

Intrusion prevention A preemptive approach to network security used to identify potential threats and respond to them swiftly.

Matching The comparison of biometric templates to determine their degree of similarity or correlation.

Palm scanning The measurement of palms is performed by techniques similar to those for finger scanning. A scanner shaped to accommodate the palm scans the ridges, valleys, and minutiae found on the palm.

Pattern matching The act of checking for the presence of the constituents of a given pattern.

Pattern recognition A subtopic of machine learning. It can be defined as the act of inputting raw data and taking an action based on the category of the data.

Physiology The branch of biology that deals with the functions and vital processes of their parts and organs.

Retina scanning Provides very high accuracy, provided the user's eye is properly focused.

Screening Involves a reasonable examination of persons, cargo, vehicles, or baggage for the protection of the vessel, its passengers, and crew.

Signature scanning This system examines the dynamics of the signing process, rather than the signature.

Template Data that represents the biometric measurement of an enrollee, used by a biometric system for comparison against subsequently submitted biometric samples.

Three-factor authentication Consists of some method of determining "something you are," "something you have," and "something you know."

Token A physical device that carries an individual's credentials.

Triple-A Authentication, authorization, and accounting.

Trust The composite of availability, performance, and security, which includes the ability to execute processes with integrity, secrecy, and privacy.

Voice recognition A biometric modality that uses an individual's speech—a feature influenced by both the physical structure of an individual's vocal tract and the behavioral characteristics of the individual—for recognition purposes.

Review Questions

1. ________________ is the study of biological data and biometric-based authentication, which is an application that uses specific personal traits for access control.

2. Provide a list and description of the biometric methods discussed.

3. ________________ is the composite of availability, performance, and security, which includes the ability to execute processes with integrity, secrecy, and privacy.

4. CIA is the acronym for ________________, ________________, and ________________.

5. An access control program must be in place that includes ________________, ________________, and ________________.

6. ________________ is the process of validating the claimed identity of an end computer or network user or a device such as a host computer, router, or some other physical element.

7. ________________ is the act of granting access rights to a user, groups of users, system, or program.

8. ________________ is defined as consisting of procedures and mechanisms which allow an entity external to a system resource to notify that system of its identity.

9. Describe three-factor authentication.
10. ________________ identification is the most popular biometrics technique and ________________ is the most accurate.
11. Behavior is described as one of a recurring or characteristic pattern of observable actions of physiological responses. (True or False?)
12. The term biometrics has two distinct meanings: "bio" meaning living creature, and "metrics" meaning the ability to measure an object qualitatively. (True or False?)
13. An algorithm is a limited sequence of instructions that tell a computer system how to solve a particular problem. (True or False?)
14. Identification verification is mainly performed via possession-based or knowledge-based ways. (True or False?)
15. The finger-scanning biometric modality has a higher comparative accuracy than does iris imaging. (True or False?)
16. Confidentiality implies protection against unauthorized modification or destruction of information. (True or False?)
17. A biometric cannot be easily transferred between individuals and represents a possible unique identifier. (True or False?)
18. Intrusion detection is a preemptive approach to network security used to identify potential attacks. (True or False?)
19. Biometric identification and verification processes both operate in a one-to-many manner when validating an individual. (True or False?)
20. Extraction is the process by which a captured biometric sample is transformed into an electronic representation. (True or False?)

Discussion Exercises

1. Develop a matrix that shows the differences between the various biometric methods.
2. Provide a list of devices that can be used to support a biometric system installation. Describe each device.
3. Provide a list of software that can be utilized to implement a biometric solution. Describe each product.
4. Develop a list of biometric hardware, software, and system providers.
5. Identify commercial, education, and government organizations that are currently utilizing biometric systems for security and access control. Provide an overview of each system.
6. Provide an overview of the AIDC categories.

7. Identify those issues and concerns employees and employers have with biometric applications.
8. Identify biometrics that fit in the categories of citizen-facing, employee-facing, and customer-facing.
9. Provide an overview of the use of PINs.
10. Identify and describe an algorithm that is used in a biometric process.
11. Arrange for some vendor presentation of biometric products.
12. Identify an organization that uses a biometric method for access control. Provide an overview of the components and areas controlled.
13. Research and develop a report that reflects the current perceptions of the use of biometrics in the public arena. This report should include personal interviews with friends, relatives, etc.

Hands-On Projects

1-1: Learning More About Biometrics

Time required—30 minutes

Objective—Become familiar with various concepts and terms involving the biometrics environment.

Description—Use various Web tools and Web sites to create a working knowledge of the biometric industry and environment.

1. Use Google to search for the terms that have been introduced in this chapter. Use the *Define: term* for the search.
2. Use pcwebopedia.com, whatis.com, and Wikipedia.com to identify biometric terms.
3. Produce a report of the results.

1-2: Establish a Progressive Case Project

Time required—45 minutes

Objective—Establish a shell company called OnLine Access, LLC, that will be a candidate for a biometric security solution. Use this company as the basis for a migration from the current security environment that will progress across all chapters of this book in incremental steps.

Description—Create an organization that currently relies on nonexistent security or outdated techniques and systems for the protection of its information systems and assets. The security systems in this organization will be modified and enhanced to migrate it from old security technologies to a state-of-the-art biometric system protection.

Online Access LLC provides for a centralized, Web-based, e-commerce site for sellers to connect with buyers who wish to purchase a wide variety of products online. Online Access is primarily a portal for personal and commercial sales activities between a buyer and seller. The database includes information about all products and services offered by the sellers.

A contract programmer/analyst maintains the Web site. A database includes records for all sellers. All financial transactions are processed by the sellers. The sellers are also responsible for all logistic functions, including shipping and returns. Income is generated from access established over the network and advertising on the Web page.

Online Access is located in the suburbs of a large Midwestern city. There are no physical security provisions at the site except for manual locks on the entrance doors. Simple network devices provide for network access to the telephone system. The only network device with a security capability is a router that has default settings. The computing devices and workstations do not require passwords. The systems do not have virus and malware protection. The owners are recent marketing majors and have not been exposed to security programs and issues.

1. Use Google to search for activities involving information security issues, particularly e-commerce activities.
2. Identify publications, books, and articles that provide information concerning biometric topics.
3. Describe the results of this research.

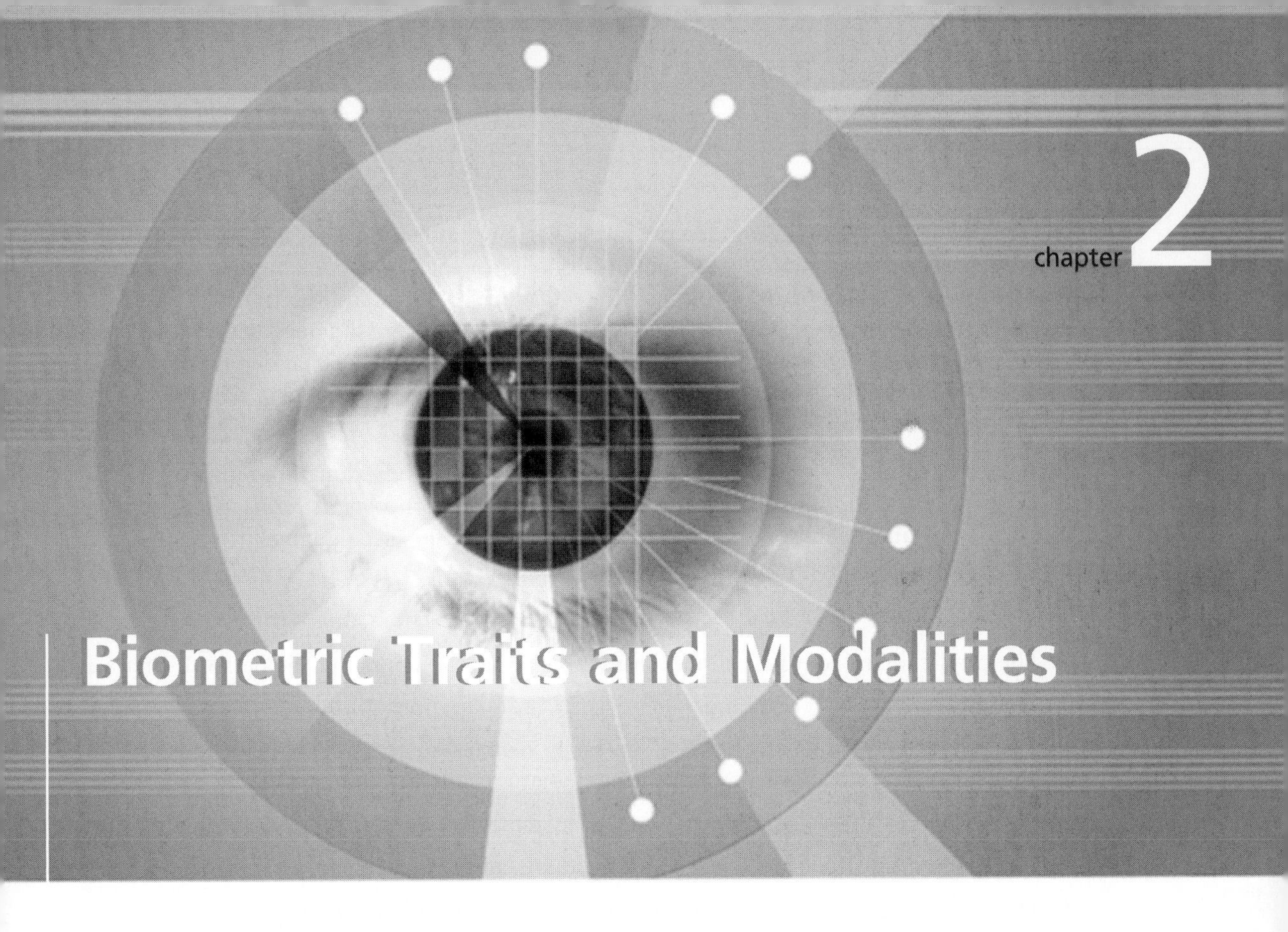

Biometric Traits and Modalities

After completing this chapter, you should be able to do the following:

- Identify the characteristics of the biometric technologies
- Understand the differences between physiological and behavioral human characteristics
- Become familiar with the various biometric methodologies
- Describe the attributes and features of the common biometric technologies
- Look at the advantages and disadvantages of the various biometric methods
- Understand the differences between verification and identification

Biometrics is in the early stage of development; most of the research and development is being conducted by computer scientists and engineers. Numerous papers have been published concerning the use of biometrics for some type of access control. Each of the biometric methods identified offer a benefit for some security issue, but some may be difficult or expensive to implement.

Two issues arise—expense and user resistance. Biometric devices can be expensive, and many can result in an invasion of privacy. Managers must look at the cost-benefit of legacy security solutions versus an untried biometric solution. A system that interacts with the human body might not be acceptable to an organization's employees or the general public, and a lack of acceptance can spell doom for a biometric access control solution.

Introduction

This chapter contains information describing the various biometric technologies and methods. Detailed descriptions are provided for the functionality of each method. This chapter presents the most technical details of the book, as the reader must understand the features, functions, and relevance for each method. Technologies introduced include scanning, imaging, recognition, and verification. Particular methods include finger scanning, iris and retina scanning/imaging, voice and speech recognition, palm and hand scanning/geometry, facial scanning, and signature verification.

This chapter identifies various characteristics of biometric methods, describes numerous physiological and behavioral biometric methods, and explains the concepts of verification and identification. Terms and definitions are presented that are used in specific biometric technologies and others that span the entire biometric environment. [1]

The following factors are needed for a successful biometric identification method:

- The physical characteristic should not change over the course of the person's lifetime.
- The physical characteristic must identify the individual person uniquely.
- The physical characteristic needs to be easily scanned or read in the field, preferably with inexpensive equipment, with an immediate result.
- The data must be easily checked against the actual person in a simple, automated way.
- The method must be easy to use by individuals and system operators.
- The willing or knowing participation of the subject is not required.
- Legacy data such as face recognition or voice analysis is used.

Biometric Characteristics

Biometrics is primarily an automated method of recognizing an individual based on a physiological or behavioral characteristic. A **characteristic** is a distinguishing feature or attribute. Biometric technologies are becoming the foundation of an extensive array of highly secure identification and personal verification solutions. The features measured include face, fingerprints, hand geometry, handwriting, iris, retina, vein, and voice. As the level of security breaches and transaction fraud increases, the need for highly secure identification and personal verification technologies is becoming a security requirement for individuals, businesses, and government operations.

Biometric technologies should be considered and evaluated addressing the following characteristics: [2]

- *Acceptability*: The general public must accept the sample collection routines. Nonintrusive methods are more acceptable to this population.
- *Circumvention*: The technology should be difficult to deceive.
- *Collectibility*: The characteristics must be easily collectible and measurable.
- *Performance*: The method must deliver accurate results under varied environmental circumstances.

- *Permanence*: The characteristics should not vary with time. A person's face, for example, may change with age.
- *Universality*: Every person should have the characteristic. Mutes, the visually impaired, or individuals without a fingerprint will need to be accommodated in some way.
- *Uniqueness*: Generally, no two people have identical characteristics. However, identical twins are hard to distinguish.

Physiological vs. Behavioral Characteristics

When referring to a biometric technology, it is important to distinguish between physiological and behavioral human characteristics. A **physiological** characteristic is a relatively stable human physical characteristic, such as a fingerprint, hand silhouette, iris pattern, or blood vessel pattern on the back of the eye. This type of measurement is unchanging and unalterable without significant duress. Alternately, a **behavioral** characteristic is a reflection of an individual's psychological makeup, although physical traits, such as size and gender, have a major influence. Examples of behavioral traits used to identify individuals include a person's typing patterns at a keyboard—commonly referred to as keystroke dynamics—signature dynamics, and the unique characteristics of how one speaks, or speech identification and/or verification.

Identification and Verification Concepts

Identity is the common sense notion of personal identity. Identification is the process of being recognized by some attribute. These attributes include a person's name, personality, physical body, and history, including such details as nationality, educational achievements, employer, security clearances, financial and credit history, etc. In a biometric system, identity is typically established when the person is registered in the system through the use of so-called "**breeder documents**" such as a birth certificate, passport, Social Security number, driver's license, etc. Identity is a subset of physical and/or behavioral characteristics by which an individual is uniquely recognizable. Identity is information concerning the person—not the actual person.

An **identification card (ID)** identifies its holder and issuer, which may carry data required as input for the intended use of the card and for transactions based thereon. As an example, new passports contain an identification feature.

Verification or Identification

Biometric identification is the one-to-many process of comparing a submitted biometric sample against all biometric reference templates on file to determine whether it matches any of the templates and, if so, the identity of the enrollee whose template was matched. A template contains data that represents the biometric measurement of an enrollee, used by a biometric system for comparison against subsequently submitted biometric samples.

Biometric data exists after it has been processed from its original representation into a form that can be used for automated matching purposes. This process involves using a biometric feature extraction algorithm and a biometric matching algorithm. Biometric data stored in a template format cannot be reconstructed into the original output image. The biometric system that uses the one-to-many approach seeks to find an identity among a database rather than authenticate a claimed identity. Note that **one-to-many** is a synonym for identification, whereas **one-to-one** is a synonym for verification.

Identity **verification** is the process of confirming or denying that a claimed identity is correct by comparing the credentials (something known, something possessed, something biological) of a person requesting access with those previously proven and stored in an ID card or system and associated with the identity being claimed. Note the difference between identification and verification.

In verification, an image is matched to only one image in the database (1:1). For example, an image taken of a subject may be matched to an image in a driver's license database to verify the subject portrayed. If identification is the goal, then the image is compared to all images in the database, resulting in a score for each potential match (1:N). In this instance, administrators may take an image and compare it to a database of mug shots to identify the subject.

Sometimes verification and identification are interpreted as similar terms; however, they have two distinct meanings. Identification occurs when an individual's characteristic is being selected from a group of stored images. Identification is the way the human brain performs most day-to-day identifications. For example, if a person encounters a familiar individual, the brain processes the information by comparing what the person is seeing to what is stored in memory. Biometric devices that implement identification techniques can be very time-consuming. Often five to 15 seconds or more are required to identify the appropriate individual. If a reasonable response time is expected, this means large computers with large databases and fast transmission circuits are required.

In many cases, verification is used to authenticate a user's identity. The individual must identify himself by presenting a code or a card. The matching formula or algorithm then needs only to compare the live, enrolled images of the user's characteristic. The question put to the machine is, "Are you who you say you are?" instead of, "Do I know who you are?" Verification can be viewed as adding another level of security. A good analogy is when a person adds a dead bolt to a door to increase the security of the entrance. [3]

There are other performance factors to consider in addition to the terms used to define biometrics, including:

- Accuracy
- Speed
- Reliability
- Acceptability
- Cost
- Resistance to counterfeiting
- Enrollment time
- Database storage requirements
- Intrusiveness

Terms and Definitions

Some of the many terms and definitions used in biometric technologies are described in this section. Most are interrelated. Other terms that are specific to one particular type of technology are included in the section about that type. For example, an *arch* is specific to the fingerprint method, so its definition is found in that section.

Liveness detection is a technique used to ensure that the biometric sample submitted is from an end user. This can help protect the system against some types of spoofing attacks. A *mimic* is the presentation of a live biometric measure in an attempt to fraudulently impersonate someone other than the submitter. *Spoofing* is the ability to fool a biometric sensor into recognizing an illegitimate user as a legitimate user (verification) or into misidentifying someone who is in the database.

Capture or submission is the process of collecting a biometric sample from an individual via a sensor. **Live capture** typically refers to a fingerprint capture device that electronically captures fingerprint images using a sensor (rather than scanning ink-based fingerprint images on a card or lifting latent fingerprints from a surface). A *sensor* consists of the hardware in a biometric device that converts biometric input into electrical signals and conveys this information to the attached computer, such as a fingerprint sensor. A **biometric sample** includes the information or computer data obtained from a biometric sensor device. Examples are images of a face or fingerprint.

Feature(s) are distinctive mathematical characteristic(s) derived from a biometric sample used to generate a reference. **Extraction** or **feature extraction** is the process of converting a captured biometric sample into biometric data so it can be compared to a reference. A template is a digital representation of an individual's distinct characteristics, representing information extracted from a biometric sample. Templates are used during biometric authentication as the basis for comparison. A **model** is a representation used to characterize an individual. Because of their inherently dynamic characteristics, behavioral-based biometric systems use models rather than static templates.

All biometric characteristics depend somewhat on both behavioral and biological characteristics. A behavioral biometric characteristic is learned and acquired over time rather than based primarily on biology. Examples of biometric modalities for which behavioral characteristics may dominate include signature recognition and keystroke dynamics. A biological biometric characteristic is based primarily on an anatomical or physiological characteristic rather than on a learned behavior. Examples of biometric modalities in which biological characteristics may dominate include fingerprint and hand geometry.

Recognition is the generic term used in the description of biometric systems (e.g., face recognition or iris recognition) relating to their fundamental function. The term "recognition" does not inherently imply verification, closed-set identification, or open-set identification. A watch list, sometimes referred to as open-set identification, describes one of the three tasks biometric systems perform. It answers the questions: Is this person in the database? If so, who is she? The biometric system determines if the individual's biometric template matches a biometric template of someone on the watch list. The individual does not make an identity claim, and in some cases does not personally interact with the system whatsoever. An **open-set identification** is a biometric task that more closely follows operational biometric system conditions to (1) determine if someone is in a database and (2) find the record of the individual in the database. This is sometimes referred to as the "watch list" task to differentiate it from the more commonly referenced closed-set identification. A **closed-set identification** is a biometric task where an unidentified individual is known to be in the database and the system attempts to determine his identity. Performance is measured by the frequency with which the individual appears in the system's top rank (or top 5, 10, etc.).

A **comparison** is the process of comparing a biometric reference with a previously stored reference or references in order to make an identification or verification decision.

A match is a decision that a biometric sample and a stored template originates from the same human source, based on their high level of similarity (difference or hamming distance). Relevant terms, discussed later, include false match rate and false non-match rate. An example is a loop fingerprint pattern. Matching, as previously stated, is the process of comparing a biometric sample against a previously stored template and scoring the level of similarity (difference or hamming distance).

A *hamming distance* is the number of noncorresponding digits in a string of binary digits; it is used to measure dissimilarity. Hamming distances are used in many Daugman* iris recognition algorithms. The *difference score* is a value returned by a biometric algorithm indicating the degree of difference between a biometric sample and a reference. The *similarity score* is a value returned by a biometric algorithm that indicates the degree of similarity or correlation between a biometric sample and a reference. [4]

Systems then make decisions based on this score and its relationship to a predetermined threshold (whether it is above or below). A *threshold* is a user setting for biometric systems operating in the verification or open-set identification (watch list) tasks. The acceptance or rejection of biometric data is dependent on the match score falling above or below the threshold. The threshold is adjustable so that the biometric system can be more or less strict, depending on the requirements of any given biometric application. [5]

Biometric Methodologies

Readers will see references to a number of biometric methods in the sections that follow. Some methods are rather impractical, even if they are technically interesting. The general public may reject a particular method for many reasons; these issues will also be addressed. *Note: Methods are described as procedures and techniques characteristic of a particular discipline or field of knowledge.*

Biometrics refers to the automatic identification or identity verification of living persons using their enduring physical or behavioral characteristics, including body parts, personal characteristics, and imaging methods such as these:

- Fingers
- Hands
- Feet

*John Daugman is a physicist and computer-vision expert at the University of Cambridge Computer Laboratory. He is best known for his pioneering work in biometric identification, and in particular the development of the Gabor wavelet-based iris recognition algorithm that is the basis of all commercially available iris recognition systems.

- Faces
- Eyes
- Ears
- Teeth
- Veins
- Voices
- Signatures
- Typing styles
- Gaits
- Odors

A proper assessment is built not only on a general understanding of biometrics, but also on an understanding of specific technologies. An understanding of both biometrics in general and specific biometric technologies is a necessary condition for a solid understanding of the larger social implications of biometrics. Numerous terms will be introduced in this chapter. Understanding the various modalities is a necessity. A **modality** is a type or class of biometric system, such as face recognition, fingerprint recognition, iris recognition, etc. [6]

Physiological identifiers include touch, speech, vision, tactile, and writing elements, such as fingerprint, face, speaker, retina, and iris recognition; hand and finger geometry; and signature verification. The following sections offer details of each of these identifiers.

Sections have been divided according to the part of the body being described for biometric usage. Each section includes information relevant to recognition, verification, and identification of the biometric element. The first section describes the biometrics associated with the finger, palm, and hand. The next section addresses the head, which includes face, iris, retina, and voice components. The last section looks at the biometrics of a signature. Included in another section are other biometric candidates currently under study.

Finger, Palm, and Hand Biometrics

Numerous hardware and software products and systems have been developed that address these methods. (See Appendix A for a list of biometric products.) Additional details are presented in Chapter 3.

Finger Biometrics

Fingerprinting is a highly familiar and well-established biometric science. Fingerprinting has traditionally been used as a forensic criminological technique used to identify perpetrators by the fingerprints they leave behind at crime scenes. Scientists compare a latent sample left at a crime scene against a known sample taken from a suspect. This comparison uses the unique features of any given fingerprint—including its overall shape, and the pattern of ridges and valleys and their bifurcations and terminations—to establish the identity of the perpetrator.

In the context of modern biometrics, these features, called fingerprint minutiae, can be captured, analyzed, and compared electronically, with correlations drawn between a live sample and a reference sample, as with other biometric technologies. Fingerprints offer tremendous invariability: they change only in size with age; they are highly resistant to modification or injury, and they very difficult to forge in any useful way. Although the development of some sort of surreptitious sensor is not inconceivable, in reality, sensors remain obtrusive, requiring a willful finger pressure to gather a useful sample. Unlike other systems based on cameras and high-tech sensors, fingerprint sampling units are compact, rugged, and inexpensive; commercially available systems from multiple vendors offer very good accuracy. Next-generation scanners can analyze below the surface of the skin, and can add pore-pattern recognition in addition to the more obvious minutia of the fingerprint.

Fingerprint recognition is a biometric modality that uses the physical structure of an individual's fingerprint for identification purposes. Important features used in most fingerprint recognition systems are minutiae points that include bifurcations and ridge endings. A **bifurcation** is the point in a fingerprint where a friction ridge divides or splits to form two ridges.

Friction ridges are the ridges present on the skin of the fingers, toes, palms, and soles of the feet which make contact with an incident surface under normal touch. On the fingers, distinctive patterns are formed by the friction ridges that make up the fingerprints. A *ridge ending* is a minutiae point at the ending of a friction ridge.

Minutiae points are friction ridge characteristics that are used to individualize a fingerprint image. Minutiae are the points where friction ridges begin, terminate, or split into two or more ridges. In many fingerprint systems, the minutiae (as opposed to the images) are compared for recognition purposes.

An *arch* fingerprint pattern occurs when the friction ridges enter from one side, make a rise in the center, and exit on the opposite side. The pattern will contain no true delta point. A *loop* fingerprint pattern occurs when the friction ridges enter from either side, curve sharply and pass out near the same side they entered as illustrated below. This pattern will contain one core and one delta. A *whorl* fingerprint pattern occurs when the ridges are circular or nearly circular, as illustrated below. The pattern will contain two or more deltas.

The *delta point* is part of a fingerprint pattern that looks similar to the Greek letter delta. Technically, it is the point on a friction ridge that is at or nearest to the point of divergence of two type lines, and is located at or directly in front of the point of divergence. The *core point* is the "center(s)" of a fingerprint. In a whorl pattern, the core point is found in the middle of the spiral/circles. In a loop pattern, the core point is found in the top region of the innermost loop. Speaking technically, a core point is defined as the topmost point on the innermost upwardly curving friction ridgeline. A fingerprint may have multiple cores or no cores.

Among the different biometrics that can be employed to authenticate an individual, fingerprint recognition is the most widely implemented. This is because of the following factors:

- Easily separates an individual from another
- Best resists aging
- Possesses a high degree of robustness, speed, and accuracy in capturing the fingerprint
- Extremely hard to fool a system when using fingerprint biometrics

Fingerprint Identification

A **latent fingerprint** is a fingerprint "image" left on a surface that was touched by an individual. The transferred impression is left by the surface contact with the friction ridges, usually caused by the oily residues produced by the sweat glands in the finger.

Slap fingerprints are those taken by simultaneously pressing the four fingers of one hand onto a scanner or a fingerprint card. "Slaps" are known as four-finger, simultaneous plain impressions. *Rolled fingerprints* obtain an image that includes fingerprint data from nail to nail by "rolling" the finger across a sensor.

A paper form called a **10-print card** is used to collect both an individual's personal and demographic information along with flat and rolled ink impression fingerprint images. It is mainly used in conjunction with an automated fingerprint identification system (AFIS). A *10-print match* or identification is an absolute positive identification of an individual by matching all 10 fingerprints to those in a system of record. This is usually performed by an AFIS system and verified by a human fingerprint examiner.

The patterns of friction ridges and valleys on an individual's fingertips are unique to that individual. Friction ridges occur on the skin of the fingers, toes, palms, and soles of the feet that make contact with an incident surface under normal touch. On the fingers, the unique patterns formed by friction ridges make up fingerprints. For decades, law enforcement has been classifying and determining identity by matching key points of ridge endings and bifurcations. Fingerprints are unique for each finger of a person, including identical twins.

On the palmar surface of the hands and feet are raised surfaces called friction ridges. The scientific basis behind friction ridge analysis is the fact that friction ridges are persistent and unique. Friction ridges are formed during fetal development, when their unique characteristics emerge due to genetic and epigenetic factors. Even identical twins do not have the same fingerprints. Uniqueness among even identical twins is due to random, or stochastic, effects during fetal development. *Stochastic* effects have widespread scientific acceptance as a source of uniqueness and have been observed in several animal studies. Such studies examined fingerprint and other unique traits, such as hair patterning. Friction ridges also persist throughout life in their permanent arrangement barring scarring or injury or until decomposition of the skin following death. Scarring occurs from damage to the basal layer of the epidermis. Like friction ridges, scars are also persistent throughout life and are regenerated in new layers of skin.

A known print is the intentional recording of the friction ridges, usually with black printer's ink rolled across a contrasting white background, typically a white card. Friction ridges can also be recorded digitally using a technique called live scan. A **live scan** occurs when a fingerprint or palm print is taken directly from a subject's hand. **Live capture** is the process of capturing a biometric sample by an interaction between an end user and a biometric system. A latent print is the chance reproduction of the friction ridges deposited on the surface of an item. Latent prints are often fragmentary and may require chemical methods, powder, or alternative light sources in order to be visualized.

When friction ridges come in contact with a surface that is receptive to a print, material on the ridges—such as perspiration, oil, grease, or ink—can be transferred to the item. The factors which affect friction ridge impressions are numerous, which means competent examiners must undergo extensive and objective study in their training. Pliability of the skin, deposition pressure, slippage, the matrix, the surface, and the development medium are just some of the various

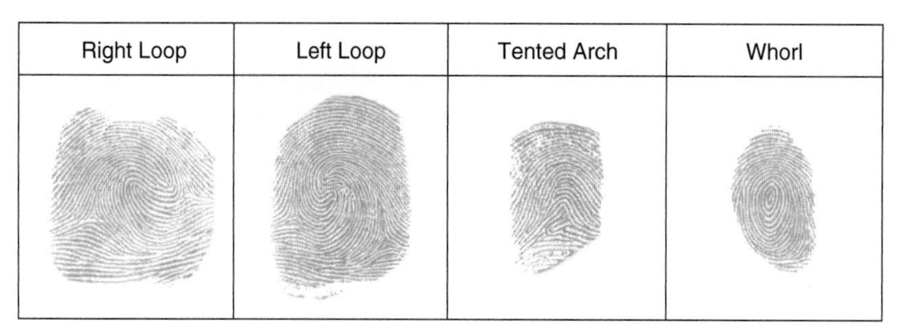

Figure 2-1 Fingerprint Categories

Course Technology/Cengage Learning

factors which can cause a latent print to appear differently from the known recording of the same friction ridges. The conditions of friction ridge deposition are unique and never duplicated.

Fingerprints remain constant throughout life. In over 140 years of fingerprint comparison worldwide, no two fingerprints have ever been found to be alike, including those of identical twins. Scanner technology is also technically simple to implement: good fingerprint scanners have been installed in PDAs like the iPaq Pocket PC. This method might not work in industrial applications since it requires clean hands. Fingerprint identification involves comparing the pattern of ridges and furrows on the fingertips, as well as the minutiae points of a specimen print with a database of prints on file. *Remember, minutiae are friction ridge characteristics that are used to individualize a fingerprint image; minutiae points are the ridge characteristics that occur when a ridge splits into two or ends.* Figure 2-1 shows four categories of fingerprints.

There are several ways to identify a fingerprint. The most common method involves recording and comparing the fingerprint's minutiae points. Minutiae points can be defined as the uniqueness of an individual's fingerprint. They are called minutiae "points" because the fingerprint scanner assigns locations (points) to the minutiae using location and directional variables. Minutiae points are and can be made up of the following characteristics:

- Bifurcation: The point at which a ridge splits into multiple ridges, called branches
- Divergence: The point where parallel ridges either spread apart or come together
- Enclosure: Where a ridge splits into two branches and then comes together again shortly thereafter
- Ending: Where a ridge terminates
- Valley: Spaces or gaps on either side of a ridge

Identification by fingerprints relies on pattern matching followed by the detection of certain ridge characteristics—also known as Galton details*, points of identity, or minutiae—and the comparison of the relative positions of these minutiae points with a reference print, usually an inked impression of a suspect's print.

There are three basic ridge characteristics, the ridge ending, the bifurcation, and the dot (or island). Figure 2-2 depicts these ridge characteristics. Bifurcation is the point in a fingerprint where a ridge divides or splits to form two ridges that continue past the point of division

*Galton details refer to friction ridge characteristics attributed to the research of English fingerprint pioneer, Sir Francis Galton.

Island
Dot
Bifurcation
Ending Ridge

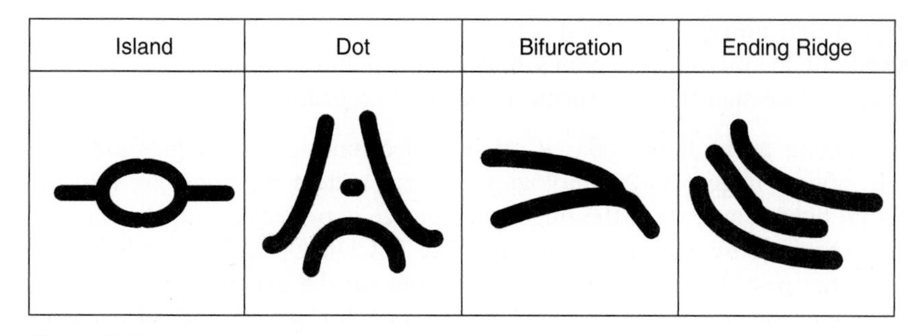

Figure 2-2 Ridge Characteristics

Course Technology/Cengage Learning

Basic and Composite Ridge Characteristics (Minutiae)
Minutiae
Example
Minutiae
Example
Ridge Ending
Bridge
Bifurcation
Double Bifurcation
Dot
Trifurcation
Island (Short Ridge)
Opposed Bifurcations
Lake (Enclosure)
Ridge Crossing
Hook (Spur)
Opposed Bifurcation/ Ridge Ending

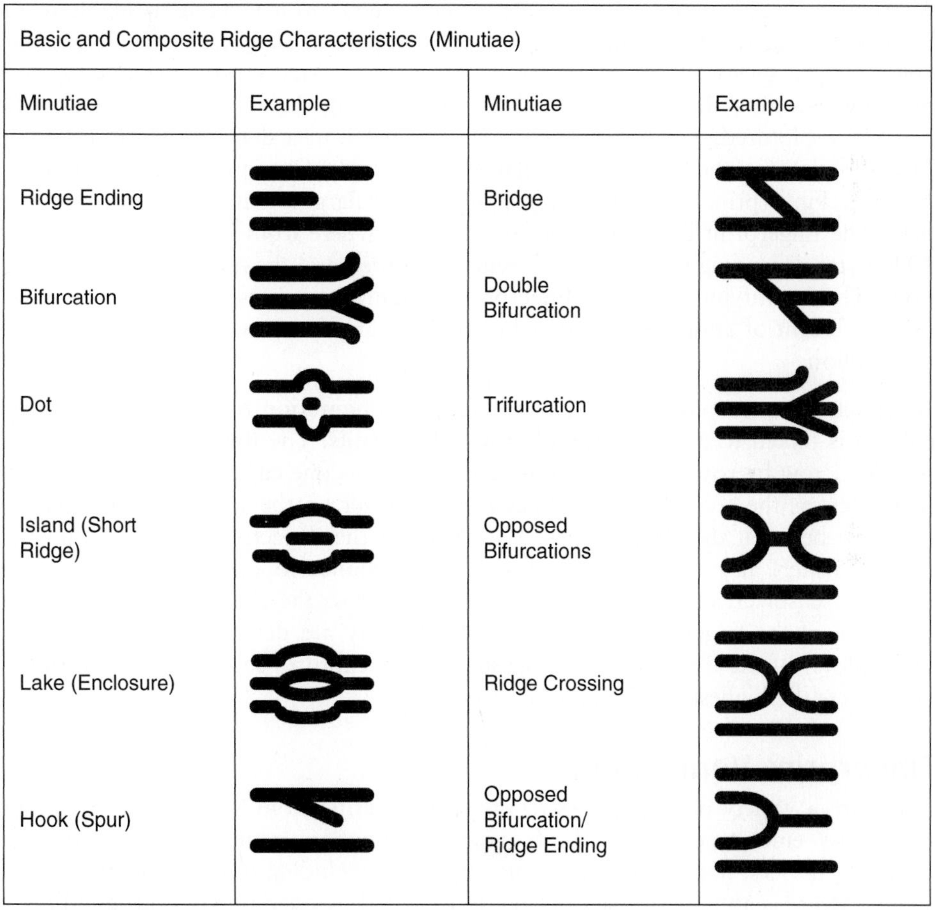

Figure 2-3 Minutiae Possibilities

Course Technology/Cengage Learning

for a distance that is at least equal to the spacing between adjacent ridges at the point of bifurcation. Figure 2-3 describes minutiae possibilities.

In a typical fingerprint that has been scanned by a fingerprint identification system, there are generally between 30 and 40 minutiae. Furthermore, the Federal Bureau of Investigation

(FBI) has found that no two individuals can have more than eight common minutiae points. Other methods of identifying a person's fingerprint include counting the number of ridges between points, processing the fingerprint image, and recording the print's sound waves.

Fingerprint imaging technology is based on two electronic capturing methods: optical and capacitive. In optical fingerprint technologies, the user places the finger on a glass substrate and an internal light source from the fingerprint device is projected onto the fingerprint. The image is then captured by a charge-coupled device (CCD). Optical methods have been used extensively for the past decade. They are proven, but on the expensive side, and not always reliable due to environmental conditions. A buildup of dirt, grime, and oil from one's finger can leave a "ghost" image, which is referred to as a "latent image." As a result, their use has been confined to specific criminal justice and military installations.

Alternatively, capacitive imaging looks to make fingerprint imaging available to the masses by making fingerprint imaging devices more compact in size, less expensive, and more reliable. Using a sensor chip and an array of circuits, capacitive systems analyze one's fingerprint by detecting the electrical field around the fingerprint. When a person's fingerprint is initially captured, a template is created and stored in a data storage system or database. This template is then compared against a person's fingerprint each time the finger is scanned. Fingerprint imaging requires one of the largest data templates in the biometric field. The fingerprint data template can range anywhere from several hundred bytes to over 1000 bytes depending on the level of security required and the method used to scan the fingerprint. The identifying power of fingerprint imaging systems shows that they tend to reject over 3 percent of authorized users, while maintaining false acceptance rates of less than one in a million.

A **capacitive sensor** consists of a row/column configuration of tiny metal electrodes. Every column is linked to a pair of sample-and-hold circuits. The fingerprint image is recorded in sequence, row by row, as each metal electrode acts as one capacitor plate and the contacting finger acts as the second plate. A "passivation" layer on the surface of the device forms the dielectric between these two plates. Variations in the dielectric between a fingerprint ridge (mainly water) and a valley (air) cause the capacitance to vary locally. Pressing the finger onto the sensor creates varying capacitor values across the array that is then converted into an image of the fingerprint. The values of the array are determined by the contour (ridges and valleys) of the fingerprint. The sensor quickly captures several images of the fingerprint and selects the highest-quality image.

Fingerprint Verification

There are a variety of approaches to fingerprint verification. Some try to emulate the traditional law enforcement method of matching minutiae, while others are straight pattern matching devices. Some adopt a unique approach, including *moiré fringe* patterns and ultrasonics. Moiré is an undesirable optical effect created by overlapping grids and lines due to an undersampling of the image data. Ultrasonics can be applied to thickness, density, flow, and level sensing, which are also used for imaging. Other methods can detect when a live finger is presented. Today, there is a greater variety of fingerprint devices available than there is for any other biometric. Devices are potentially capable of good accuracy (low instances of false acceptance), but can suffer from usage errors among insufficiently disciplined users (higher instances of false rejection). This situation may arise with large user bases.

False acceptance and false rejection are two concepts that must be addressed. When a biometric system incorrectly identifies an individual or incorrectly authenticates an imposter against a claimed identity, a false acceptance situation occurs. The *false acceptance rate (FAR)* includes the probability that a biometric system will incorrectly identify an individual or will fail to reject an imposter.

A false rejection occurs when a biometric system fails to identify an enrollee or fails to verify the legitimate claimed identity of an enrollee. The *false rejection rate (FRR)* is the probability that a biometric system will fail to identify an enrollee, or verify the legitimate claimed identity of an enrollee. *False non-match rate (FNMR)* is an alternative term for false rejection rate. These processes are used in the context of a negative claim of identity. These integrity concepts will be presented in Chapter 9.

One must also consider the transducer–user interface and how this would be affected by large-scale usage in a variety of environments. *Note: A transducer converts energy from one system to another.* Fingerprint verification may be a good choice for in-house systems where adequate explanation and training are provided to users and where the system is operated within a controlled environment. It is not surprising that the workstation access application area seems to be based almost exclusively on fingerprints. This is primarily due to the relatively low cost, small size, and ease of integration.

Autocorrelation is a proprietary finger scanning technique. Two identical finger images are overlaid in the autocorrelation process, so that light and dark areas, known as moiré fringes, are created. The *automated fingerprint identification system (AFIS)* is an example of a highly specialized biometric system that compares a single finger image with a database of finger images. AFIS is predominantly used for law enforcement, but is also used in civil applications. AFIS will be described in Chapter 12. In law enforcement, finger images are typically collected from crime scenes or directly from criminal suspects after they are arrested. These fingerprints are known as "latents." In civilian applications, finger images may be captured by placing a finger on a scanner or by electronically scanning inked impressions on paper.

Hand and Palm Biometrics

Palm print recognition is a biometric modality that uses the physical structure of an individual's palm print for recognition purposes; *hand geometry recognition* uses the physical structure of an individual's hand for recognition purposes.

These methods of personal authentication are well established. Hand recognition technology has been available for over 20 years. To achieve personal authentication, a system may measure either physical characteristics of the fingers or the hands. These include length, width, thickness, and surface area of the hand. One interesting characteristic is that some systems require a small biometric sample. Hand geometry has gained acceptance in a range of applications. It can frequently be found in physical access control in commercial and residential applications, in time and attendance systems, and in general personal authentication applications. Figure 2-4 provides an example of hand geometry measurements.

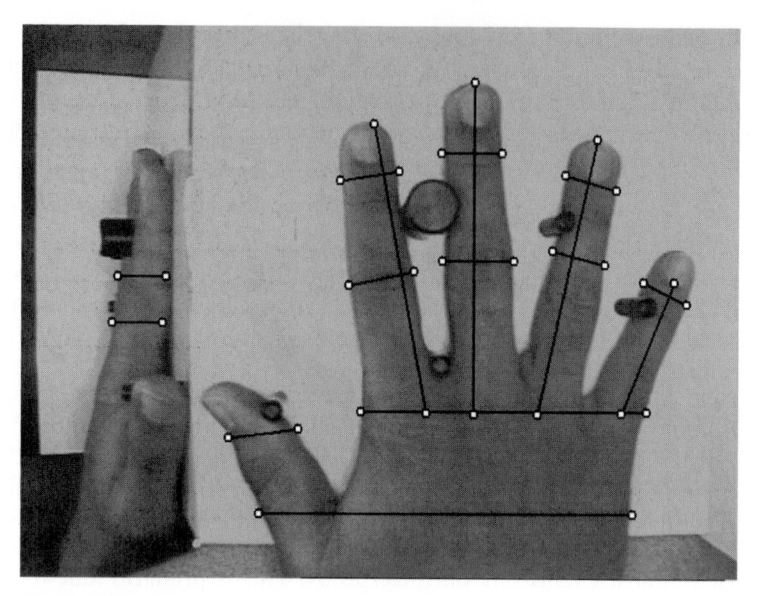

Figure 2-4 Hand Geometry Measurements

Course Technology/Cengage Learning

Hand Geometry

Perhaps the most ubiquitous electronic biometric systems are hand-geometry-based. Hand-geometry-based systems require the subject to place the hand (usually the right hand) on a plate, where it is photographically captured and measured. The human hand consists of 27 bones and a complex web of interconnected joints, muscles, and tendons. It presents a sufficiently peculiar conformation of anatomical features to enable authentication, but is not considered sufficiently unique to provide full identification. Further, the geometry of the hand is variable over time, as hand shape may be altered due to injury, disease, aging, or dramatic weight swings. A simple hand geometry system will measure length and thickness of digits, width of the palm at various points, and the radius of the palm. This results in a relatively simple identification that can be expressed in a very simple, compact string of data. Efforts have been made to improve the accuracy of hand geometry, including three-dimensional sampling (i.e., a second camera measuring the thickness of the hand from the side), and a patented system that measures the pattern of the veins of the hand. Providers have claimed that their systems provide a high degree of accuracy, and that the hand vein feature is unique and relatively invariable, changing little over a person's lifespan.

Traditional hand geometry systems have found acceptance in applications requiring verification of an identity, rather than full proof or establishment of an identity. Airports, prisons, and factories have successfully employed hand-geometry-based systems to restrict access to runways, to prevent walkout escapes during visits, and to ensure that time cards are being punched only by the worker, and not by that worker's pal on their behalf. In all these instances, the subject is attempting to prove or disprove his membership in a relatively small group of people, such as authorized runway personnel, prisoners/visiting family, and factory workers. When stakes are high, these systems are not relied on exclusively to confirm identity; rather, they are used to provide an additional layer of security above and beyond that provided by existing security systems.

Since they must accommodate the largest of hands, any hand geometry or hand vein system must be somewhat bulky, and requires the user to perform the obtrusive task of placing the hand on the platen for sampling. Because of this, hand-based biometrics represents less of a privacy threat than some other systems: subjects cannot have their biometric features sampled without their knowledge, and the sampling method is unambiguous in its intent.

Hand Geometry Readers

Hand geometry is concerned with measuring the physical characteristics of the individual's hand and fingers from a three-dimensional perspective. One of the most established methodologies, hand geometry, offers a good balance of performance characteristics and is relatively easy to use. This methodology may be suitable for larger user bases or users who access the system infrequently and may, therefore, be less disciplined in their approach to the system. Accuracy can be very high, while flexible performance tuning and configuration can accommodate a wide range of applications. Hand geometry readers are deployed in a wide range of scenarios, including time and attendance recording, where they have proved to be extremely popular. Ease of integration into other systems and processes, coupled with ease of use, make hand geometry an obvious first step for many biometric projects.

Hand geometry readers work in harsh environments, do not require clean conditions, and form a very small dataset. It is not regarded as an intrusive kind of test and is often the authentication method of choice in industrial environments.

Face, Iris, Retina, and Voice Biometrics

Certain biometric methods involve the head, including the face, the eye, and the mouth. The human eye offers two features with excellent properties for identification. Both the iris (the colored part visible at the front of the eye) and the veins of the retina (the thin film of nerve endings inside the eyeball that capture light and send it back to the brain) provide patterns that can uniquely identify an individual.

Iris Biometrics

Iris scanning is a method of biometric identification and pattern recognition that is used to determine the identity of the subject. Iris scans create high-resolution images of the irides of the eye; infrared (IR) illumination is used to reduce specular reflection from the cornea. Irides (the plural of iris) are the pigmented, round, contractile membranes of the eye situated between the cornea and the lens, and perforated by the pupil. The iris itself is a good subject for biometric identification because it is an internal organ that is well protected, is mostly flat, and has a fine texture that is unique even for identical twins. Iris scans can be done regardless of whether the subject is wearing contact lenses or glasses. However, the system must take eyelids and eyelashes into account, as both can obscure the necessary parts of the eye and cause false information to be added into automated systems. Iris scans are extremely accurate.

Iris scanning is undoubtedly the least intrusive of the eye-related biometrics. It uses a fairly conventional *charge-coupled device (CCD)* camera element and requires no intimate contact between the user and reader. It also has the potential for higher-than-average template matching performance. As a technology, it has attracted the attention of various third-party integrators, and one would expect to see additional products launched as a result. It has been demonstrated to work with vision enhancers in place and with a variety of ethnic groups. It is also one of the few devices which can work well in identification mode. Ease of use and system integration have not traditionally been strong points with iris scanning devices, but improvements can be expected in these areas as new products are introduced.

Like a retinal scan, an iris scan also provides unique biometric data that is very difficult to duplicate and remains the same for a lifetime. The scan is similarly difficult to make and may be difficult to perform on children or the infirm. However, there are ways of encoding the iris scan biometric data in a way that it can be transported securely in a "barcode" format.

Iris recognition is a biometric modality that uses an image of the physical structure of an individual's iris for recognition purposes. The iris muscle is the colored portion of the eye surrounding the pupil. It is a method of biometric authentication that uses pattern recognition techniques based on high-resolution images of the irides of an individual's eyes. Not to be confused with retina scanning, which is another, less-prevalent ocular-based technology, iris recognition uses camera technology, and subtle infrared illumination to reduce specular reflection from the convex cornea to create images of the detail-rich, intricate structures of the iris. When these unique structures are converted into digital templates, they provide mathematical representations of the iris that yield an unambiguous positive identification of an individual.

Iris recognition efficacy is rarely impeded by glasses or contact lenses. Iris technology has the smallest outlier (those who cannot use/enroll) group of all biometric technologies.

It is the only biometric authentication technology designed for use in a one-to-many search environment. A key advantage of iris recognition is its stability—or template longevity—as, barring trauma, a single enrollment can last a lifetime.

This recognition method uses the iris of the eye, which is the colored area that surrounds the pupil (Figure 2-5). Iris patterns are considered to be unique and are obtained through a

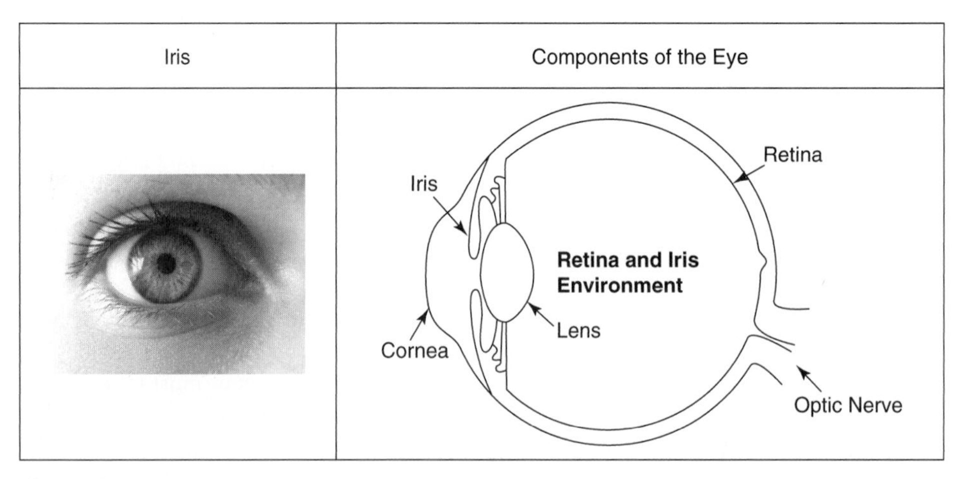

Figure 2-5 Components of the Eye and the Iris

video-based image acquisition system. Iris scanning devices have been used in personal authentication applications for several years. Systems based on iris recognition have substantially decreased in price; this trend is expected to continue. The technology works well in both verification and identification modes where systems perform one-to-many searches in a database.

Current systems can be used even in the presence of eyeglasses and contact lenses. The technology is not intrusive, as it does not require physical contact with a scanner. Iris recognition has been demonstrated to work with individuals from different ethnic groups and nationalities. Iris recognition is a biometric identification technology that uses high-resolution images of the irides of the eye. The iris of the eye is well suited for authentication purposes as it is an internal organ that is protected from most damage and wear.

Iris recognition is accomplished by applying proprietary algorithms for image acquisition and subsequent one-to-many matching. Iris recognition algorithms produce remarkable results. Daugman's algorithms have produced accuracy rates in authentication that are better than those of any other method. IrisCode, a commercial system derived from Daugman's work, has been used in the United Arab Emirates as a part of their immigration process. After more than 200 billion comparisons, there has never been a false match. [7]

Iris recognition is easier to do than retina scanning, in that it merely requires the subject to look at a camera from a distance of three to 10 inches. The iris scanner illuminates the iris with invisible infrared light, which shows details on darker-colored eyes that are not visible to the naked eye. The pattern of lines and colors on the eye are, as with other biometrics, analyzed, digitized, and compared against a reference sample for verification.

Iridian Technologies, who holds the patents on iris recognition, claims that the iris is the most accurate and invariable of biometrics, and that its system is the most accurate form of biometric technology. Iridian's system also has the benefit of extremely swift comparisons. The company claims it can match an iris against a database of 100,000 reference samples in two to three seconds, whereas a fingerprint search against a comparable database might take 15 minutes.

Retina Biometrics

The human retina is a thin tissue composed of neural cells located in the posterior portion of the eye. The blood vessels within the retina absorb light more readily than the surrounding tissue, and are easily identified with appropriate lighting. A retinal scan is performed by casting an undetectable ray of low-energy infrared light into a person's eye as she looks through the scanner's eyepiece. This beam of light outlines a circular path on the retina. Because retinal blood vessels are more sensitive to light than the rest of the eye, the amount of reflection fluctuates. The results of the scan are converted to computer code and stored in a database.

Figure 2-6 show the various components of the eye and a retina. Because of the complex structure of the capillaries that supply the retina with blood, each person's retina is unique. The network of blood vessels in the retina is so complex that identical twins do not even share a similar pattern. Although retinal patterns may be altered in cases of diabetes, glaucoma, retinal degenerative disorders or cataracts, the retina typically remains unchanged from birth until death. Due to its unique and unchanging nature, the retina appears to be the most precise and reliable biometric.

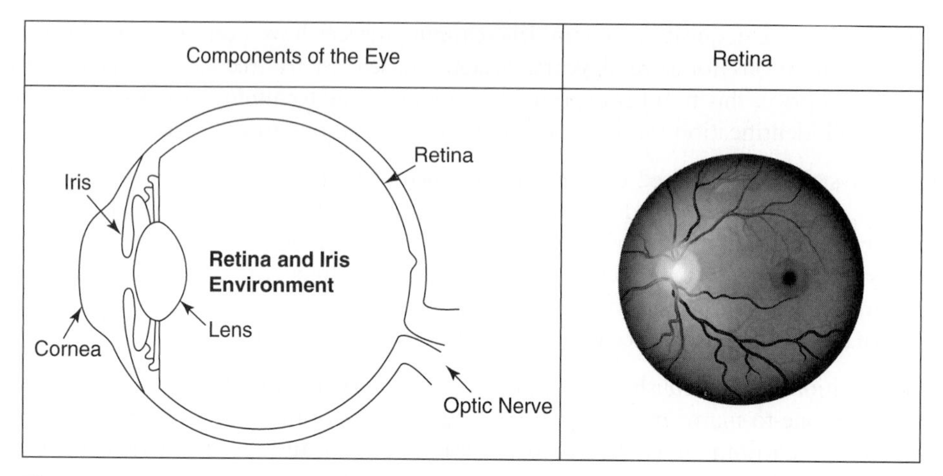

Figure 2-6 Components of the Eye and Retina

Course Technology/Cengage Learning

Retinal Scanning

Retinal scanning requires the subject to look into a reticle and focus on a visible target while the scan is completed. A reticle consists of a glass plate that contains a pattern of transparent and opaque areas. It contains the pattern for one or more die, but is not large enough to transfer a wafer-sized pattern all at once. Retina scans require that the person remove any glasses, place his eye close to the scanner, stare at a specific point, remain still, and focus on a specified location for approximately 10 to 15 seconds while the scan is completed. A retinal scan involves the use of a low-intensity coherent light source, which is projected onto the retina to illuminate the blood vessels, which are then photographed and analyzed. An optical coupler is used to read the blood vessel patterns.

This is definitely one of the more intrusive biometric technologies, with some subjects reporting discomfort in the scanning process. A retina scan cannot be faked as it is currently impossible to forge a human retina. Furthermore, the retina of a deceased person decays too rapidly to be used to deceive a retinal scan. A retinal scan has an error rate of 1 in 10,000,000, compared to fingerprint identification error which can be as high as 1 in 500. The pattern of the blood vessels at the back of the eye is supposedly unique and remains the same for a lifetime. Retinal scanning remains a standard in military and government installations.

This method has proved to be quite accurate in use, but does require the user to look into a receptacle and focus on a given point. This is not particularly convenient if vision enhancement is used or if the individual has concerns about intimate contact with the reading device. For these reasons, retinal scanning has user acceptance problems, although the technology can function well. The leading product was redesigned in the mid-1990s, providing enhanced connectivity and an improved user interface; however, this is still a relatively marginal biometric technology.

Retina Recognition

Retinal recognition technology is used to measure the unique configuration of blood vessels contained in the retina. A low-intensity laser light source is used to illuminate the retina.

A 360-degree circular scan is taken while approximately 400 readings are captured. Up to 192 reference points are identified and reduced to a 96 byte template.

Biometric retina recognition systems are among the most accurate of all biometric technologies and are used at military installations and other high-risk facilities. They are also quite expensive due to the hardware required. Precautions are not needed to prevent fooling the system because it is not possible to make a "fake" retina, and a retina from a cadaver would deteriorate too fast.

Retina scan–based *personal identification (PI)* exploits the uniqueness of the vascular pattern of the retina in defining the biometric. To date, the only vendor producing retina scan systems is *EyeDentify, Inc.* A retina scanner illuminates, through the pupil, an annular region of the retina with infrared (IR) light and records the reflected vasculature contrast information. Retina scanning is considered an exceptionally accurate and invulnerable biometric technology and is established as an effective solution for very high security environments. Even so, the technology is difficult to use because it requires well-trained, cooperative, and patient users for proper performance to be achieved. Because of this, and the fact that many people shy away from devices that interact with their eyes, retina scanning is thought to be fairly intrusive.

Face Biometrics

Face biometrics includes sections that explain face recognition and face verification techniques.

Face Recognition

Face recognition technology uses an image of the visible physical structure of an individual's face for recognition purposes. The identification of a person by her facial image can be done in a number of different ways, such as by capturing an image of the face in the visible spectrum using an inexpensive camera or by using the infrared patterns of facial heat emission. Facial recognition in visible light typically models key features from the central portion of a facial image. Using a wide assortment of cameras, the visible light systems extract features from the captured image(s) that do not change over time while avoiding superficial features such as facial expressions or hair. Several approaches to modeling facial images in the visible spectrum are:

- Principal component analysis
- Local feature analysis
- neural networks
- elastic graph theory
- multiresolution analysis

Some of the challenges of facial recognition in the visual spectrum include reducing the impact of variable lighting and detecting a mask or photograph. Some facial recognition systems may require a stationary or posed user in order to capture the image, though many systems use a real-time process to detect a person's head and locate the face automatically. Major benefits of facial recognition are that it is nonintrusive, hands-free, continuous, and accepted by most users.

Of the various biometric identification methods, face recognition is one of the most flexible, working even when the subject is unaware of being scanned. It also shows promise as a way to search through masses of people who spent only seconds in front of a "scanner" that can be an ordinary digital camera. Face recognition systems work by systematically analyzing specific features that are common to everyone's face.

- Distance between the eyes
- Width of the nose
- Position of cheekbones
- Jaw line
- Chin

These numerical quantities are then combined in a single code that uniquely identifies each person.

Facial characteristics include the size and shape of facial characteristics and their relationship to each other. Although this method is the one that human beings have always used with each other, it is not easy to automate. Typically, this method uses relative distances between common landmarks on the face to generate a unique "face print."

Facial recognition is a technique which has attracted considerable interest and whose capabilities have often been misunderstood. Extravagant claims have sometimes been made for facial recognition devices which have been difficult, if not impossible, to substantiate in practice. It is one thing to match two static images and another to unobtrusively detect and verify the identity of an individual within a group. It is easy to understand the attractiveness of facial recognition from the user perspective, but one needs to be realistic in expectations of the technology. To date, facial recognition systems have had limited success in practical applications. Progress, however, continues to be made in this area and it will be interesting to see how future implementations perform. If technical obstacles can be overcome, facial recognition may eventually become a primary biometric methodology.

Facial recognition remains one of the more controversial biometric technologies because of its unobtrusiveness. With good cameras and good lighting, a facial recognition system can sample faces from tremendous distances without the subject's knowledge or consent.

Most facial recognition technology works by one of two methods: facial geometry or eigenface comparison. *Facial geometry analysis* works by taking a known reference point, such as the distance from eye to eye, and measuring the various features of the face in their distance and angles from this reference point. **Eigenface comparison** uses a palette of about 150 facial abstractions, and compares the captured face with these archetypal abstract faces. In laboratory settings, facial recognition results are excellent, but critics have questioned the effectiveness of the technology in real-world circumstances. Nevertheless, the accuracy of facial recognition has been good enough for casinos to have put the technology to use since the late 1990s as a means to spot banned players. Facial recognition technology proponents claim good performance even against disguises, weight changes, aging, or changes in hairstyle or facial hair.

Face Verification

Every face has numerous, distinguishable landmarks, the different peaks and valleys that make up facial features. FaceIt® defines these landmarks as **nodal points**. Each human face has approximately 80 nodal points. Some of these measured by the software are:

- Distance between the eyes
- Width of the nose
- Depth of the eye sockets
- The shape of the cheekbones
- The length of the jaw line

These nodal points are measured, creating a numerical code called a **faceprint**, which represents a face in the database. Using 3D software, the system goes through a series of steps to verify the identity of an individual. FaceIt®, described in Chapter 3, uses five steps to accomplish the facial recognition process. [8]

- Step 1: Detection: Acquiring an image by digitally scanning an existing photograph
- Step 2: Alignment: System determines the head's position, size, and pose
- Step 3: Measurement: System measures the curves of the face and creates a template
- Step 4: Representation: System translates the template into a unique code
- Step 5: Matching: Matching takes place without any changes being made to the image

Voice Biometrics

Like face recognition, voice biometrics provides a way to authenticate identity without the subject's knowledge. It is easier to fake a voice using a tape recording; however, it is not possible to fool an analyst by imitating another person's voice. Voice recognition involves the analysis of the pitch, tone, cadence, and frequency of a person's voice.

Speech/Voice Recognition Speaker recognition or *voice recognition* is a biometric modality that uses an individual's speech—a feature influenced by both the physical structure of an individual's vocal tract and the behavioral characteristics of the individual—for recognition purposes. Speech recognition recognizes the words being said, and is not a biometric technology; it is a technology that enables a machine to recognize spoken words.

Speaker recognition has a history dating back four decades, where the output of several analog filters was averaged over time for matching. Speaker recognition uses the acoustic features of speech that have been found to differ between individuals. These acoustic patterns reflect anatomy, which includes the size and shape of the throat and mouth, and learned behavioral patterns, which includes voice pitch and speaking style. This incorporation of learned patterns into the voice templates (voiceprints) has earned speaker recognition its classification as a "behavioral biometric." Speaker recognition systems employ three styles of spoken input: text dependent, text prompted and text independent. Most speaker verification applications use text-dependent input, which involves the selection and enrollment of one or more voice passwords. Text-prompted input is used whenever there is concern of imposters. The various technologies used to process and store voiceprints include:

- Hidden Markov models
- Pattern matching algorithms
- Neural networks
- Matrix representation
- Decision trees

Some systems also use "antispeaker" techniques, such as cohort models and world models.

Ambient noise levels can impede the collection of the initial and subsequent voice samples. Performance degradation can result from changes in behavioral attributes of the voice and from enrollment using one telephone and verification on another telephone. Voice changes due to aging also need to be addressed by recognition systems. Many companies market speaker recognition engines, often as part of large voice processing, control, and switching systems. Capture of the biometric is seen as noninvasive. The technology needs little additional hardware by using existing microphones and voice-transmission technology that allows recognition over long distances via ordinary wired or wireless telephones.

The difference in vocal cord size between men and women means they have differently pitched voices. Genetics also causes variances among the same sex, with men and women's singing voices being categorized into types. For example, among men, there are basses, baritones, and tenors, and altos, mezzo-sopranos, and sopranos among women. There are additional categories for operatic voices.

In most cases, the human voice can be a complex instrument. Humans have vocal folds which can loosen or tighten or change their thickness and over which breath can be transferred at varying pressures. The shape of chest and neck, the position of the tongue, and the tightness of otherwise unrelated muscles can be altered. Any one of these actions results in a change in pitch, volume, timbre, or tone of the sound produced. One important categorization that can be applied to the sounds singers make relates to the register or the "voice" that is used. Singers refer to these registers according to the part of the body in which the sound most generally resonates, and which have correspondingly different tonal qualities. There are widely differing opinions and theories about what a register is, how they are produced and how many there are. The distinct change or break between registers is called a *passaggio*, a term used in classical singing to describe the pitch ranges in which vocal registration events occur.

Nearly all untrained voices have noticeable differences in timbre at different pitches. This is most easily seen in males, who are generally aware that they can speak in both their usual voice, which sounds "male," and in another, higher voice that is lighter, breathier, and sounds "female." Each of these timbres is referred to as a "register" (hence "vocal registration").

Beyond these facts, it should be noted that registration is somewhat controversial, with disagreement on how many registers there are and indeed whether they exist at all. Further complication arises from the fact that registration is, broadly, unique to a particular person and is to some extent subjective—a head note for a bass is a chest note for a soprano.

But voice recognition technology is still not good enough to be used as a front-line biometric technology. Voice verification systems have to account for a lot more variables than do other systems, starting with the inevitable compression of a voice captured by cheap microphones and phone handsets, discriminating a voice from background noise and other sonic artifacts, and the tremendous variability of the human voice due to colds, aging, and simple tiredness. Also, just as a voice can be surreptitiously recorded over the telephone or face-to-face, a person's voice can be captured surreptitiously by a third party, either by tapping or bugging, and replayed, or might be biometrically sampled remotely without consent during a fake door-to-door or telephone sales call. Because of these difficulties, commercial deployments of voice verification have been limited to backup status, systems in which there are other token-based methods of identification, with voice verification providing an added layer of protection.

Voice Verification

It is very difficult for computers to positively identify someone. The prospect of accurate voice verification offers one great advantage: it would allow a remote identification using the phone system, an infrastructure that's already been built and thus has zero client-side cost, as no special reader needs to be installed in a home. Even without the phone system, the sampling apparatus, a microphone, remains far cheaper than competing, largely optically based biometric technologies.

Voice verification is a potentially interesting technique because considerable voice communication takes place with regard to everyday business transactions. Some designs have concentrated on wall-mounted readers, while others have sought to integrate voice verification into conventional telephone handsets. A number of voice verification products have been introduced into the market, however many have not been successful. This is because of the variability of both transducers and local acoustics. In addition, the enrollment procedure has often been more complicated than with other biometric methods. This has led to the perception that voice verification is unfriendly in some quarters. However, much work has been and continues to be undertaken in this context and it will be useful to monitor progress.

Other biometric methodologies include the use of scent, ear lobes, and various other parameters. While these may be technically interesting, at this stage they are not considered to be workable solutions in practical applications. The methodologies outlined above represent the majority interest, and are be a good starting place for designers to consider within a biometric project.

Signature Biometrics

The biometric most familiar to citizens is the signature. The ability to judge by sight if one signature matches another has made this a time-proven and legally binding biometric. However, most people cannot recognize the pressure of the pen on the paper or the speed and rhythms of its traverse of the page by sight alone. Computers can do this, as well as quantify, analyze, and compare each of these properties to make signature recognition a viable biometric technology. Being based on things that are not visible, such as pressure and velocity, signature-based biometric technology offers a distinct advantage over regular signature verification. In addition to mimicking the letter forms, any potential forger has to fabricate a signature at the same speed, and with the same pen weight, as the victim.

Signature biometrics poses a couple of unique problems. The first is the comfort with which people are already willing to use their signature as a form of identification. While this high level of consumer acceptance is viewed as a strength by vendors of such systems, this bears with it a strong downside. Without proper notification, a person may sign an electronic signature pad and unwittingly also be surrendering a reference or live biometric sample. Since the custom of leaving a signature as one's "official mark" is based on the presumption of irreproducibility, people are willing to provide a signature without giving its potential for reproduction a second thought. However, electronic data is easy to copy and transmit. Thus, a forger posing as a delivery man might fraudulently secure a signature biometric by presenting a victim with a "gift" box, requesting a signature to confirm delivery, and making off with the victim's biometric data.

The second unique property of signature biometrics is that unlike all other biometrics, which either establish an identity (identification) or confirm an identity (authentication), a signature can convey *intent* (authorization). In other words, a traditional signature on paper is taken both to authenticate the signer and convey the signer's legal authority. An electronic system that solicits a user's nonsignature biometric must provide a separate step to convey the user's legal authorization for any binding transaction. A signature-based biometric system could mimic current legally customary acceptance of a signature to simultaneously convey both identity and authority.

Signature dynamics is a behavioral biometric modality that analyzes dynamic characteristics of an individual's signature—such as shape of signature, speed of signing, pen pressure when signing, and pen-in-air movements—for recognition.

Signature Scanning

A signature is another example of biometric data that is easy to gather and is not physically intrusive. Digitized signatures are sometimes used, but usually have insufficient resolution to ensure authentication. Although an individual's signature does change over time, and can be consciously changed to some extent, it provides a basic means of identification. The pen pressure and duration of the signing process, which is done on a digital-based pen tablet, is recorded as an algorithm that is compared against future signatures. A signature scan is implemented in situations where a signature or written input processes are already in place. These applications include contract execution, formal agreements, acknowledgement of services received, access to controlled documents, etc.

Signature Verification

This technology uses the dynamic analysis of a signature to authenticate a person. The technology is based on measuring the speed, pressure, and angle used when a person is signing something. One focus for this technology has been e-business applications and other applications where signature is an accepted method of personal authentication.

Signature verification is the process used to recognize an individual's handwritten signature. Dynamic signature verification technology uses the behavioral biometrics of a handwritten signature to confirm the identity of a computer user. This is done by analyzing the shape, speed, stroke, pen pressure, and timing information during the act of signing. Natural and intuitive, the technology is easy to explain and an effective replacement for a password or a PIN number.

Signature verification technology utilizes the distinctive aspects of the signature to verify the identity of individuals. The technology examines the behavioral components of the signature as opposed to comparing visual images of signatures. Unlike traditional signature comparison technologies, signature verification measures the physical activity of signing. While a system may also leverage a comparison of the visual appearance of a signature, or "static signature," the primary components of signature verification are behavioral.

Signature verification enjoys a synergy with existing processes that other biometrics do not. People are used to signatures as a means of transaction-related identity verification and most would see nothing unusual in extending this to encompass biometrics. Signature verification devices have proved to be reasonably accurate in operation and obviously lend themselves to applications where the signature is an accepted identifier. There have been, however, relatively few significant applications to date in comparison with other biometric methodologies. If the application fits, it is a technology worth considering, although signature verification vendors have tended to have a somewhat checkered history. The following steps are generally required in a signature verification process:

- Software identifies the unique characteristics of the way the signature is signed. This process includes its acceleration, speed, pressure, and time, and not the way it looks.
- Everyone on the network who needs to authorize an action in the workflow process initially establishes their signature with the software product.
- When a signature is required, users can either sign directly with a stylus onto a form presented on a PC Tablet, or use a separate signature reader attached to a PC or computing device.
- There is a verification process of the signature by a software product.
- The user signature is either verified or rejected.

Signature verification has several strengths. Because of the large amount of data present in a signature verification template, as well as the difficulty in mimicking the behavior of signing, signature verification technology is highly resistant to imposter attempts. As a result of the low false acceptance rates (FAR), a measure of the likelihood that a user claiming a false identity will be accepted, implementers can have a high confidence level that successfully matched users are who they claim to be.

There is an important distinction between simple signature comparisons and dynamic signature verification. Both can be computerized, but a *simple signature comparison* only takes into account what the signature looks like. *Dynamic signature verification (DSV)* takes into account how the signature was made. With DSV it is not the shape or look of the signature that is meaningful; it is the changes in speed, pressure, and timing that occur during the act of signing. Only the original signer can re-create the changes in timing and pressure.

The global features extracted, listed below, compose the feature vector for DSV comparison. The advantage to using these features is that little preprocessing must be performed on the raw signature data. As such, these features could be extracted in real time as the signature is captured, eliminating the need to store the raw data.

- Average pressure: The average pen-tip pressure over the entire signature
- Pen tilt: The average tilt of the pen while writing over the entire signature
- Number of strokes: The number of velocity-based strokes over the entire signature

- Number of pen-ups: The number of times the pen was lifted over the entire signature
- Average velocity: The average x, y velocity over all sample points

A pasted bitmap, a copy machine, or an expert forger may be able to duplicate what a signature looks like, but it is virtually impossible to duplicate the timing changes in pressure. The practiced and natural motion of the original signer would be required to repeat the patterns shown. There will always be slight variations in a person's handwritten signature, but the consistency created by natural motion and practice over time creates a recognizable pattern that makes the handwritten signature a natural for biometric identification.

Signature verification is natural and intuitive. The technology is easy to explain and trust. The primary advantage of these systems over other biometric technologies is that signatures are already accepted as the common method of identity verification. This history of trust means that people are very willing to accept a signature-based verification system.

Unlike the older technologies of passwords and keycards, which are often shared or easily forgotten, lost, and stolen, dynamic signature verification provides a simple and natural method for increased computer security and trusted.

Emerging Biometric Technologies

Newer biometric technologies using diverse physiological and behavioral characteristics are in various stages of development. Some are commercially available, some may emerge over the next two to four years, and others are many years from implementation. Each technique's performance can vary widely, depending on how it is used and the environment in which it is used. Table 2-1 provides a functional comparison of emerging biometric technologies. This section is divided into keystroke dynamics and all others that are being researched. [9]

Technology	How it Works
Blood pulse measurement	Infrared sensors measure blood pulse on a finger
Skin pattern recognition	Extracts distinct optical patterns by spectroscopic measurement of light scattered by the skin
Ear shape recognition	Based on distinctive ear shape and the structure of the cartilaginous projecting portion of the outer ear
Facial thermography	Infrared camera detects heat patterns created by the branching of blood vessels and emitted from the skin
Gait recognition	Captures a sequence of images to derive and analyze motion characteristics
Nailbed identification	An interferometer detects phase changes in back-scattered light shone on the fingernail
DNA matching	Compares accrual samples of DNA rather than templates generated
Odor sensing	Captures the volatile chemicals that the skin's pores emit
Vein scan	Captures images of blood vessel patterns

Table 2-1 Functional Comparison of Emerging Biometric Technologies

Keystroke Dynamics

Keystroke dynamics is a biometric modality that uses the cadence of an individual's typing pattern for recognition. The rhythms with which one types at a keyboard are sufficiently distinctive to form the basis of this biometric technology. While distinct, keystroke dynamics are not sufficiently unique to provide identification, but can be used to confirm a user's identity.

Keystroke dynamics, unlike other biometric technologies, are 100% software-based, requiring no sensor more sophisticated than a home computer. Because of this, deployment is occurring in fairly low-stakes, computer-centric applications, such as content filtering (*Net Nanny owns BioPassword, the leading keystroke dynamics vendor*) and digital rights management, in which passwords to download music are bolstered by keystroke dynamic verification to prevent password-sharing. As a general rule, any method involving home or office computers is inherently insecure, as these devices leave a lot more room for experimentation than devices like ATMs or entry systems, and the information they use tends to travel over unsecured communication lines.

Other Biometric Technologies

Included in this section are those biometric technologies that are in the development process. These include:

- Vein scan
- Facial thermography
- DNA matching
- Body odor
- Blood pulse
- Gait
- Ear shape

Vein scan biometric technology can automatically identify a person from the patterns of the blood vessels in the back of the hand. The technology uses near-infrared light to detect vein vessel patterns. Vein patterns are distinctive between twins and even between a person's left and right hand. Developed before birth, they are highly stable and robust, changing throughout one's life only in overall size. The technology is not intrusive, and works even if the hand is not clean. It is commercially available.

Facial thermography detects heat patterns created by the branching of blood vessels and emitted from the skin. These patterns, called thermograms, are highly distinctive. A **thermogram** is an image of an object taken with an infrared camera that shows surface temperature variations. Even identical twins have different thermograms. Developed in the mid-1990s, thermography works much like facial recognition, except that an infrared camera is used to capture the images. The advantages of facial thermography over other biometric technologies are that it is not intrusive—no physical contact is required—every living person presents a usable image, and the image can be collected on the fly. Also, unlike visible light systems, infrared systems work accurately even in dim light or total darkness. Although identification systems using facial thermograms were undertaken in 1997, the effort was suspended because of the cost of manufacturing the system.

DNA matching is a type of biometric in the sense that it uses a physiological characteristic for personal identification. It is considered to be the "ultimate" biometric technology in that it can produce proof-positive identification of a person, except in the case of identical twins. However, DNA differs from standard biometrics in several ways. It compares actual samples rather than templates generated from samples. Also, because not all stages of DNA comparison are automated, the comparison cannot be made in real time. DNA's use for identification is currently limited to forensic applications. The technology is many years away from any other kind of implementation and will be very intrusive.

Researchers are investigating a biometric technology that can distinguish and measure body odor. This technology would use an odor-sensing instrument (an electronic "nose") to capture the volatile chemicals that skin pores all over the body emit to make up a person's smell. Although distinguishing one person from another by odor may eventually be feasible, the fact that personal habits such as the use of deodorants and perfumes, diet, and medication influence human body odor renders the development of this technology quite complex.

Blood pulse biometrics measure the blood pulse on a finger with infrared sensors. This technology is still experimental and has a high false match rate, making it impractical for personal identification.

The exact composition of all the skin elements is distinctive to each person. For example, skin layers differ in thickness, the interfaces between the layers have different undulations, pigmentation differs, collagen fibers and other proteins differ in density, and capillary beds have distinct densities and locations beneath the skin. Skin pattern recognition technology measures the characteristic spectrum of an individual's skin. A light sensor illuminates a small patch of skin with a beam of visible and near-infrared light. The light is measured with a spectroscope after being scattered by the skin. The measurements are analyzed, and a distinct optical pattern can be extracted.

Nailbed identification technology is based on the distinct longitudinal, tongue-in-groove spatial arrangement of the epidermal structure directly beneath the fingernail. This structure is mimicked in the ridges on the outer surface of the nail. When an interferometer is used to detect phase changes in back-scattered light shone on the fingernail, the distinct dimensions of the nailbed can be reconstructed and a one-dimensional map can be generated.

Gait recognition, or recognizing individuals by their distinctive walk, captures a sequence of images to derive and analyze motion characteristics. A person's gait can be hard to disguise because a person's musculature essentially limits the variation of motion, and measuring it requires no contact with the person. However, gait can be obscured or disguised if the individual, for example, is wearing loose-fitting clothes. Preliminary results have confirmed its potential, but further development is necessary before its performance, limitations, and advantages can be fully assessed.

Ear shape recognition is still a research topic. It is based on the distinctive shape of each person's ears and the structure of the largely cartilaginous, projecting portion of the outer ear. Although ear biometrics appears to be promising, no commercial systems are available.

Biometric Method Advantages and Disadvantages

There does not appear to be one method of biometric data gathering and reading that produces the "best" result of ensuring secure authentication. Each of the different methods of biometric identification has positive elements. Some are less invasive, some can be done without the knowledge of the subject, and some are very difficult to fake.

There are a number of advantages to biometric identification technology:

- Biometric identification can provide extremely accurate, secured access to information; fingerprints and retinal and iris scans produce absolutely unique data sets when properly constructed.
- Current methods like password verification have many problems as people tend to write them down, forget them, and make up easy-to-hack passwords.
- Automated biometric identification can be accomplished rapidly and uniformly, with a minimum of training.
- Identities can be verified without resorting to documents that may be stolen, lost, or altered.

Table 2-2 provides a comparison of the major biometric methods and their respective performance objective categories. Each biometric method is ranked as high, medium, or low. By assigning a 1 to low, a 2 to medium, and a 3 to high, readers can arrive at a ranking. An average across the performance objective categories shows that the iris biometric is the most effective method.

Biometrics	Acceptability	Collectability	Circumvention	Performance	Permanence	Uniqueness	Universality
DNA	L	L	L	H	H	H	H
Face	H	H	L	L	M	L	H
Facial thermogram	H	H	H	M	L	H	H
Fingerprint	M	M	H	H	H	H	M
Hand geometry	M	H	M	M	M	M	M
Hand vein	M	M	H	M	M	M	M
Iris	L	M	H	H	H	H	H
Keystroke dynamics	M	M	M	L	L	L	L
Retina	L	L	H	H	M	H	H
Signature	H	H	L	L	L	L	L
Voice	H	M	L	L	L	L	M

Table 2-2 Comparison of Major Biometric Methods

Face Recognition

Of the various biometric identification methods, face recognition is one of the most flexible, working even when the subject is unaware of being scanned. It also shows promise as a way to search through masses of people who spent only seconds in front of a "scanner," which is an ordinary digital camera.

Fingerprint Identification

Fingerprints remain constant throughout life. In over 140 years of fingerprint comparisons worldwide, no two fingerprints have ever been found to be alike, not even those of identical twins. Good fingerprint scanners have been installed in PDAs like the iPaq Pocket PC, so scanner technology is also a trivial matter. This technology might not work in industrial applications since it requires clean hands.

Hand Geometry Biometrics

Hand geometry readers work in harsh environments, do not require clean conditions, and form a very small dataset. It is not regarded as an intrusive kind of test. It is often the authentication method of choice in industrial environments.

Retina Scan

There is no known way to replicate a retina. As far as anyone knows, the pattern of the blood vessels at the back of the eye is unique and stays the same for a lifetime. However, it requires about 15 seconds of careful concentration to take a good scan. Retina scan remains a standard in military and government installations.

Iris Scan

Like a retina scan, an iris scan also provides unique biometric data that is very difficult to duplicate and remains the same for a lifetime. The scan is similarly difficult to make and may be difficult to perform on children or the infirm. However, there are ways of encoding the iris scan biometric data in a way that it can be transported securely in a "barcode" format.

Signature

A signature is another example of biometric data that is easy to gather and is not physically intrusive. Digitized signatures are sometimes used, but usually have insufficient resolution to ensure authentication.

Voice Analysis

Like face recognition, voice biometrics provides a way to authenticate identity without the subject's knowledge. It is easier to fake (using a tape recording); however, it is not possible to fool an analyst by imitating another person's voice.

Summary

- Biometrics is primarily an automated method of recognizing an individual based on a physiological or behavioral characteristic. Biometric technologies are considered and evaluated using a number of characteristics, including acceptability, circumvention, collectability, performance, permanence, universality, and uniqueness.
- Biometric identification is the one-to-many process of comparing a submitted biometric sample against all biometric reference templates on file to determine whether it matches any of the templates. Identity verification is the process of confirming or denying that a claimed identity is correct by comparing the credentials of a person requesting access with those previously proven.
- Biometric performance factors that need to be thoroughly investigated include accuracy, speed, reliability, acceptability, and cost. Additional factors include resistance to counterfeiting, enrollment time, database storage requirements, and intrusiveness.
- The definition of "biometrics modality" is a class of biometric system. These modalities include face, hand, voice, signature, and eye variations.
- There does not appear to be one method of biometric data gathering and reading that produces the "best" result of ensuring secure authentication. Each of the different methods of biometric identification has positive elements.

Key Terms

10-print card Used to collect both an individual's personal and demographic information along with flat and rolled ink impression fingerprint images.

Autocorrelation A proprietary finger scanning technique.

Behavioral (characteristic) A reflection of an individual's psychological makeup, although physical traits, such as size and gender, have a major influence.

Bifurcation The point in a fingerprint where a ridge divides or splits to form two ridges that continue past the point of division for a distance that is at least equal to the spacing between adjacent ridges at the point of bifurcation.

Biometric identification The one-to-many process of comparing a submitted biometric sample against all biometric reference templates on file to determine whether it matches any of the templates.

Biometric sample The identifiable, unprocessed image or recording of a physiological or behavioral characteristic, acquired during submission, used to generate biometric templates. Also referred to as biometric data.

Breeder document A document used as an original source of identity to apply for (or breed) other forms of identity credentials.

Capacitive sensor Consists of a row/column configuration of tiny metal electrodes. Every column is linked to a pair of sample-and-hold circuits. The fingerprint image is recorded in sequence, row by row, as each metal electrode acts as one capacitor plate and the contacting finger acts as the second plate.

Capture The process of taking a biometric sample from the user. In the capture, a sample of the user's biometric is acquired using the required sensor (i.e., camera, microphone, fingerprint scanner).

Characteristic A distinguishing feature or attribute.

Closed-set identification A biometric task where an unidentified individual is known to be in the database and the system attempts to determine his/her identity.

Comparison The process of comparing biometric data with a previously stored reference template or templates (biometric data).

Eigenface comparison Uses a palette of about 150 facial abstractions, and compares the captured face with these archetypal abstract faces.

Faceprint Represents a face in the database.

Feature extraction The automated process of locating and encoding distinctive characteristics from a biometric sample in order to generate a template.

Fingerprint recognition A biometric modality that uses the physical structure of an individual's fingerprint for identification purposes.

Friction ridge The ridges on the skin of the fingers, toes, palms, and soles of the feet that make contact with an incident surface under normal touch. On the fingers, the unique patterns formed by friction ridges make up fingerprints.

Identification card A card identifying its holder and issuer which may carry data required as input for the intended use of the card and for transactions based thereon.

Iris recognition A biometric modality that uses an image of the physical structure of an individual's iris for recognition purposes.

Iris scanning A method of biometric identification and pattern recognition that is used to determine the identity of the subject.

Keystroke dynamics A biometric modality that uses the cadence of an individual's typing pattern for recognition.

Latent fingerprint The transferred impression of a friction ridge detail that is not readily visible; a generic term used for a questioned friction ridge detail.

Live capture The process of capturing a biometric sample by an interaction between an end user and a biometric system.

Liveness detection A technique used to ensure that the biometric sample submitted is from an end user.

Live scan A scan of a fingerprint or palm print taken directly from a subject's hand.

Minutiae points The point where a friction ridge begins, terminates, or splits into two or more ridges. Minutiae are friction ridge characteristics that are used to individualize a fingerprint image.

Modality A type or class of biometric system, such as face recognition, fingerprint recognition, iris recognition, etc.

Model A detailed description or scaled representation of one component of a larger system that can be created, operated, and analyzed to predict actual operational characteristics of the final produced component.

Nodal points Every face has numerous, distinguishable landmarks, the different peaks and valleys that make up facial features.

One-to-many (2-N) Synonym for identification.

One-to-one (2-1) Synonym for verification.

Open-set identification A biometric task that more closely follows operational biometric system conditions to 1) determine if someone is in a database and 2) find the record of the individual in the database. See watch-list.

Palm print recognition A biometric modality that uses the physical structure of an individual's palm print for recognition purposes.

Physiological (characteristic) A relatively stable human physical characteristic, such as a fingerprint, hand silhouette, iris pattern, or blood vessel pattern on the back of the eye. This type of measurement is unchanging and unalterable without significant duress.

Recognition The generic term used in the description of biometric systems (e.g., face recognition or iris recognition) relating to their fundamental function.

Signature dynamics A behavioral biometric modality that analyzes dynamic characteristics of an individual's signature.

Signature verification The process used to recognize an individual's handwritten signature.

Speech recognition Recognizes the words being said, and is not a biometric technology; it is a technology that enables a machine to recognize spoken words.

Thermogram An image of an object taken with an infrared camera that shows surface temperature variations.

Verification The process of confirming or denying that a claimed identity is correct by comparing the credentials.

Voice verification A potentially interesting technique because considerable voice communication takes place with regard to everyday business transactions.

Review Questions

1. A ________________ is a distinguishing feature or attribute.
2. A ________________ characteristic is a relatively stable human physical characteristic, such as a fingerprint, hand silhouette, iris pattern, or blood vessel pattern on the back of the eye.
3. A ________________ characteristic is a reflection of an individual's psychological makeup, although physical traits, such as size and gender, have a major influence.
4. ________________ is the process of being recognized by some attribute.
5. A ________________ is a user setting for biometric systems operating in the verification or open-set identification (watchlist) tasks.

6. Identity ________________ is the process of confirming or denying that a claimed identity is correct by comparing the credentials of a person requesting access with those previously proven and stored in an ID card or system and associated with the identity being claimed.
7. In a biometric system, identity is typically established when the person is registered in the system through the use of so-called "________________," such as birth certificate, passport, Social Security number, driver's license, etc.
8. ________________ is a synonym for identification, whereas one-to-one is a synonym for verification.
9. ________________ is the generic term used in the description of biometric systems (e.g., face recognition or iris recognition) relating to their fundamental function.
10. Which biometric has the most effective performance objective?
11. A ________________ contains data that represents the biometric measurement of an enrollee.
12. ________________ is a synonym for identification, whereas ________________ is a synonym for verification.
13. Live capture and liveness detection are two techniques used to ensure that a biometric sample is from an end user. (True or False?)
14. A ________________ is a term referring to an open-set identification task that a biometric system can perform.
15. A biometric task that determines if someone is in a database and finds that individual is termed a/an ________________.
16. A ________________ is a user setting for biometric systems operating in verification tasks.
17. Face and fingerprint recognition is a ________________ or type of biometric system.
18. Friction ridges and minutiae points are limited to characteristics on an individual's fingerprint image. (True or False?)
19. Based on industry specifications, an iris match can be conducted more quickly than a fingerprint search. (True or False?)
20. The ________________ comparison uses a palette of 150 facial abstractions and compares the captured face with these archetypal abstract faces.

Discussion Exercises

1. Describe the extraction process of converting a captured biometric sample into biometric data.
2. Describe the differences between a model and a template.

3. What is the difference between an open-set and closed-set identification?
4. Provide an overview of hamming distance, difference, and similarity score.
5. Describe the features and characteristics of fingerprint biometrics.
6. Describe the various fingerprint patterns.
7. Provide an overview of all fingerprint minutiae.
8. Provide an overview of moiré fringe patterns.
9. Describe the differences between false acceptance and false rejection concepts.
10. Produce a comparison matrix of emerging biometric technologies.

Hands-On Projects

2-1: Identifying Biometric Modalities

Time required—30 minutes

Objective—Identify the various biometric modalities being researched, developed, and implemented.

Description—Use various Web tools, Web sites, and other tools to identify the potential for biometrics in industry and government environments.

1. Research the biometric modalities that use the following body parts and create a matrix that provides the pros and cons for each modality: Fingers, hands, eyes, voices, and signatures. Indicate the potential for error in scanning and matching for each modality.
2. Research the Web, magazines, and news articles that provide information on the subject of biometrics. Prepare a summary report and slide show reflecting the research conducted.
3. Describe a situation where a biometric modality would be useful. Identify the flow of a user interacting with this biometric system. Identify issues that would impact the success or failure of using this biometric system.

2-2: Researching Access Control Alternatives

Time required—45 minutes

Objective—Identify those biometrics options that can provide some level of access control and security for OnLine Access assets and resources.

Description—The owners and managers of OnLine Access have identified several instances of some compromise to the organization's database entries. Some data has been compromised and/or stolen. They have decided to research the access control aspects and capabilities of using some biometric solution.

1. Research the topic of biometric access control and security. Develop a comparison of the various biometric methods that could provide a solution for OnLine Access.
2. Produce a matrix describing each method's features and benefits.
3. Rank order these solutions that would best provide the benefits desired by OnLine Access management.

chapter 3

Biometric Applications and Solutions

After completing this chapter, you should be able to do the following:

- Identify biometric applications and solutions in commercial, government, military, and e-commerce and e-business environments
- Look at the various biometrics modalities where applications are being developed for access control
- See how biometric solutions are utilized in citizen-facing, employee-facing, and customer-facing security environments
- Become familiar with biometric applications in justice/law enforcement, border control/airports, physical access control, logical access control, laptops, PDAs, and locks
- Identify vendors and applications that provide biometric security access control using identification and verification techniques

Numerous software marketing organizations have jumped on the bandwagon for security product sales. Buyers should, however, be aware as many of these biometric applications and solutions are expensive and may not have been successfully field tested. Diligence on the part of an implementer is required. Not only is a cost–benefit analysis required, but also a feasibility analysis.

The latest gizmo may not fit the security situation and may not be acceptable to the organization's members. It is also essential that top management supports the solution. Awareness and training sessions for those impacted are required to sell the benefits of any access control and security solution as the options and features are numerous. Many of the latest versions of biometric solutions have only been implemented in government environments. This type of environment may not be viable for a security solution in a commercial or educational setting.

Introduction

Since the beginning of civilization, identifying fellow human beings has been crucial to the fabric of human society. Consequently, individual identification is an integral part of the infrastructure needed for diverse business sectors such as finance, health care, transportation, entertainment, law enforcement, security, access control, border control, government, and communication.

Legacy security applications include technical, hardware, software, and protocol solutions to ensure the protection of an organization's computer, database, and network resources. A system is a set of elements that acts as a single, goal-oriented entity, whereas an application usually consists of software programs that provide some functionality to a user or entity. Controls must be implemented to protect specific applications.

Various biometric methods can provide access control and security for legacy applications. Application solutions are classified into citizen-facing, customer-facing, and employee-facing. Each of these solution categories can be used to solve some access control issue.

Many biometric applications are the product of government and big business. This situation arises because of development and implementation expenses. Several of these large applications will be discussed in this chapter. *Note: A common attribute of software is change, and these biometric applications are no different. Readers should endeavor to obtain the most current information for the products being discussed and evaluated.*

Legacy Applications Requirements

Information resources are distributed throughout an organization. Employees travel and take home sensitive electronic data on personal computers; information is transmitted to and from the organization and among organizational components; and numerous secure facilities are accessed. Understanding potential threats and vulnerabilities of information system applications is a critical task, and organizing an appropriate defense system is one of the major activities of administrative and security managers. Defense safeguards include application controls for protecting specific software applications. Controls for software applications are usually built into the programs and take the form of validation rules. These can be classified into three major categories of:

- Input controls: Designed to prevent data alteration or loss
- Processing controls: Ensures data are complete, valid, and accurate
- Output controls: Includes complete, accurate, consistent, and valid results.

Information security is concerned with protecting data from accidental or intentional disclosure to unauthorized persons, or from unauthorized modification or destruction. Data security functions are implemented through security access control programs, database products, backup/recovery procedures, applications, and external control procedures. Data security must address the following issues:

- Access control
- Confidentiality of data
- Integrity of data
- Critical nature of data

The Need for Secure Transactions

As society becomes electronically connected to form one large global community, it has become necessary to conduct reliable person recognition, often remotely and through automatic means. Surrogate representations of identity such as passwords and cards are no longer sufficient. Further, passwords and cards can be shared and thus cannot provide nonrepudiation. Biometrics—the automatic recognition of people based on their distinctive anatomical and behavioral characteristics—could become an essential component of effective person identification solutions because biometric identifiers cannot be shared or misplaced, and they intrinsically represent the individual's bodily identity. Recognition of a person by the physical body, then linking that body to an externally established "identity," forms a very powerful tool with tremendous potential consequences, both positive and negative. Consequently, biometrics is not only a fascinating pattern recognition research problem but, if carefully used, could also be an enabling technology with the potential to make society safer, reduce fraud, and lead to user convenience.

Reliable person recognition is an important problem in diverse organizations such as commercial markets, government, and e-commerce and e-business networks. Biometrics has the potential to become an irreplaceable part of many identification systems. While successful in some niche markets, the biometrics technology has not yet delivered its promise of foolproof automatic human recognition. With the availability of inexpensive biometric sensors and computing power, it is becoming increasingly clear that broader usage of biometric technologies is being stymied by a lack of understanding of four fundamental problems:

- Accurately and efficiently represent and recognize biometric patterns
- Guarantee that the sensed measurements are not fraudulent
- Ensure that the application is exclusively using pattern recognition for the expressed purpose (function creep)
- Acquire repeatable and distinctive patterns from a broad population

Solving these core problems is required to move biometrics into mainstream applications, and may also stimulate the adoption of other pattern-recognition applications which provide effective automation of sensitive tasks without jeopardizing individual freedoms.

As previously stated, biometrics consists of automated methods of recognizing a person based on a physiological or behavioral characteristic. Among the features measured include facial, fingerprints, hand geometry, handwriting, iris, retinal, vein, and voice. Biometric technologies are becoming the foundation of an extensive array of highly secure identification and personal verification solutions. As the level of security breaches and transaction fraud increases, the need for highly secure identification and personal verification technologies is becoming apparent.

Biometric-based solutions can enable confidential financial transactions and personal data privacy. The need for biometrics can be found in federal, state, and local governments, in the military, and in commercial applications. Enterprise-wide network security infrastructures, government IDs, secure electronic banking, investing and other financial transactions, retail sales, law enforcement, and health and social services are already benefiting from these technologies.

Biometric Technology on the Leading Edge

Biometric-based authentication applications include workstation, network, and domain access, single sign-on, application logon, data protection, remote access to resources, transaction security, and Web security. Trust in these electronic transactions is essential to the healthy growth of the global economy. Utilized alone or integrated with other technologies such as smart cards, encryption keys, and digital signatures, biometrics is set to pervade nearly all aspects of the economy and daily life. More information about biometrics, standards activities, government and industry organizations, and research initiatives on biometrics can be found throughout this book. Some leading edge applications include:

- Fingerprint scanners and the necessary software to store and compare fingerprints have already been installed in laptop computers and PDAs.
- Sensors installed in automobiles can identify the driver, and adjust mirrors, seat positions, and climate controls.
- Special readers can measure various elements of hand geometry, comparing the result with data on file for each person.
- Surveillance cameras can search crowds for missing persons or criminal suspects.
- Face recognition software can be modified to recognize gestures, leading to improved assistive technologies for quadriplegic patients.

Authentication by biometric verification is becoming increasingly common in corporate and public security systems, consumer electronics, and point of sale (POS) applications. In addition to security, the driving force behind biometric verification has been convenience. Biometric devices, such as finger scanners, consist of:

- A reader or scanning device
- Software that converts the scanned information into digital form and compares match points
- A database that stores the biometric data for comparison

To prevent identity theft, biometric data is usually encrypted when it's gathered. To convert the biometric input, a software application is used to identify specific points of data as *match points*. The match points in the database are processed using an algorithm that translates that information into a numeric value. The database value is compared with the biometric input the end user has entered into the scanner and authentication is either approved or denied.

The **application programming interface (API)** provides formatting instructions or tools used by an application developer to link and build hardware or software applications. The **biometrics application programming interface (BioAPI)** defines the application programming interface and service provider interface for a standard biometric technology interface. The BioAPI enables biometric devices to be easily installed, integrated, or swapped within the overall system architecture. Additional details concerning BioAPI will be presented in Chapter 12.

Biometric Application Environment

Biometrics is expected to be incorporated into applications providing for security and access control solutions in:

- Commercial markets
- Education
- Government
- Military
- E-commerce and e-business (network)

A number of biometrics initiatives are ongoing in federal, state, and local governments. Congressional offices and a large number of organizations are addressing the important role that biometrics will play in identifying and verifying the identity of individuals and protecting national assets.

There is a great potential need for biometrics beyond government security applications. A range of new applications have been identified in such diverse environments as:

- Amusement parks
- Banks
- Credit unions
- Other financial organizations
- Enterprise and government networks
- Passport programs
- Driver's licenses
- Colleges
- Physical access to multiple facilities (e.g., nightclubs)
- School lunch programs

Utilizing biometrics for personal authentication is becoming convenient and considerably more accurate than current methods, such as the utilization of passwords or PINs. This is because biometrics:

- Links the event to a particular individual—a password or token may be used by someone other than the authorized user
- Is convenient—there's nothing to carry or remember

- Is accurate, and provides for positive authentication
- Can provide an audit trail
- Is becoming socially acceptable and inexpensive

Biometric authentication requires comparing a registered or enrolled biometric sample (biometric template or identifier) against a newly captured biometric sample (for example, a fingerprint captured during a login). During *enrollment*, a sample of the biometric trait is captured, processed by a computer, and stored for later comparison. Figure 3-1 depicts a logical view of a biometric enrollment and verification flow.

Biometric recognition can be used in **identification mode**, where the biometric system identifies a person from the entire enrolled population by searching a database for a match based solely on the biometric. For example, an entire database can be searched to verify that a person has not applied for entitlement benefits under two different names. This is sometimes called "one-to-many" matching. A system can also be used in **verification mode**, where the biometric system authenticates a person's claimed identity from his previously enrolled pattern. This is also called "one-to-one" matching. In most computer access, physical access, or network access environments, verification mode is used. A user enters an account and user name, or inserts a token such as a smart card, but instead of entering a password, a simple touch with a finger or a glance at a camera is enough to authenticate the user.

As previously stated, biometric-based authentication applications include workstation and network access, single sign-on, application logon, data protection, remote access to resources, transaction security, and Web security. The promises of e-commerce and e-government can be achieved through the utilization of strong personal authentication procedures. Biometric technologies are expected to play a key role in personal authentication for large-scale enterprise network authentication environments, POS, and for the protection of all types of digital content, such as in digital rights management and health care applications. Used alone or

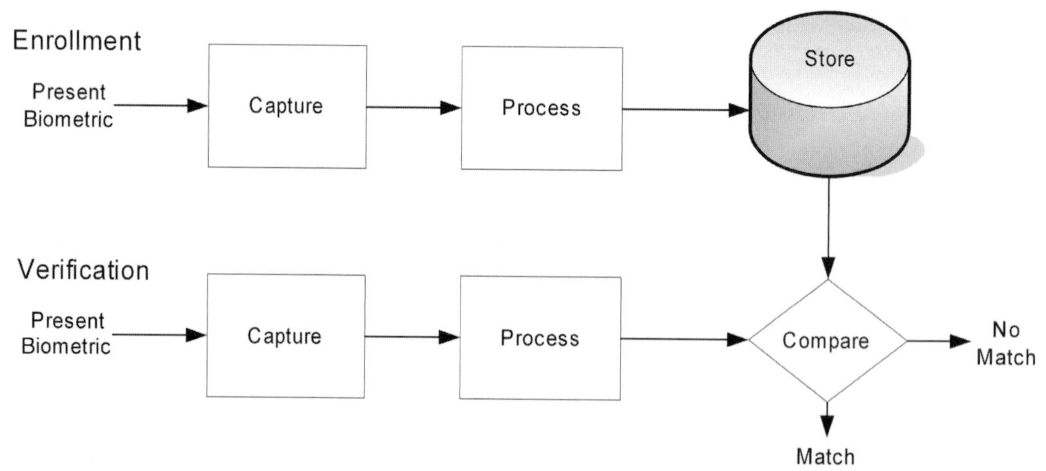

Figure 3-1 Biometric Enrollment and Verification Flow

Course Technology/Cengage Learning

integrated with other technologies such as smart cards, encryption keys, and digital signatures, biometrics is anticipated to pervade nearly all aspects of the economy and daily life. For example, biometrics is used in schools, such as in lunch programs in Pennsylvania and school libraries in Georgia. Examples of other current applications include verification of annual pass holders in an amusement park, speaker verification for television home shopping, Internet banking, and user authentication in a variety of social services.

Using biometrics to indentify human beings offers some unique advantages. Biometrics can be used to positively identify individuals. Tokens, such as smart cards, magnetic stripe cards, photo ID cards, and physical keys, can be lost, stolen, duplicated, or left at home. Passwords can be forgotten, shared, or observed. Moreover, today's fast-paced electronic world means citizens are asked to remember a multitude of passwords and PINs for computer accounts, bank ATMs, e-mail accounts, wireless phones, and Web sites. Biometrics holds the promise of fast, easy-to-use, accurate, reliable, and less-expensive authentication for a variety of applications.

The advantage of biometric authentication is its ability to allow more instances of authentication in such a quick and easy manner that users are not bothered by the additional requirements. As biometric technologies mature and come into wide-scale commercial use, dealing with multiple levels of authentication or multiple instances of authentication will become less of a burden for users.

An Indication of Biometric Activities

As of March 2005, NIST and NSA have cosponsored and spearheaded a number of biometric-related activities, including the development of a **common biometric exchange file format** (CBEFF), NIST Biometric Interoperability, Performance, and Assurance Working Group, a BioAPI users' and developers' seminar, and the NIST BioAPI interoperability testbed. CBEFF describes a set of data elements necessary to support biometric technologies in a common way which is independent of the application and the domain of use. These would include mobile devices, smart cards, protection of digital data, and biometric data storage. This documentation will be described further in Chapter 12. CBEFF facilitates the:

- Biometric data interchange between different system components or between systems
- Interoperability of biometric-based application programs and systems
- Forward compatibility for technology improvements
- Simplification of the software and hardware integration process

The NIST Biometric Interoperability, Performance and Assurance Working Group supports the advancement of technically efficient and compatible biometric technology solutions on a national and international basis. It promotes and encourages an exchange of information and collaborative efforts between users and private industry in all things biometric. The Working Group consists of numerous organizations representing biometric vendors, system developers, information assurance organizations, commercial end users, universities, government agencies, national labs, and industry organizations. The Working Group is currently addressing the development of a simple testing methodology for biometric systems as well as addressing issues on biometric assurance. It is also addressing the utilization of biometric data in smart card applications by developing a smart card format compliant to the CBEFF. [1]

Using Biometric Methods

This section provides a brief review of the most prevalent biometric methods previously described.

Fingerprint Recognition

The patterns of friction ridges and valleys on an individual's fingertips are unique to that individual. For decades, law enforcement has been classifying and determining identity by matching key points of ridge endings and bifurcations. Each fingerprint is unique, including the prints of identical twins. Fingerprint recognition is one of the most commercially available biometric technologies; fingerprint recognition devices for desktop and laptop access are now widely available from many different vendors at a low cost. With these devices, users no longer need to type passwords because a touch provides instant access. Fingerprint systems can also be used in identification mode. Several states check fingerprints for new applicants to social services benefits to ensure recipients do not fraudulently obtain benefits under fake names.

Fingerprint ridges are formed in the womb; babies have fingerprints by the fourth month of fetal development. Once formed, fingerprint ridges are like a picture on the surface of a balloon. As a person ages, the fingers get larger; however, the relationship between the ridges stays the same, just as the picture on a balloon is still recognizable as the balloon is inflated.

Among the different biometrics used to authenticate an individual, fingerprint recognition is the most widely used in the market because it:

- Consistently separates an individual from another
- Best resists aging
- Employs a system with a high degree of robustness, speed, and accuracy in capturing the fingerprint
- Is extremely hard to fool

Face Recognition

This biometric method uses facial characteristics, such as the size and shape of facial characteristics and their relationship to each other. Although human beings have always used facial recognition, it is not easy to automate. Typically, relative distances between common landmarks on the face are used to generate a unique "face print."

The identification of a person by her facial image can be done in many different ways, such as by capturing an image of the face in the visible spectrum using an inexpensive camera or by using the infrared patterns of facial heat emission. Facial recognition in visible light typically models key features from the central portion of a facial image. Using a wide assortment of cameras, the visible light systems extract features from the captured image(s) that do not change over time, while avoiding superficial features, such as facial expressions or hair. Several approaches to modeling facial images in the visible spectrum are:

- Principal component analysis
- Local feature analysis

- Neural networks
- Elastic graph theory
- Multiresolution analysis

Some of the challenges of facial recognition in the visual spectrum include reducing the impact of variable lighting and detecting a mask or photograph. Some facial recognition systems may require a stationary or posed user in order to capture the image, although many systems use a real-time process to detect a person's head and locate the face automatically. The major advantages of facial recognition are that it is nonintrusive, hands-free, continuous, and accepted by most users.

Speaker Recognition

Speaker recognition dates back four decades, where the outputs of several analog filters were averaged, over time, for matching. Speaker recognition uses the acoustic features of speech (pitch, tone, cadence, and frequency of a person's voice) that have been found to differ between individuals. These acoustic patterns reflect both anatomy (e.g., size and shape of the throat and mouth) and learned behavioral patterns (e.g., voice pitch and speaking style). This incorporation of learned patterns into the voice templates (the latter called "voice prints") has earned speaker recognition its classification as a behavioral biometric. Speaker recognition systems employ three styles of spoken input:

- Text dependent
- Text prompted
- Text independent

Most speaker verification applications use text-dependent input, which involves selection and enrollment of one or more voice passwords. Text-prompted input is used whenever there is concern of imposters. The various technologies used to process and store voiceprints include:

- Hidden Markov models
- Pattern matching algorithms
- Neural networks
- Matrix representation
- Decision trees

Some systems also use "antispeaker" techniques, such as cohort models and world models.

Ambient noise levels can impede collection of the initial and subsequent voice samples. Performance degradation can result from changes in behavioral attributes of the voice and from enrollment using one telephone and verification on another telephone. Voice changes due to aging also need to be addressed by recognition systems. Many companies market speaker recognition engines, often as part of large voice processing, control, and switching systems. Capture of the biometric is seen as noninvasive. The technology needs little additional hardware by using existing microphones and voice-transmission technology allowing recognition over long distances via ordinary wire line or wireless telephones.

Speaker recognition applications have been developed for banking, medical, and customer support. Several examples are:

- Securing online transactions: To enhance e-banking and phone banking services
- Securing critical medical records: As a means of verifying the identity of an individual who needs to access personal medical records via phone or Internet
- Preventing benefit fraud: So governments can track individuals who claim benefits
- Resetting passwords: Fulfilling help desk requests
- Voice indexing: Creates indexes for broadcast news (reporter model)

Iris Recognition

This recognition method uses the iris of the eye, which is the colored area that surrounds the pupil. Iris patterns are thought to be unique and are obtained through a video-based image acquisition system. Iris scanning devices have been used in personal authentication applications for several years. Systems based on iris recognition have substantially decreased in price and this trend is expected to continue. The technology works well in both verification and identification modes (applications performing one-to-many searches in a database). Current applications can be used even in the presence of eyeglasses and contact lenses. The technology is not intrusive as it does not require physical contact with a scanner. Iris recognition has been demonstrated to work with individuals from different ethnic groups and nationalities.

The iris scan provides an analysis of the rings, furrows, and freckles in the colored ring that surrounds the pupil of the eye. More than 200 points are used for comparison. Iris scans were proposed in 1936, but it was not until the early 1990s that algorithms for iris recognition were created and patented. All current iris recognition systems use these basic patents, held by Iridian Technologies (Viisage Corporation). The iris of the eye is well suited for authentication purposes. It is an internal organ protected from most damage and wear, it is practically flat and uniform under most conditions and it has a texture that is unique even to genetically identical twins. [2]

Iris recognition is accomplished by applying proprietary algorithms for image acquisition and subsequent one-to-many matching. Iris recognition algorithms produce remarkable results. These algorithms have produced accuracy rates in authentication that are better than those of any other method. The picture of an eye is first processed by software that localizes the inner and outer boundaries of the iris and the eyelid contours in order to extract just the iris portion. Eyelashes and reflections that may cover parts of the iris are detected and discounted. Sophisticated mathematical software then encodes the iris pattern by a process called "demodulation." This creates a phase code for the texture sequence in the iris, similar to a DNA sequence code. The demodulation process uses functions called 2-D wavelets that make a very compact, yet complete, description of the iris pattern, regardless of its size and pupil dilation, in just 512 bytes.

This phase sequence is called an IrisCode® template, and it captures the unique features of an iris in a robust way that allows easy and very rapid comparisons against large databases of other templates. The IrisCode template is immediately encrypted to eliminate the possibility of identity theft and to maximize security. This is a purely passive process achieved by using CCD video cameras. This image is then processed and encoded into an IrisCode® record, which is stored in an IrisCode® database. This stored record is then used for identification in any live transaction when an iris is presented for comparison. [3]

Retinal Scanning

A retina scan provides an analysis of the capillary blood vessels located in the back of the eye; the pattern remains the same throughout life. A scan uses a low-intensity light to take an image of the pattern formed by the blood vessels. Retina scans were first suggested in the 1930s.

Retinal scanners are typically used for authentication and identification purposes. Retinal scanning has been utilized by several government agencies including the FBI, CIA, and NASA. However, in recent years, retinal scanning has become more commercially popular. Retinal scanning has been used in prisons, for ATM identity verification, and the prevention of welfare fraud.

Some states require mandatory retinal scans for truck and bus drivers. Scan results are used by state agencies to stop incompetent drivers from obtaining licenses in numerous states in order to conceal their driving records. A controversial proposal for the use of retina scans is the formation of a worker registry. In such a registry, every U.S. citizen would have to undergo a retinal scan. Scans would be stored in a database and would be used to ensure that each individual was a legal resident who qualified for employment within the United States. Opponents of this application of retinal scanning are concerned that it is a major invasion of privacy and a serious threat to personal liberties.

Retinal scanning also has medical applications. Communicable illnesses such as AIDS, syphilis, malaria, chicken pox, and Lyme disease as well as hereditary diseases like leukemia, lymphoma, and sickle cell anemia impact the eyes. Pregnancy also affects the eyes. Likewise, indications of chronic health conditions such as congestive heart failure, atherosclerosis, and cholesterol issues first appear in the eyes. Advantages and disadvantages are summarized as follows:

Advantages

- Low occurrence of false negatives
- Extremely low false positive rates
- Highly reliable because no two people have the same retinal pattern
- Speedy results: Identity of the subject is verified very quickly

Disadvantages

- Vulnerability to diseases such as cataracts and glaucoma
- Scanning procedure is highly invasive
- Not very user-friendly
- Limited government, corporate, and other funding
- Subject being scanned must focus on the scanner from about three inches away
- High equipment costs
- Poor lighting can affect results

Hand and Finger Geometry

Hand geometry is defined as the measurement and comparison of the different physical characteristics of the hand. Although hand geometry does not have the same degree of permanence or individuality as some other characteristics, it is still a popular means of biometric authentication.

These methods of personal authentication are well established as hand recognition has been available for over 20 years. To achieve personal authentication, a system may measure either physical characteristics of the fingers or the hands, which includes length, width, thickness, and surface area of the hand. Hand geometry has gained acceptance in a range of applications and can frequently be found in physical access control in commercial and residential applications, in time and attendance systems, and in general personal authentication applications.

The Palm vein system uses an infrared beam to penetrate the user's hand as it is waved over the system; the veins within the palm of the user are returned as black lines. Palm vein authentication has a high level of authentication accuracy due to the complexity of vein patterns of the palm. Because the palm vein patterns are internal to the body, this is a difficult system to counterfeit. Also, the system is contactless and therefore hygienic for use in public areas.

Signature Verification

This technology uses the dynamic analysis of a signature to authenticate a person, based on measuring speed, pressure and angle used by the person when a signature is produced. One focus for this technology has been e-business applications and other applications where signature is an accepted method of personal authentication. [4] Although the way individuals sign their name does change over time, and can be consciously changed to some extent, it provides a basic means of identification.

Biometric Application Solutions

Numerous solutions are available to address access and security requirements. A search of the Web reveals a tremendous list of products that promise to solve some problem. The reader will need to determine which product or service addresses a particular security issue. There is a diverse range of ways in which biometrics can be used, from network access control and employee screening, to verifying customer identity at an ATM machine. This section highlights the main drivers and detractors in the market and assesses the applications that are proving to be successful endeavors for biometric suppliers. Applications are generally classified according to solutions that are categorized as citizen facing, customer facing, or employee facing.

Citizen Facing

Millions of citizens around the world are being issued new identity credentials to cross borders, enter buildings, board aircraft, use computers and networks, and access employee benefits or government-provided services. The United States is rolling out new identity credentials for government employees, transportation workers, and first responders, as well as defining *citizen-facing* programs for enhanced driver's licenses and border crossing cards. Many countries are also developing programs for implementing national ID or national health cards; such programs include the British national ID card, the electronic European identity and travel document, the Chinese national ID card, and the Australian health and welfare services access card.

As organizations consider putting new secure credentialing systems into place, privacy and the protection of personal information quickly emerge as key issues that must be balanced against the need for more accurate identity verification. This requirement of how to identify people unequivocally, while also protecting their privacy, shapes every discussion of how to design, build, or implement secure credentialing systems.

Citizen-facing biometric solutions can be used to identify and verify individuals when they interact with government agencies and other organizations that require a high level of access security. Sample application areas include:

- Driver's license
- Background checks
- Government benefits
- Immigration
- Voting and voter registration
- Surveillance
- Employment
- Medical/HIPAA
- Counterterrorism

Customer Facing

Retailers can identify employees and customers accurately by scanning a portion of a fingerprint and matching it with another piece of data on record, such as a name. The technology can be used in the same way to identify a group of customers, such as members of a loyalty scheme.

To allay privacy concerns, retailers have the option to put biometric information on a swipe card, which the customer possesses. This is then presented at the POS for each transaction, and the identity of the card holder can be confirmed with a fingerprint. When the only record of a customer's information is retained by the customer, biometrics becomes much less threatening.

Customer-facing biometric solutions also play an important role in the future of retail stores. Key applications will center on reducing fraud associated with bad checks, credit cards, and debit cards. Biometric solutions could become a competitive differentiator for retailers. In the right environment, biometric solutions can be positioned as something that makes the shopping experience more convenient.

Cost savings from reduced fraud can be passed on to the customer. The retail POS is a strong environment for biometrics as a replacement for PINs or cards. Because verifying the identity of customers and employees through biometrics has great potential to reduce fraud and enrich the shopping experience, many believe that the technology will grow significantly in the near future.

Employee Facing

Biometrics can be used to verify the identity of employees at the POS and other locations, such as restricted stockrooms. The technology can reduce internal fraud and practices such as "buddy punching" (when employees clock in for their absent friends).

Since employees are responsible for a very significant percentage of retail fraud, the importance of an *employee-facing* solution is clear. And while security guards and technologies exist to minimize customer fraud, no such measures are effective against crimes perpetrated by employees.

All the current methods of preventing employee fraud have inherent security risks. Retailers can choose to use binary or traditional keys, magnetic cards, or buttons on POS devices to identify the store associate on duty. However, in each case, keys and cards can be replicated or swapped between members of staff.

Retailers are interested in biometrics because the technology eliminates all of these risks. It depends on fingerprints and other biological information unique to the person involved. In addition, retailers that implement solutions on the employee side do not need to invest in keys and cards or replace those that are lost or damaged. They can also easily protect access to restricted areas of the store or to sensitive data stored in the back office.

Biometric solutions that reduce employee fraud are already big business for commercial merchandisers. The turnover of employees is so great in large stores that retailers need secure solutions to protect against fraud, and be able to accurately determine which cashiers and other staff are working, as well as when and where.

Application Types and Use

Biometric application of leading industry products by use include: [5]

- Physical access
- Time and attendance
- Financial/transactional
- Logical access
- HIPAA
- Integrators/resellers
- Border control/airports
- Laptops, PDAs
- Locks
- Civil ID
- Criminal ID
- PC/network access
- POS authentication
- Surveillance
- Middleware/software
- Consultants, other

Application solutions of leading industry products by type include: [6]

- Fingerprint
- Facial recognition

- Smart cards/multimodal
- Iris recognition
- Voice/speaker
- Hand and finger
- Vascular pattern
- Fingerprint recognition
- Voice verification
- Retina scan
- Hand geometry
- Signature verification
- Gait recognition
- Earlobe recognition
- DNA recognition

Characterizing Biometric Applications

Characterizing biometric applications can be expressed in terms of seven variables:

- Cooperative
- Overt
- Attended
- Habituated
- Standard environment
- Public
- Open application

Cooperative vs. **noncooperative** refers to the behavior of the "threat" or would-be deceptive user. Is the "threat" trying to cooperate with the system? If the threat is trying to enter a restricted area, the subject either cooperates with the positive ID system to fool it into thinking entrance is allowed, or is deceptively uncooperative with a negative ID system and does not trigger the alarm. One implication of this variable is the scope of the database search. In cooperative applications, users may first identify themselves with a card or PIN so that the system need only match against the claimed identity's template. In noncooperative applications, users can't be trusted to identify themselves correctly, so the entire database may need to be searched.

Overt vs. *covert* describes the awareness that biometric sampling and identification is occurring. Covert methods usually include some element that is secret or hidden and not openly practiced or engaged in or shown or avowed.

Attended vs. *nonattended* addresses the supervision level required or expected of the intended user when using the system.

Habituated vs. **nonhabituated** describes the level of experience and expertise that the intended user possesses in using the system. *Habituation* is a key factor in biometric systems that will be used daily or routinely. The user's behavior could potentially change as a result of acclimatization.

Standard vs. *nonstandard environment* describes the level of environmental control of the conditions for operation.

Public vs. *private* depicts whether users are customers (public) or employees (private).

Open vs. *closed* states whether the application is required to exchange biometric data with other systems.

An example of an application that uses these seven variables is the positive hand geometry biometric identification of users of the Immigration and Naturalization Service's Passenger Accelerated Service System (INSPASS). This application can be used to rapidly admit frequent travelers into the United States.

The system is *cooperative* because those wishing to defeat the system will attempt to be identified as someone already holding a pass. It will be *overt* because all will be aware that they are required to give a biometric measure as a condition of enrollment into this system. It will be *nonattended* and in a *standard environment* because collection of the biometric will occur near the passport inspection counter inside an airport, but not under the direct observation of an INS employee. It will be *nonhabituated* because most international travelers use the system less than once per month. The system is *public* because enrollment is open to any frequent travelers into the United States. It is *closed* because INSPASS does not exchange biometric information with any other system.

A Comparison of Biometrics

Another evaluation, described in Chapter 2, compares several biometrics with each other against seven categories [7]. These categories include the degree of approval of a technology by the public in everyday life, the ease it is to fool the authentication system, and how well a biometric resists aging. In addition, they explain how easy it is to acquire a biometric for measurement, how well the biometric separates one individual from another, and how commonly a biometric is found in each individual. A performance category indicates the accuracy, speed, and robustness of the system capturing the biometric.

Each biometric category is ranked as low, medium, or high. A low ranking indicates poor performance in the evaluation criterion, whereas a high ranking indicates a very good performance.

Using Biometric Applications

Biometric technology was inevitable and will probably be commonplace in the future. The technology offers considerable benefits when compared to conventional personal identification techniques, such as access tokens and photo ID cards. It is difficult to distinguish individuals, especially if they have a vested interest in remaining incognito. Biometric verification techniques provided a potential solution to a long-standing problem in this respect.

Why, in this fast-moving world of technological innovation, has it taken so long for biometric verification to be used in mainstream applications? Early implementations of biometric technology were far from the perfect solution many expected. With hindsight, this was partly because expectations had been raised too high by over-enthusiastic industry claims and partly because, like any emerging technology, there was a learning curve, especially in the areas of ergonomics, consistency, and user response.

Administrators have the benefit of practical experience and continued refinement over time, and of several products that have turned in excellent performance figures, which are relatively easy to use and integrate. Costs are also a lot more realistic. In many instances, biometric readers represent a viable alternative to token technology.

There are benefits to be realized across a variety of application areas in a cost-effective and relatively straightforward manner. Existing systems in physical and logical access control, time monitoring, immigration, law enforcement, and other areas have proven the concept of biometrics. While it is still a relatively low-profile technology outside of the immediate industry, public awareness is considerably greater now than it was a decade ago and many initially perceived user objections have weakened with the passage of time. Users are often intrigued to discover that biometrics technology is easy to use and functions properly.

Different technologies may suggest themselves for different applications depending upon perceived user profile, the need to interface with other systems or databases, environmental conditions, and a host of other parameters specific to the application at hand. While there are some obvious areas of application for certain technologies, it is wise to seek specialist advice in this context. Independent bodies, such as the Association for Biometrics, represent a good starting place for the newcomer to learn this technology. Similarly, there are existing publications and software tutorials to guide the potential user or systems integrator. Biometric parameters such as fingerprint scanning, hand geometry, iris scanning, voice verification, retinal scanning, signature verification, and others are well established with their own particular characteristics, which suit different circumstances and applications accordingly. Their practical performance might, in many instances, be better than supposed, with the cost of implementation realistic in comparison with conventional methods, such as token technology.

By sampling an individual's physiological and/or behavioral characteristic, such as hand geometry, voice pattern, or signature dynamics, it is possible to create a unique template to verify the user in subsequent transactions. Typically, three or more samples are taken at the time of enrollment and averaged to produce a reliable template for an individual. The storage requirement for this template varies between 9 bytes—in the case of hand geometry—to up to 1000 bytes, making it feasible to store large numbers of templates in standard memory media or to incorporate the template on a token. The majority of available products are verification systems, which means that the user is verified against a single template claimed to be his. Biometric identification requires that the system identifies an individual from a list of templates held in a database.

This is an important distinction with interesting implications, and only a few systems currently offer this facility. However, the potential biometric user must identify the task that is achieved when considering this point. In many cases, biometric verification will probably be appropriate, except in a search for multiple enrollments of the same individual or a wish to use the template as search criteria for other reasons. Several questions arise:

- When designing a system, should the emphasis be placed on the front end, or should the system be considered as a whole?

- What happens behind the identity verification?
- Should planning consider the immediate requirements only, or should an open systems approach be allowed to accommodate future requirements and advances in technology?

A criticism sometimes leveled at the security industry is that of developing closed-loop proprietary systems which, for all intents and purposes, remain static in their functionality, especially regarding potential interfaces to other systems. It is unlikely that such a software program from a security vendor would answer all requirements across all applications, and yet administrators continue to seek this approach. The alternative has often been solutions created at a significant cost to tackle a particular requirement, often with their own set of startup difficulties.

This is unfortunate, as advances in parallel areas of technology are beginning to provide for a modular open-systems approach, utilizing standard tools found in the majority of corporate technology departments. It is logical to use current standards whenever possible, providing a familiar environment for the end user. If standard data formats and links are incorporated, then there is a basis for a flexible, open-ended approach to security and related applications. This is becoming easier, as administrators witness the trend toward strategic alliances in the IT industry aimed at promoting compatibility at software, hardware, and network levels. This is good news for the systems developer, who can utilize these standard tool sets to good effect. These standards will be outlined in Chapter 12.

How does this bode for the future of biometric and related applications? Consider the constituent parts of a typical system, which include:

- The user interface at the reader or data acquisition point
- The communications subsystem, which will probably include a degree of distributed intelligence and processing
- The control and programming interface, which may be at a central, or possibly multiple points
- The main software engine

Given the importance of the last two parameters, it would seem logical to incorporate as much commonality and flexibility as possible, providing for a familiar and intuitive interface, while allowing interaction with other systems and formats where applicable. The software is the heart and brain of the system, utilizing a variety of peripheral devices and technologies as necessary to suit the application. It is in this area where future advances and benefits may occur, particularly in areas of integration. It is likely that digital imaging technology will sit side-by-side with biometric verification technology in many instances. In turn, both of these might require interaction with existing databases and report-generating methodology. They might all be subject to standard communications protocols. Peripheral components, such as biometric readers, imaging devices, and in/out controllers, will be developed in their own right and at their own pace. The user may choose different technologies in different areas of application in this respect, all interfacing to a common master system, providing, of course, that there is a suitable degree of compatibility. With a communications network, input/output devices, image processing, and standard software controls, other initiatives can be achieved utilizing these elements within a typical corporate environment.

Security related or otherwise, visual verification of events might be usefully incorporated. This could include point monitoring for both security and building management purposes. For the ability to program automated responses to certain exceptional conditions, such as energy management or personnel management, a low-cost corporate visual communications system might be provided via a simple tool set that uses standard computer technology. Management might blur the lines a little between traditional security-system methodology and the broader-based IT requirement, using the best of both worlds to provide scaleable, open systems. These applications can be tailored as needed without incurring high costs of obsolescence or redundancy.

Biometric technology viewed in isolation is interesting enough; however, when seen as a complimentary part of a larger system tool set, administrators can foresee a considerable upsurge in its implementation. Attention might have been focused too closely on the individual reading technologies, and not closely enough on the broader application picture. Standard, individually addressable devices would control everything in the corporate or government environment via intuitive user interfaces and a common data bus. Visual communication would be the norm and the Internet (Web) would contain useful information. [8]

Issues and Concerns

As with many interesting and powerful developments of technology, there are concerns about biometrics. The biggest concern is the fact that once a fingerprint or other biometric source has been compromised, it is compromised for life because users can never change their fingerprints. Theoretically, a stolen biometric can haunt a victim for decades.

Identity Theft and Privacy Issues

Concerns about identity theft through biometrics use have not been resolved. If a person's credit card number is stolen, for example, it can cause her great difficulty, since this information can be used in situations where the security system requires only "single-factor" authentication; just knowing the credit card number and its expiration date can sometimes be enough to use a stolen credit card successfully. "Two-factor" security solutions require something known plus something owned; for example, a debit card and a PIN or a biometric. Some argue that if a person's biometric data is stolen it might allow someone else to access personal information or financial accounts, in which case the damage could be irreversible. But this argument ignores a key operational factor intrinsic to all biometrics-based security solutions; biometric solutions are based on matching, at the point of transaction, the information obtained by the scan of a "live" biometric sample to a pre-stored, static "match template" that was created when the user originally enrolled in the security system. Most of the commercially available biometric systems ensure that the static enrollment sample has not been tampered with (by using hash codes and encryption), so the problem is effectively limited to cases where the scanned "live" biometric data is hacked. Even then, most competently designed solutions contain antihacking routines. For example, the scanned "live" image is virtually never the same from scan to scan owing to the inherent plasticity of biometrics; ironically, a "replay" attack using the stored biometric is easily detected because it is too perfect a match. Computer system and network privacy concerns will be described in Chapter 10.

Sociological Concerns

As technology advances and time progresses, additional private companies and public utilities will use biometrics for safe, accurate identification. However, these advances will raise many concerns throughout society, where many may not be educated on the methods. Here are some examples of society's concerns with biometrics:

- Physical: Some believe this technology can cause physical harm to an individual using the methods, or that the instruments used are unsanitary. For example, there are concerns that retina scanners might not always be clean.
- Personal Information: There are concerns about whether personal information obtained through biometric methods can be misused, tampered with, or sold by criminals stealing, rearranging, or copying the data. Also, the data obtained using biometrics might be used in unauthorized ways without the individual's consent.

Uses and Initiatives

The U.S. government has become a strong advocate of biometrics with the increase in security concerns in recent years. Starting in 2005, U.S. passports with facial biometric data have been produced. Privacy activists in many countries have criticized the use of this technology for the potential harm to civil liberties, privacy, and the risk of identity theft. Currently, there is some apprehension in the United States and other countries that the information can be "skimmed," identifying people's citizenship remotely for criminal intent, such as kidnapping and extortion. In addition, technical difficulties are currently delaying biometric integration into passports. These difficulties include compatibility of reading devices, information formatting, and nature of content. Some organizations use only image data, whereas others use fingerprint and image data in their passport *radio frequency ID* (*RFID*) biometric chip(s) or transponder (*transmitter-responder*). An RFID tag is an object that can be applied to or incorporated into a product, animal, or person for the purpose of identification using radio waves. Some tags can be read from several meters away and beyond the line-of-sight of the reader.

The U.S. government stated in 2006 that anyone willing to enter the United States legally to work would be card-indexed and must provide fingerprints upon entry. Many foreigners will be subjected to these procedures, which were formerly imposed only on criminals and spies and not to immigrants and visitors—even less to citizens. In his address on immigration reform in May 2006, President George W. Bush said, "A key part of that system should be a new identification card for every legal foreign worker. This card should use biometric technology, such as digital fingerprints, to make it tamper-proof."

The U.S. Department of Defense's (DoD) common access card is an ID card issued to all U.S. service personnel and contractors on U.S. military sites. This card contains biometric data and digitized photographs, and has laser-etched photographs and holograms to add security and reduce the risk of falsification.

Activities and Research

Commercial applications such as computer network logon, electronic data security, ATMs, credit card purchases, physical access control, cellular phones, PDAs, medical records management, and distance learning are examples of authentication applications. These are typically cost-sensitive with a strong incentive for being user-friendly. Given an input biometric sample, a large-scale identification determines if the pattern is associated with any of a large number of enrolled identities. Typical large-scale identification applications include:

- Driver's license
- Criminal investigation
- Welfare-disbursement
- Voter ID cards
- Missing children identification
- National ID cards
- Border control
- Corpse identification
- Parenthood determination

These large-scale identification applications require a large sustainable throughput with as little human supervision as possible. Screening applications covertly and unobtrusively determine whether a person belongs to a watch list of identities. Examples of screening applications could include airport security, security at public events, and other surveillance applications. The screening watch list consists of a moderate number of identities. By their very nature, the screening applications:

- Do not have a well-defined "user" enrollment phase
- Can expect only minimal control over their subjects and imaging conditions
- Require large sustainable throughput with as little human supervision as possible

Neither large-scale identification nor screening can be accomplished without using token-based or knowledge-based identification. Biometric supporters everywhere agree it is not a mere matter of a superficial system tuning or clever system improvisation. For example, almost a century after the fingerprints were observed to be distinctive, a 2004 fingerprint contest revealed that fingerprint matching algorithms have a false nonmatch error rate of 2 percent.

While the error rate of the fingerprint system can be significantly reduced by using multiple fingers, the point being emphasized is that the error rate is nonzero. Similarly, even though the first paper on automatic face recognition appeared in the early 1970s, state-of-the-art face recognition systems have been known to be fragile in recent operational tests. Speaker recognition field awaits good solutions to many of its critical problems. More recent biometric identifiers (such as the iris) have low error rates. The technology test data may not be representative of a target application population, but the performance is certainly representative of the order-of-magnitude estimate of the best-of-the-breed matcher capability. Operational test performance is expected to be significantly lower than the technology test performance.

The biometric recognition problem appears to be more difficult than had been perceived by the pattern-recognition research community. The complexity of designing a biometric system is based on the following three main factors:

- Accuracy
- Database size
- Usability

Many application domains require a biometric system to operate on the extreme (tolerance) for only one of the factors; such systems have been successfully deployed. The challenge is to design a system that will operate on the extremes of all these three axes simultaneously. This entails overcoming the fundamental barriers that have been cleverly avoided in designing the currently successful niche biometric solutions. Addressing these core research problems may significantly advance the state of the art and make biometric systems more secure, robust, and cost-effective. This could promote the adoption of biometric systems, resulting in potentially broad economic and social impact.

Smart Card Personalization and Distribution

The process of making the smart card ready for the user is called personalization. The procedure includes printing the user's picture, installing applets and PKI (X.509) certificates, and binding the smart card to both the user and the physical access system.

In most cases, a smart **card management system** (CMS) is a deployment requirement. Important considerations when evaluating a CMS include:

- Administrative delegation and scoping capabilities: These components enable the secure management of smart cards across the organizational hierarchy.
- Platform support: This is necessary for vendor physical access systems, the smart card, and printer products.
- Key escrow and recovery capabilities: Such features can re-create the user's PKI credentials on a new smart card in the event that the previous card is destroyed or lost.
- Provisioning system integration: Integration provides a single authoritative source of identity information and consistent access rights.
- Remote applet distribution capabilities: Java card applets provide most smart card functionality, and the CMS can deliver new applets after the smart cards have been distributed to the users.

It's easy to see why the distribution of smart cards to an organization's employees is considered "involved." The process can take months or even years, and the many important details require careful planning. The distribution of smart cards to "virtual" employees, those who rarely visit a campus, requires special attention. The PKI standard will be described in Chapter 12.

Upgrade of Physical Access Control Systems

It's a toss-up as to which activity causes more organizational grief: the distribution of smart cards, or the upgrading of the physical access systems across an organization's campuses. Organizations may have a wide spectrum of physical access technologies throughout their environments, from keys to magnetic stripes to biometric authenticators to contactless tools. As part of the planning process, an organization should take an inventory its campus-wide physical access system and determine what upgrades are necessary for the implementation of a common authenticator.

While there is wide variability, the typical components of modern physical access systems include readers, controllers, security servers (hosts), and cards. Even if some of these components are present, the organization may still be required to upgrade them. During the gradual migration to a common authenticator, multitechnology readers, or even controllers, may be required to enable the use of various card types. Controllers may also need to be upgraded to work over Ethernet (LAN) instead of a serial, RS-232 protocol. Such improvements can better support a modern architecture and an organization's physical–logical convergence goals.

Emergency Access Procedures

It is a fact of life: users will leave their smart card at home. Without the card, they cannot access applications, workstations, buildings, and maybe the parking lot or the bathroom. With proper emergency access measures, such an error should only be temporary. The organizational challenge is to implement emergency access procedures that give forgetful, cardless individuals a timely access to resources. The access processes must also occur in a cost-effective manner. Some tricks of the trade include:

- Self-service kiosks in the building entrance where employees can authenticate and get a temporary smart card
- Physical access readers with PIN pads that enable the user to temporarily authenticate with an identification number
- IT software management tools that temporarily allow the user to authenticate with a password instead of a smart card

Even in the face of the details mentioned above, planning for a common authenticator appears more daunting than real. If the organization defines achievable milestones and exercises vigilance against the temptation of expanding and redefining the objective of the project, implementation is possible. [9]

Device authentication doesn't replace user authentication. It augments it. Hence the idea behind using identity-enabled network devices is to authenticate a device, rather than a user, before it can access a network. This emerging authentication paradigm is intended to add an extra layer of security to any kind of network device, whether a workstation, desktop or laptop, PDA, cell phone, or even a wireless access point. A user would still have to use a user ID and password, or some other authentication mechanism, to log onto a network, but the device itself would also have to authenticate.

Traditional access management is meant to allow authorized users while blocking unauthorized or malicious ones. Hardware authentication does the same thing for devices. An authorized device, like an authorized user, is trusted: it's confirmed to be virus and malware free, is patched with up-to-date software, and won't bring anything harmful in, or take anything unauthorized out, like data, from the network. [10]

Biometric Applications and Vendors

Reliable human recognition is an important problem in diverse businesses and other organizations. Biometric applications that can be used to recognize distinctive personal traits have the potential to become an irreplaceable part of many identification systems.

Numerous vendors supply system, hardware, and software solutions utilizing biometric products and services. Typical categories of systems that benefit from a biometric application include:

- Physical access control
- Logical access control

- Justice/law enforcement
- Time and attendance
- Border control/airports
- HIPAA
- Financial/transactional
- Integrators/resellers
- Laptops, PDAs
- Locks
- Other

Physical Access Control Solutions

Web references to access control involving fingerprint biometric solutions and systems appeared frequently, offering personal, convenient, and portable identity verification devices. One device replaced multiple passwords and access cards to offer secure authentication for logical and physical access (including vehicle gates from 300 feet), working with existing security infrastructure, and performing all fingerprint processing, including matching, on-device. Some devices provided a biometric terminal for access control and, additionally, time and attendance applications. Some devices could operate as a stand-alone device or in a network with other devices. Some could be equipped with *radio frequency* or a smart card reader, thus allowing storage of PIN and/or templates on cards.

One product line offered three different biometric options, fingerprint in both "one-to-one" and "one-to-many" configurations as well as a finger geometry reader for those environments that are not conducive to fingerprints. A hand *vascular pattern recognition (VPR)* biometrics device, ideally suited for physical access control and time and attendance solutions was also offered. A number of fingerprint biometric solutions addressed both access control and time and attendance applications.

Providers offered automated fingerprint identification system (AFIS), and other fingerprint biometric solutions to governments, corporations, law enforcement agencies, and other organizations worldwide. These solutions enabled customers to capture fingerprint images electronically, encode fingerprints into searchable files, and accurately compare a set of fingerprints to a database containing potentially millions of fingerprints in seconds. Manufacturers offered fingerprint recognition hardware and software products. A stated mission was to provide advanced biometric authentication products and solutions to customers who need enhanced security in identity management. With proprietary fingerprint recognition algorithms, fingerprint authentication products and solutions were available for:

- Physical access control
- Computer and network logon
- Identity management
- E-commerce

Another manufacturer advertised a full range of security devices for proximity readers and cards, finger biometrics, smart card readers, and access control panels and user-friendly application software. Biometric identification replacement for swipe-card readers, barcode readers, and PIN pads were also available.

The AFIS is a system originally developed for use by law enforcement agencies, which compares a single fingerprint with a database of fingerprint images. Subsequent developments have seen its use in commercial applications, where a client or customer has a finger image compared with existing personal data by placing a finger on a scanner, or by the scanning of inked paper impressions.

Today, vendors are beginning to offer multibiometric solutions involving the development, manufacture, and integration of multiple biometric technologies, including fingerprint, facial, and iris recognition products and services. Biometric products are being offered addressing markets including:

- Law enforcement
- Federal Government
- Homeland Security
- Driver's licenses
- Civil identification
- Applicant background checks
- Consumer and commercial applications

One vendor was identified that offered an iris recognition platform. A user interface provided sophisticated countermeasures and unique security features, integrating with contact and contactless smart cards in a variety of public and private sector applications.

Time and Attendance Biometrics Solutions

Time and attendance systems (T & A) are utilized to easily and efficiently track employee time and work activities. These systems can manage employee profiles and enhance employee scheduling, and provide data to payroll systems.

T & A solutions enable customers to capture fingerprint images electronically, encode fingerprints into searchable files, and accurately compare a set of fingerprints to a database containing employee fingerprints in seconds. Systems provide individuals with conclusive identity credentials and organizations with the ability to undeniably identify and manage their assets. Security devices include proximity readers and cards, finger biometrics, smart card readers, as well access control panels.

As mentioned in the previous section, fingerprint biometrics is often used in time and attendance applications. Time and attendance hardware is available in three different biometric options: "one-to-one" and "one-to-many" configurations as well as a finger geometry reader for those environments not conducive to fingerprint. Fingerprint biometrics solutions provide template storage and matching ability to enroll and verify in the industry. This solution eliminates "buddy punching" and has an immediate return on investment.

A number of vendors offer complete time and attendance systems designed for small- to mid-sized businesses. A series of stand-alone time attendance, F series of access control, rotation identification, and stand-alone SDK for customer application are also offered. The hand VPR biometric device is suited for time and attendance solutions.

Secure workforce management solutions provide advanced fingerprint biometrics and an intelligent JAVA programming environment with the power of a PC. This provides fast, accurate, and reliable data collection for any application, including time and attendance, scheduling, employee self-service, and access control.

Multibiometric solutions integrate multiple biometric technologies including fingerprint, facial, and iris recognition products and services for biometric identification.

Multibiometric enrollment and management platforms are available that enables the biometric searching and matching of unlimited population sizes. Multibiometrics will be described in Chapter 6. Biometric devices and algorithms meet users' identity management requirements for T & A applications.

Biometric identification replacement of swipe-card readers, barcode readers, and PIN pads reduce problems and costs created by cards, and PINs are eliminated. This results in irrefutable proof of identification for security, T & A, and record keeping.

Iris recognition for T & A applications is under development, providing sophisticated countermeasures and unique security features. Various models integrate contact and contactless smart cards with accurate noncontact authentication.

Financial/Transactional Biometrics Solutions

Financial transactions generally fit in one of these categories:

- Purchase
- Loan
- Mortgage
- Account
- Credit-card purchase
- Debit-card purchase

Security in the financial and transactional industry is essential in the e-commerce environment. Biometric hardware readers are employed to take advantage of a three-form (factor) authentication: what is owned (ID card), and what is known (added security by using a PIN) and physical (fingerprint for the final security advantage, something that cannot be stolen or misplaced).

A major concern is to protect critical business assets. Secure authentication, verification, and management of users are employed in enterprise, commercial, educational, and governmental entities to verify their identities and control their access to networks and applications. Advanced voice biometrics systems can be used to provide a more convenient, practical, and secure alternative to using PINs, passwords, or security tokens over the phone or Internet. Identity verification can be effected by a voice-verification platform certified and accredited to issue advanced electronic signatures (voice digital certificates) and to be ISO 27001 rated.

Logical access using identity authentication and file security can improve employee productivity and reduce information technology administrative costs. Prevention of data and identity theft, while preventing false aliases by a finger identification algorithm that has been certified by the ICSA, is effective. One-to-many finger and one-to-one identification solutions provide security and positive identification for both government and enterprise applications.

Solutions for digital signatures (biometrically enabled) include single-sign-on (or "password bank") applications. Fingerprint biometrics can be applied to everyday life with personal, convenient, and portable identity verification devices. AFIS and other fingerprint biometric solutions can be applied to numerous financial and banking applications. Solutions enable customers to capture fingerprint images electronically, encode fingerprints into searchable files, and accurately compare a set of fingerprints to a database containing potentially millions of fingerprints in seconds.

Identity management provides multibiometric, secure credential, and investigative solutions. Multibiometric enrollment and management platform enables the biometric searching and matching of unlimited population sizes. It supports any biometric device and algorithm for identity management requirements. Systems can store templates, files, and passwords in a safe area and are able to perform cryptographic operations. Scanners are capable of matching fingerprints on board computing devices, increasing security and privacy.

Fingerprint authentication eliminates the security risks and costs of password management. Users and customers can log on to networked and Internet applications with a touch of their finger; reducing help desk calls, improving network security, and accelerating regulatory compliance.

Similar to other biometric applications, multibiometric solutions are available for integration into multiple biometric technologies including fingerprint, iris, and facial recognition products and services. Biometric identification using fingerprint technology is useful for processing financial transactions. Security hardware devices for proximity readers and cards, finger biometrics, smart card readers, as well as access control panels are available for protection of financial assets. An iris recognition platform integrates contact and contactless smart cards for accurate, noncontact authentication.

Laptops, PDAs, Locks, and Other Biometrics Solutions

Solid-state fingerprint sensors and world-class minutia-based matching algorithms improve the security and convenience of user authentication. Sensors provide a quick, reliable alternative to passwords, PINs, keys, and other methods of user authentication.

Fingerprint technology includes three categories: optical fingerprint reader technology, fingerprint verification algorithm technology which obtains a state invention patent, and build-in application technology. Fingerprint technology products include fingerprint locks, fingerprint safes, and fingerprint Bio-Guard or Bio-Pass to PC.

Biometric finger identification technologies provide applications for prevention of data and identity theft while preventing false aliases. Finger identification algorithms have been certified by the ICSA. One-to-many finger and one-to-one identification solutions provide security and positive identification for both government and enterprise applications involving computer devices.

Biometric hardware including drivers' license and government-issued ID card readers utilize the *common access card (CAC)*. It reads 2-D magnetic strip, and contact smart card data. Multiple lock control applications are implemented involving law enforcement, casinos, access control, bars and nightclubs, convenient stores, and the health care industry.

A common biometric system solution involves AFIS solutions to governments, corporations, law enforcement agencies, and other organizations worldwide. These systems capture fingerprint images electronically, encode fingerprints into searchable files and accurately and quickly compare a set of fingerprints to a large database containing millions of fingerprints.

Logical Access Control Biometrics Solutions

Biometric security software products for logical access, identity authentication, and file security improve employee productivity and reduce information technology administrative costs. Many solutions offer secure authentication, verification, and management of identities. Products allow enterprise, commercial, or governmental entities to verify the identity of users and control their access to networks and applications. Payment verification adds "speak on the dotted line" digital signatures to credit or debit card payments over the phone or Web. Web authentication dramatically improves the security of Web-based services.

Breed identity management infrastructure products are available to support customers, integrators, and device and algorithm vendors in offering solutions with the highest levels of privacy, security, availability, throughput, and timeline performance. Biometric middleware minimizes implementation and total ownership costs and risk while maximizing business benefits by integrating with universal IT platforms and managing the long-term identity life cycle.

Voice biometrics systems can provide a more convenient, practical, and secure alternative to using PINs, passwords, or security tokens over the phone or the Internet. Banks, insurers, and government agencies already use VoiceVault to deliver enhanced levels of service and increased security, and to optimize identity verification costs. Voice verification platforms are certified and accredited to issue Advanced Electronic Signatures (voice digital certificates) and be ISO 27001 rated.

VoiceVault uses spoken words to calculate vocal measurements of an individual's vocal tract. Sophisticated algorithms convert these measurements into a voice print; creating a unique digital representation of an individual's voice. Numerous different calculations and tests are used to compare a previously enrolled voice print with a voice print generated at the time of verification. (*Note that a voice print is not a recording, set of words, or a wave pattern of a voice. It cannot be played back or used for any other purpose than a comparison with subsequent voice prints*). Special filters and algorithms are used to eliminate background noise as well as to detect and reject any attempt to use voice recordings.

Because verification is based on vocal physiology, accuracy is not affected by the caller having a cold, for example. It adapts voice prints, over time, to take into account the aging effects on the voice. VoiceVault is a cross-platform system. This means the same caller can be accurately verified again and again regardless of the phone he is using, whether it is a mobile phone, fixed line, speakerphone, or even the Internet. Tackling identity-theft risks has never been more important to protect a business and retain customer confidence in phone and Internet channels to market. Caller authentication verifies the true identity of callers through their voice rather than easily lost or stolen passwords or PINS.

Biometric verification technology supports a range of fingerprint capture devices. A single enrollment can be used to distribute the user's biometric information throughout all of the physical access points, directories, and applications that are designated to utilize fingerprint verification. It enables each user to create a biometric template that is interoperable with a multitude of supported fingerprint sensors, including those in fingerprint readers used for securing building access. Solutions enable organizations to utilize a user's biometric record throughout all applications within an enterprise, without the need for costly reenrollment.

Solid-state fingerprint sensors and world-class minutia-based matching algorithms improve the security and convenience of user authentication. These solutions offer an alternative to passwords, PINs, keys, and other methods of user authentication. Additional biometric products, based on fingerprint acquisition and recognition provide image processing technologies and the use of smaller and lighter optical-mechanics modules, products and solutions for government, police forces, institutions, commercial, and international organizations.

An optical fingerprint scanner is capable of performing several useful tasks, improving security, efficiency, and usability. An optical design allows high-quality images (569 dots per inch (dpi)). Encrypted data transfer makes communications extremely secure. Solutions include identity verification/recognition. AFIS and other fingerprint authentication eliminate the security risks and costs of password management. It also allows users to log on to both networked and Internet applications with a touch of their finger; reducing help desk calls, improving network security, and accelerating regulatory compliance.

Biometric finger identification technologies provide for prevention of data and identity theft, while preventing false aliases. BIO-key pioneered the only finger identification algorithm that has been certified by the ICSA one-to-many finger and one-to one identification solutions that provide security and positive identification for both government and enterprise applications. Fingerprint biometrics is applied to everyday life with personal, convenient, and portable identity verification devices. One device replaces multiple passwords and access cards to offer secure authentication for logical and physical access (including vehicle gates from 300 feet), works with existing security infrastructure, and performs all fingerprint processing, including matching, on-device.

Numerous vendors supply fingerprint recognition hardware and software products. Advanced biometric authentication products and solutions are available to customers who need enhanced security in identity management. Fingerprint recognition algorithms provide fingerprint authentication products and solutions for physical access control, computer and network logon, identity management, and e-commerce.

Mobile handheld reader-identification terminals are available for specialized applications and interoperability into existing systems. Deployment is available into mainstream verticals such as commercial, law enforcement, government, and military markets. Vendors develop, manufactures, and integrate multiple biometric technologies including fingerprint, facial and, iris recognition products and services. Biometric identification markets include law enforcement, Federal Government, Homeland Security, driver's licenses, civil identification, and applicant background checks as well as consumer and commercial applications.

Multibiometric solutions for identity management, provides multibiometric, secure credential and investigative solutions. Multibiometric enrollment and management platforms enable the biometric searching and matching of unlimited population sizes. These solutions are supporting any biometric device and algorithm for identity management.

Integrated smart card and biometric fingerprint readers facilitate government and commercial organizations to comply with HSPD12 and HIPAA. The use of smart cards and biometrics for authentication to digital information for security and ease of use is made simple by integrating this process into a Cherry keyboard (Cherrycorp). Replacements are available for swipe card readers, barcode readers, and PIN pads. Problems and costs created by cards and PINs are eliminated resulting in irrefutable proof of identification for security and record-keeping. Single identification solutions and integrations can be seamlessly installed with different applications already in place.

Public Domain Biometrics Applications

The bulk of biometrics applications to date are most likely in areas that are unique. This is because there are a very large number of relatively small security related applications undertaken by specialist security systems suppliers. These systems account for the majority of unit sales as far as the device manufacturers are concerned and are often supplied via a third-party distribution chain. There are numerous biometric applications in the public domain. [11]

Border control is being enhanced by **NEXUS**, [12] which is a joint Canada–United States program designed to let preapproved travelers cross between that border more quickly. Members of the program can avoid long waits at border entry points by using self-serve kiosks at airports, reserved lanes at land crossings, or by phoning border officials when entering by water. This program is operated by both the Canada Border Services Agency and the United States Customs and Border Protection. A NEXUS membership card is a valid document under the Western Hemisphere Travel Initiative when used at a NEXUS kiosk.

Only citizens and permanent residents of Canada or the United States are eligible for the NEXUS program. Applicants must have lived in one of these countries continuously for the last three years. Authorities will check applicants' documentation, run background checks, and take fingerprints, photographs, and digital iris photographs. Applicants must also pay a fee, which is waived for applicants younger than 18 years old. Applicants who pass these checks are issued a NEXUS card, which is valid for five years.

While NEXUS card holders are screened more quickly, they are still subject to all the regular standard border checks and procedures. In addition, a vehicle can only use the NEXUS lane if everyone in the vehicle has a NEXUS card, including children. Likewise, NEXUS card holders declaring certain items must use the regular lane, and are prohibited from transporting certain items across the border.

Biometric technology offers countless solutions for military applications in future systems, including identity assurance, access control, and force protection, as well as new uses yet to be discovered. Several devices have been deployed in recent military operations in Afghanistan and Iraq, viewed by the entire world in real time, clearly showing the success of and the increased need for secure information and weapon systems. Through collaboration of the best-of-breed biometric technologies in government, industry, and academia, the *Biometrics Task Force (BTF)* can ensure the nation's security efforts stay technologically advanced and relevant.

Additional areas of the government that benefit from these technologies include voting systems, schools, prisons, benefits, and licensing. Specifics include:

- Voting systems: Eligible politicians are required to verify their identity during a voting process. This is intended to stop "proxy" voting, where the vote may not go as expected.
- School areas: Problems have been experienced with children being molested or assaulted. Subsidized campus meals are available to bona fide students; however, the system was being heavily abused in some areas.

- Prison visitor systems: Visitors to inmates are subject to verification procedures to prevent identities from being swapped during the visit—a familiar occurrence among prisons worldwide.
- Benefit payment systems: Several states have saved significant amounts of money by implementing biometric verification procedures. Not surprisingly, the numbers of individuals claiming benefit has dropped dramatically in the process, validating the systems as an effective deterrent against multiple claims.
- Driver's licenses: Some authorities have found that drivers (particularly truck drivers) had multiple licenses or swapped licenses among themselves when crossing state lines or national borders.

In addition, there are numerous applications in gold and diamond mines, bullion warehouses, and bank vaults, as might be expected, as well as the more commonplace physical access control applications in industry.

E-Passport Application

The defense of a nation's borders and the protection of its citizens have always been basic duties of government. Many nations now call it homeland security. To what extent does homeland security depend on an integrated approach across boundaries and organizations, and how do the emerging technologies of secure e-passports and biometrics facilitate collaboration? [13]

Homeland security has become a key issue since 9/11. Organizations that move people and goods across borders find themselves at the frontline of the conflict against new threats from criminals who are harnessing technology as never before to further their aims. High-security networks and mobile technologies are exploited equally by criminals, illegals, and the general public.

The major challenge in delivering homeland security is to operate business as usual in extraordinary circumstances. If air and seaports cannot permit swift movement of goods and people, the economy suffers. Fundamental to achieving these twin objectives of security and openness is the ability for authorities to be certain of the identity of the people crossing borders for travel or trade.

Secure electronic documents like ID cards, e-passports, and cargo manifests provide trustworthy and accurate identification of goods and travelers. With agreement on standards, reductions in costs and the heightened state of government awareness, projects to issue secure personal credentials are no longer considered too expensive.

The e-passport is a common, secure form of identification that contains a chip within the document which stores digital versions of the printed passport data, including the photograph of the bearer. In some cases, fingerprint data is also being considered for inclusion. The contents of the chip are read using a secure key. If the digital picture matches both the printed version in the passport and its owner, it is safe to assume that both are genuine.

E-passports are being issued by many countries. They are designed to be interoperable with standard technology at border control points. In this way, they are more appropriate for the secure identification of travelers than ID cards, which are issued by the nation principally for access to national services. Although there are agreements at European borders that an ID card will be accepted instead of a passport, in the future there may be technology incompatibilities where a nation's needs have taken precedence over international standards in the specification of the card. E-passports are designed to be interoperable, durable, and trustworthy identification tokens. Border control organizations looking to build trust into their systems should consider the role that e-passports can play.

The ability to insert computer chip technology within paper documents is just part of the story. Crucial, too, will be the development of infrastructures to issue, read, update, and maintain the new computers, and integrate them into business processes. Collaboration is vital, not just between police, immigration, and customs, but also between carriers, airlines, and airport authorities. Efficiency can only be achieved if all the various groups concerned with travel security know each other's roles and working procedures, and share the same vision. To implement technology on a process that does not have widespread support and common understanding is inviting problems. When there is common agreement and shared vision, there is a chance of success.

The common **Schengen visa** system uses biometrics to ensure an applicant has not previously applied for a visa at another embassy or under an assumed identity. Once issued, the visa allows for rigorous checking by comparing the fingerprints presented at the border to those held in a central database. Over the next five years, the database is expected to grow to store data on over 70 million people. The system depends on collaboration between national systems that are accepting the visa applications and the central EU system that supports the processing of the applications. Without collaboration the system is pointless. If no government installs the enrollment system, there will be no biometric data to process.

For this particular application, fingerprints provide the best balance across a range of issues including accuracy, cost, acceptability, and convenience. Each of the main methods of biometry has different characteristics, which means each may be appropriate for different uses. Cultural forces are particularly influential. Some countries associate fingerprints with crime, but for others it has been perfectly acceptable as a means of identification for decades. Iris recognition is extremely accurate, but some cultures prefer not to bow to a machine. Other methods are not quite infallible; facial biometric recognition is not yet as accurate as the human brain and voice recognition is vulnerable if the subject has a cold or is emotional.

For the new Schengen visa, it would be possible to store the fingerprint data within the e-travel document, such as a visa or passport. Indeed, many employee security access-control systems use that technique. However, to guarantee future interoperability with new biometric approaches, it has been decided that the fingerprint image will be held on a central database and not in the visa. The visa sticker will tell the checking system where to find fingerprint data in the central system to verify the identity of the visa holder. The new system will be more secure than it is now and may prevent the theft of valid visa documents, because the visa will not be usable for anyone else.

With a Schengen visa, travelers may enter one country and travel freely throughout the Schengen zone. Internal border controls have disappeared; there are no or few stops and checks. This means internal air, road, and train travel are handled as domestic trips, similar to travel from one U.S. state to another.

The emergence of new technology is always accompanied by great expectations and some exaggerated views of capabilities. Projections that new e-documents can be read as people pass through the travel process without a need for passengers to be stopped as they travel are unlikely to be realized in the near future. The contactless smart chip selected for e-passports uses a radio frequency that operates at distances of up to 10 centimeters. The potential for chips to be read at distance by powerful receivers has led to the use of secure readers and messaging to minimize the dangers of chip contents being read by unauthorized personnel.

The increasing dependence of homeland security initiatives on the secure identification of individuals is not welcomed by all. Forcing citizens to assert their identity to a higher authority using technology that can be used to track their movements is not compatible with certain perspectives on civil liberty. The potential for error (*the false rejection of the good actors and the false acceptance of the bad actors*) means that biometry is at best viewed with suspicion and at worst rejected outright. Of course, all technology can be misused, but with appropriate procedures in place for the accurate capture of biometrics in the first place, and thereafter, accurate reading through good technology and processes, the fears over false acceptance and false rejection are manageable. The problems highlighted in recent pilot projects are all addressable and do not justify dismissal of the whole concept. The reason why the technology is being tested is to identify areas for rectification rather than to prove that it doesn't work. [14]

Personal Biometric Devices

Before the mid-1990s, optical fingerprint-capture devices were bulky and expensive. Technological advances have brought the size and cost down dramatically; the new solid-state sensors cost less than $100 and occupy the surface area of a postage stamp. Previously used primarily for government applications, fingerprint authentication technology is now steadily progressing into the private sector for many applications requiring both convenience and security.

The small size and cost of these devices can provide secure access to desktop PCs, laptops, the Web, and most recently, to mobile phones and palm computers. Automobile manufacturers are building prototype cars with access and personalization (of seat position, radio channels, and so on) that are controlled by fingerprint authentication devices. When these sensors become small, inexpensive, and low power enough to build into a key fob, many of us will carry a universal key to facilitate secure. Fingerprint authentication devices will find increasing application in securing laptops and providing access to everything from front doors to car doors, computers, and bank machines.

A sensor, located either in the car door handle or in a key fob, could unlock a vehicle, and another in the dashboard could control the ignition. Reliability is a concern, however, because automobile sensors must function under extreme weather conditions on the car door and high temperature in the passenger compartment. And a key fob sensor must be scratch-, impact-, and spill-resistant. It must also be able to sustain an electrostatic discharge of greater than 25 kV, which is no small voltage for a chip. Despite these concerns, automotive parts manufacturers are forging ahead. Safeguards, such as protecting the sensor within an enclosure or placing it in a protected location on the car, are under consideration.

The companies developing this technology have used different means for fingerprint capture, including electrical, thermal, or others. For example, a capacitive-sensing chip measures the varying electrical-field strength between the ridges and valleys of a fingerprint. A thermal sensor measures temperature differences in a finger swipe, the friction of the ridges generating more heat than the nontouching valleys as they slide along the chip surface. Some companies are working on optical and hybrid optical/electrical capture devices whose optics have shrunk to about 1.5 cubic inches. One of the first widespread applications of personal authentication will be for portable computing. In terms of financial losses for corporate computing, laptop theft ranks high.

Personal authentication also can come into play in cryptography, in the form of a private-key lockbox, which provides access to a private key only to the true private-key owner via a fingerprint. The owner can then use this private key to encrypt information relayed on private networks and the Internet. Although good encryption methods are very difficult to break, the Achilles heel in many encryption schemes is ensuring secure storage of the encryption key (or private key). Frequently, a 128-bit or higher key is safeguarded only by a six-character (48-bit) password. A fingerprint provides much better security and unlike a password is never forgotten. In the same way, a fingerprint-secured lockbox can contain digital certificates or more secure passwords (ones that are much longer and more random than those commonly chosen) for safeguarding e-commerce and other Internet transactions. These schemes assure a user security of electronic transactions as well as personal privacy.

Recognizing the potential of small and inexpensive fingerprint sensors, several companies have developed technologies for this purpose. Among these are the following:

- Authentec makes a biometric identification subsystem. It uses CMOS and electric-field imaging (www.authentec.com).
- Veridicom has products that use CMOS and capacitive imaging (www.veridicom.com).
- L1 Identity Solutions makes optical fingerprint readers (www.l1id.com/).

The small size and low cost of these new fingerprint sensors make them an ideal human interface to secure systems. These and many more applications will soon include personal biometric authentication. If the current trends continue, the public sector can expect to see such devices increasingly incorporated into everyday life.

Providers of Biometric Solutions

If the list of providers below is any indication, hardware and software vendors are all trying to get on the biometric bandwagon. The number of vendors appears to be endless, with offerings in all categories of biometric modalities. A search of the Web for vendors providing biometric systems and solutions included the following references:

Accu-Time Systems, Inc

ADEL

Anviz

BIO-key

Bioscrypt

Ceelox

Cherry

Cogent

DAON

Datastrip

DigitalPersona

E-Seek

Fujitsu

Futronic
Fx2000
Fx3000
FxLock
Genus®
Green Bit
Identica Corp
identiMetrics
IdentiPHI
IDTECK
ImageWare
LG Electronics
Privaris
SAGEM Morpho
Sequiam Corporation
VoiceVault
Wasp Barcode Technologies
ZK Software

Many of these vendors offer products and services for multiple application areas. This list is a small snapshot of potential biometric solution suppliers. [15] Many others can be identified given additional time and energy.

Future Applications

There are many varying views concerning potential biometric applications. Some popular examples are:

- ATM machine use
- Workstation and network access
- Travel and tourism
- Telephone transactions
- Public identity cards

ATM Machine Use

Most of the leading banks have been experimenting with biometrics for ATM machine use and as a general means of combating card fraud. Surprisingly, these experiments have rarely consisted of carefully integrated devices into a common process, as could easily be achieved with certain biometric devices. The banks have a view concerning the level of fraud and the cost of combating it via a technology solution such as biometrics. They also express concern about potentially alienating customers with such an approach. However, it still surprises many in the biometric industry that the banks and financial institutions have failed to enthusiastically embrace this technology.

Workstation and Network Access

Historically, this was an area often discussed but rarely implemented until recent developments saw the unit price of biometric devices fall dramatically and several designs aim squarely at this application. In addition, with household names such as Sony, Compaq, KeyTronics, Samsung, and others entering the market, these devices appear almost as a standard computer peripheral. Many see this as the application which will provide critical mass for the biometric industry and create the transition between sci-fi device to regular systems component, thus raising public awareness and lowering resistance to the use of biometrics in general.

Travel and Tourism

Many in this industry foresee a multiapplication card for travelers which, incorporating a biometric, would enable them to participate in various frequent flyer and border control systems as well as pay for their air ticket, hotel room, rental car, etc., all with one convenient token. Technically this is possible, but from a political and commercial point of view there are still many issues to resolve, not the least being who would own the card, be responsible for administration, and so on. These may not be insurmountable problems and perhaps something along these lines may still emerge. A notable challenge in this respect would be to package this initiative in a way that would be truly attractive for users.

Internet Transactions

Many immediately think of online transactions as an obvious area for biometrics, although there are some significant issues to consider in this context. Assuming device cost could be brought down to a level whereby a biometric card reader could be easily incorporated into a standard PC, the problem of authenticated enrollment and template management still exist, although there are several approaches available. If a credit card already incorporated a biometric, this would simplify things considerably. It is interesting to note that certain device manufacturers have collaborated with key encryption providers to provide an enhancement to their existing services.

Telephone Transactions

No doubt many telemarketing and call-center managers have pondered the use of biometrics. It is an attractive possibility to consider, especially for automated processes. However, voice verification is a difficult area of biometrics, especially if one does not have direct control over the voice transducers. The variability of telephone handsets coupled with the variability of line quality and the variability of user environments presents a significant challenge to voice verification technology, and that is before the variability in understanding among users is considered. The technology can work well in controlled closed-loop conditions, but is extraordinarily difficult to implement on anything approaching a large scale. Designing in the necessary error correction and fallback procedures to automated systems, in a user-friendly manner, is also not a job for the fainthearted.

Public Identity Cards

A biometric incorporated into a multipurpose public ID card would be useful in a number of scenarios if one could win public support for such a scheme. Unfortunately, many individuals definitely do not want to be identified. This ensures that any such proposal would quickly become a political hot potato and a nightmare for the politicians concerned. From a dispassionate technology perspective, it represents something of a lost opportunity, but this is of course nothing new. It is interesting that companies have issued cards where named card holders can receive various benefits, including discounts at local stores and on certain services. These do not seem to have been seriously challenged, even though they are in effect an ID card.

Perhaps further developments will largely overcome these problems. Certainly there is a commercial incentive to do so and no doubt research is under way in this respect. The technology can work well in controlled closed-loop conditions but is extraordinarily difficult to implement on anything approaching a large scale. Designing in the necessary error correction and fallback procedures to automated systems in a user-friendly manner is a difficult assignment. Perhaps there will be additional developments which will largely overcome these problems. Certainly there is a commercial incentive to do so and no doubt considerable research is underway.

Summary

- Security applications include hardware, software, technical, managerial, and protocol solutions to ensure the protection of an organization's computer, database, network, and physical resources and assets. Many of these applications are oriented toward issues relating to access control, which restricts access to only authorized individuals and processes.
- Biometric-based solutions are able to provide for confidential financial transactions and personal data privacy. The need for biometrics can be found in federal, state and local governments, in the military, and in commercial applications. Applications are generally classified according to solutions that are citizen-facing, customer-facing, and employee-facing.
- Biometric applications of leading industry products by use include physical access, time and attendance, financial/transactional, border control/airports, surveillance, and many others.
- There are a number of concerns and issues with applications using biometrics. A major concern with biometric applications is the fact that once a fingerprint or other biometric source has been compromised, it is compromised for life. Obviously, users can never change their fingerprints or other biological and physiological identifiers, therefore making it possible that a stolen biometric can haunt a victim for decades.

Key Terms

Application programming interface (API) A set of services or instructions used to standardize an application. An API is computer code used by an application developer. Any biometric system that is compatible with the API can be added or interchanged by the application developer.

Biometrics application programming interface (BioAPI) specification Defines the application programming interface and service provider interface for a standard biometric technology interface. Developed by the BioAPI consortium.

Card management system (CMS) A smart card/token and digital credential management solution that is used to issue, manage, personalize, and support cryptographic smart cards and PKI certificates for identity-based applications throughout an organization.

Common biometric exchange file format (CBEFF) Describes a set of data elements necessary to support biometric technologies in a common way independently of the application and the domain of use.

Cooperative Refers to the behavior of the "threat" or would-be deceptive user.

Habituated Describes the level of experience and expertise the intended user possesses in using the system.

Identification mode The biometric system that identifies a person from the entire enrolled population by searching a database for a match based solely on the biometric.

NEXUS A joint Canada–United States program designed to let preapproved travelers cross between that border more quickly.

Noncooperative Refers to the behavior of the "threat" or would-be deceptive user.

Nonhabituated Describes the level of experience and expertise the intended user possesses in using the system.

Schengen visa A system that uses biometrics to ensure an applicant has not previously applied for a visa at another embassy or under an assumed identity.

Verification mode The biometric system that authenticates a person's claimed identity from their previously enrolled pattern.

Review Questions

1. Biometric applications are generally categorized into three classifications. These are: ________________, ________________, and ________________.

2. Biometric devices, such as finger scanners, consist of ________________, ________________, and ________________.

3. ________________ provides formatting instructions or tools used by an application developer to link and build hardware or software.

4. The ________________ defines the application programming interface and service provider interface for a standard biometric technology interface.

5. Biometric identification systems can be grouped based on the main physical characteristics. These are ________________, ________________, and ________________.

6. The common ________________ system uses biometrics to ensure an applicant has not applied previously under an assumed identity.

7. The biggest concern about biometrics is ________________.

8. The ______________ in the database are processed using an algorithm that translates that information into a numeric value.

9. Applications are generally classified according to solutions that are ______________, ______________, and ______________.

10. ______________ vs. ______________ describes the level of experience and expertise the intended user possesses in using the system.

11. ______________ vs. ______________ states whether the application is required to exchange biometric data with other systems or not.

12. Identify the four data security issues that must be addressed in legacy applications. ______________, ______________, ______________, and ______________.

13. Speaker recognition systems employ three styles of spoken input: ______________, ______________, and ______________.

14. A retina scan provides an analysis of the veins located in the front of the eye. (True or False?)

15. Driver's license, immigration, and employment are all examples of customer-facing applications. (True or False?)

16. Overt vs. covert describes the awareness that biometric sampling and identification is occurring. (True or False?)

17. Open vs. closed determines whether the biometric application requires a password to enter a sample. (True or False?)

18. Society has several categories of concerns relative to the use of biometrics. Two of these are ______________ and ______________.

19. An ______________ is an object that can be applied to or incorporated into a product, animal, or person for the purpose of identification using radio waves.

20. There are three tools for emergency access procedures. These include: ______________, ______________, and ______________.

Discussion Exercises

1. Develop a list of business processes that can benefit from a biometric application or solution.

2. Identify organizations that are involved in biometric access-control issues.

3. Research various venues for applications and solutions for each of the biometric modalities. Produce a summary for each modality.

4. Describe a number of solutions that fit in the categories of customer, employee, and citizen facing.

5. Provide an overview of biometric applications by use.

6. Provide an overview of biometric applications by type.
7. Select a biometric modality and provide a list of advantages and disadvantages for deployment and use.
8. Provide an overview of the activities of the Biometrics Task Force.
9. Provide an overview of the Schengen visa system.
10. Provide an overview of the VoiceVault system.

Hands-On Projects

3-1: Identifying Biometric Applications and Solutions

Time required—30 minutes

Objective—Identify biometric applications and solutions that have been implemented in some organization.

Description—Use various Web tools, Web sites, and other tools to identify biometric applications currently in development or production in both commercial and government organizations.

1. Identify biometric applications and solutions that have been implemented in some organization. Provide an overview of each. Develop graphics that depict the process and operation of these applications. Describe the pros and cons of the implementations.
2. Identify a biometric application or solution that would provide security and access control for some public transportation organization. Provide a detailed analysis of the security issues and describe how a biometric application would provide a solution. Produce a graphic that shows the flow of those who are impacted by the application.
3. Describe a biometric solution that could be employed to protect U.S. borders. How would this impact the public flow between Canada and Mexico? How would this impact the movement of shipping and truck traffic between the countries?
4. Develop a comparison matrix that characterizes and ranks biometric applications expressed in terms of the seven variables listed in this chapter.

3-2: Evaluating Applications and Systems

Time required—45 minutes

Objective—Project management must identify biometrics applications and systems options that can provide some level of access control and security for the OnLine Access assets and resources.

Description—The owners and managers of OnLine Access have previously identified several instances of some compromise to the organization's database entries. There is a possibility of additional security violations. Project management has decided to research the various product and service offerings that apply biometric solutions to security and access control.

1. Research the vendors offering biometric applications and systems that provide for access control and security.

2. Develop a comparison of the various biometric applications and systems that could provide a solution for OnLine Access.
3. Produce a matrix describing each solution's features and benefits. Provide costs if attainable.
4. Rank order these solutions that would best provide the benefits desired by OnLine Access management.

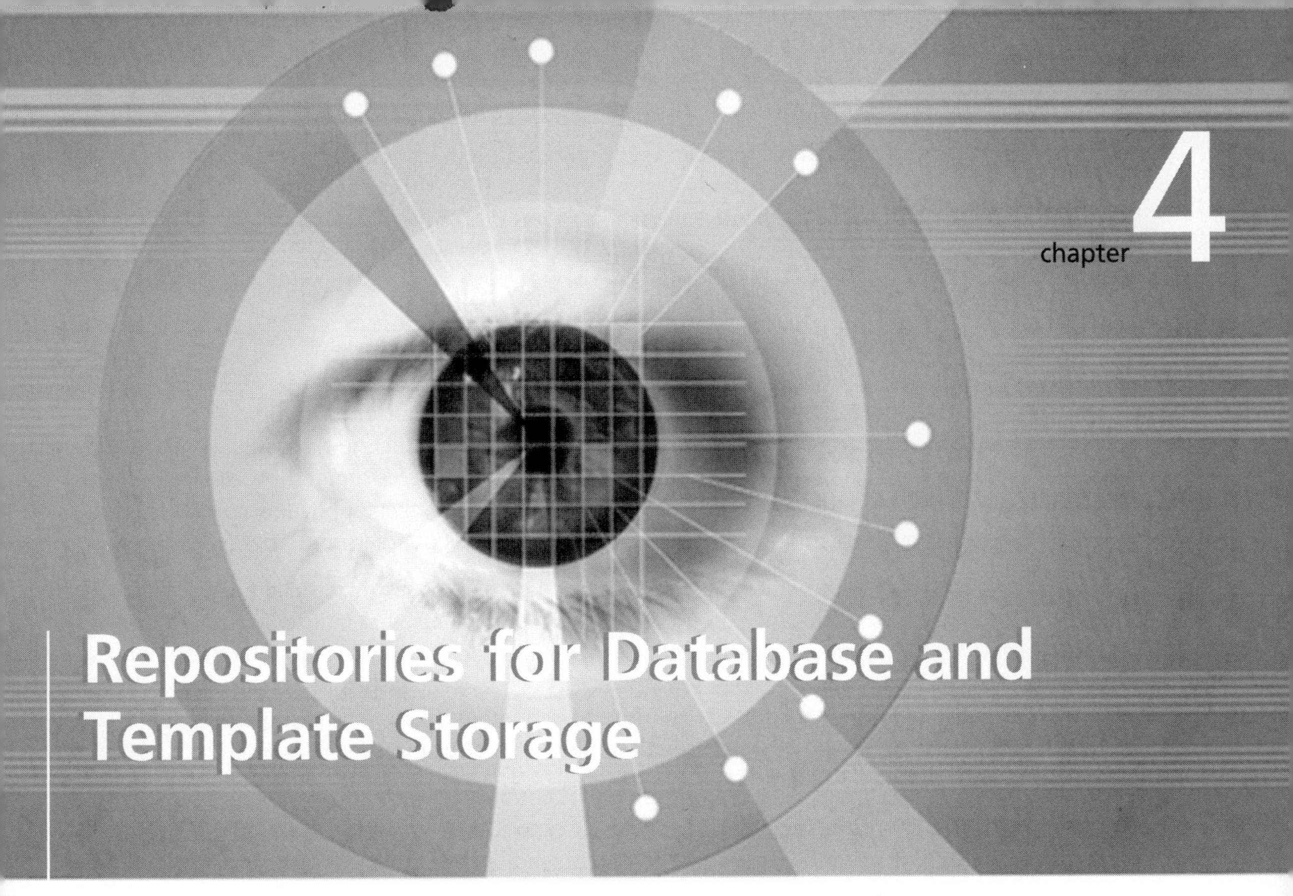

chapter 4

Repositories for Database and Template Storage

After completing this chapter, you should be able to do the following:

- Understand the basics of database and database management functions and capabilities
- Identify the various components and functions of a biometric database and repository
- See how biometric modalities use the database asset for access control
- Identify a number of biometric applications and systems that operate in the database environment
- Look at the various issues where a database operation can impact the successful operation of a biometric application
- Become familiar with the various physical and security impediments in the operation of a computer-based biometric database

Major components of any information system are the database hardware and software configuration. For any biometric solution, these resources contain data and templates. The repository associated with the biometric application and system often is an extension of a legacy computer system. A biometric security and access control solution may also require additional disk storage and computer memory.

Security and access control policies and procedures should already be in place for embedded legacy systems; however, this is often not the case. Many organizations do not have a formalized set of standards and procedures for their computer and network operations. Disaster is waiting for this kind of an opportunity. Administrators must ascertain the current level of database security and provide the necessary enhancements to support the additional requirements of a biometric solution.

Introduction

Secure access to database systems and repositories holds a high priority in today's e-commerce environment. Issues relating to this access and security is presented and discussed. Of particular interest are the response time and disk latency required when accessing and processing security requirements. The size of the database asset must be considered when applying security solutions. Security systems must be employed that provide various levels of access and "rights" control.

Physical security can be thought of as the lock and key part of security, since the primary protection mechanism is a lock of some type. Equipment rooms that house sensitive database storage devices and systems' equipment are normally locked, workstations contain locking mechanisms, and doors into various sensitive areas are locked. In lieu of actual physical locks and keys, magnetically encoded cards are used to activate workstations and open doors. These methods of entry are also used as personal employee identification.

Auditing involves the process of collecting and monitoring all aspects of the computer system that provides access to the biometric database asset to identify potential holes and flaws in the systems. This information becomes part of contingency planning and disaster recovery efforts.

Security awareness programs are necessary to educate the employee and user community as to the importance of physical and database security. Many users are not aware of the consequences of simple lapses in security, which could cost the organization millions of dollars in computer, database, and network corruption and damages.

Physical security planning includes addressing a number of emergency situations, such as natural disasters, power failures, and human-initiated incidents. Disasters can occur from both fire and water damage originating from inside and outside sources. There are a number of proactive measures that can be taken to reduce the damage caused by these natural and man-made disasters. The database hardware systems are susceptible to a number of these disastrous incidents.

Intrusion detection is the art of detecting and responding to incidents that can impact the biometric resources. The benefits include attack anticipation, deterrence, detection, response, damage assessment, and prosecution support. The security administrator must evaluate the potential risks and implement the appropriate system for maximum protection.

Legacy Data and Database Primer

As mentioned in a previous chapter, legacy systems consist of older hardware and software computer components that have become necessary for ongoing business operations, yet are still viable and capable of meeting the business requirements. Additional requirements may cause a need for additional information and a database upgrade to support these changes.

Data is defined as items representing facts, text, graphics, bit-mapped images, sound, and analog or digital live-video segments. Data is a system's raw material, supplied by data producers and used by information consumers to create information. What is data and what can be done to ensure its integrity and security? Data is basically any material that is represented in a formalized manner so that it can be stored, manipulated, and transmitted by a computer. Data is stored in a database according to a predefined format and an established set of rules, called a **schema**. A database that consolidates an organization's data is called a **data warehouse**. **Data warehousing** is a software strategy in which data is extracted from large transactional databases and other sources and stored in smaller databases, which makes analysis of the data easier. **Information** is data that has been processed to add or create meaning and knowledge for the person receiving it.

Database

A **database (DB)** is a collection of information organized in such a way that a computer program can quickly select desired pieces of data. A database is basically an electronic filing system. Traditional databases are organized by *fields*, *records*, and *files*.

These are defined as:

- A field is a single piece of information: name or address or Social Security Number (SSN)
- A record is one complete set of fields: name and address and SSN
- A file is a collection of records: multiple different SSNs and associated records

For example, a telephone book is analogous to a file. It contains a list of records, each consisting of three fields: name, address, and telephone number.

An alternative concept in database design is known as *hypertext*. In a hypertext database, any object, whether a piece of text, a picture, or a film, can be linked to any other object. Hypertext databases are particularly useful for organizing large amounts of disparate information, but they are not designed for numerical analysis.

Computer databases typically contain aggregations of data records or files, such as sales transactions, product catalogs and inventories, and customer profiles. Typically, a database manager provides users with the capabilities of controlling read/write access, specifying report generation, and analyzing usage. Databases and database managers are prevalent in large mainframe systems, but are also present in smaller distributed workstation and mid-range systems such as the AS/400 and on personal computers. Figure 4-1 depicts the environment and connectivity of a computer system and the associated database component.

SQL (Structured Query Language) is a standard language for making interactive queries from and updating a database such as IBM's DB2, Microsoft's Access, and database products from Oracle, Sybase, and Computer Associates. Using keywords and sorting commands, users can rapidly search, rearrange, group, and select the field in many records to retrieve or create reports on particular aggregates of data according to the rules of the database management system being used. To access information from a database, a database management system is required. This system is a collection of programs that enables entering, organizing, and selecting data in a database.

Microsoft SQL Server is a relational database management system (RDBMS) produced by Microsoft. Its primary query language is Transact-SQL, an implementation of the ANSI/ISO standard Structured Query Language (SQL) used by both Microsoft and Sybase.

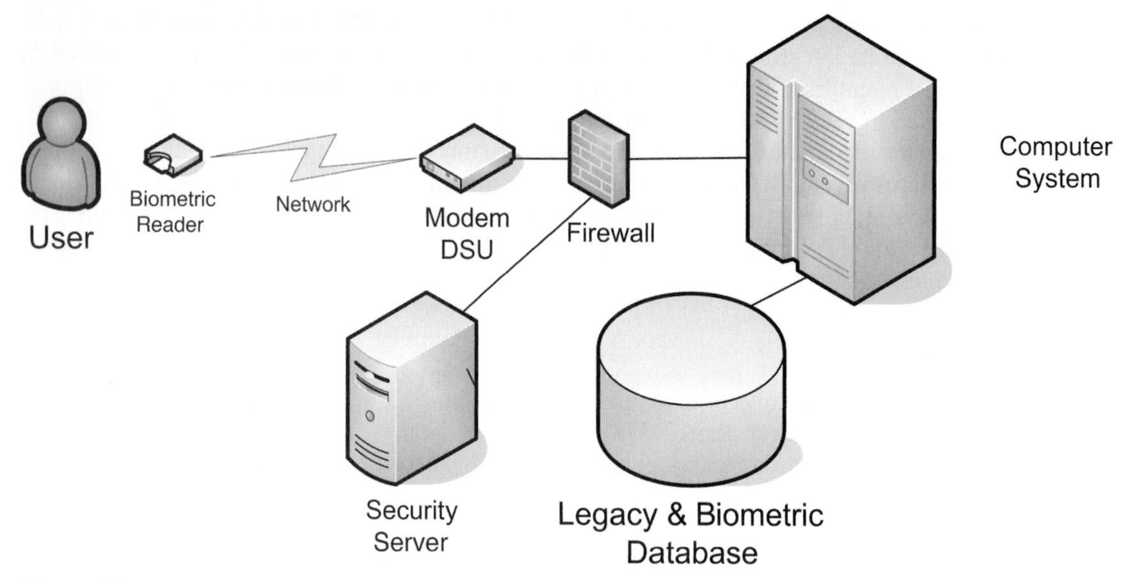

Figure 4-1 Computer and Database Environment

Course Technology/Cengage Learning

Data Storage Access

A data storage device consisting of input/output equipment may be considered data storage equipment if it writes to and reads from a data storage medium. Data storage equipment uses either:

- Portable methods which are easily replaced
- Semiportable methods requiring mechanical disassembly tools and/or opening a chassis
- Fixed methods, meaning loss of memory if disconnected from the unit

An important issue with data storage is the response time required to access the data in the database from some remote device. This issue is particularly important when someone is attempting to enter some area and a biometric scan and comparison are required. The technical term for this is called latency or delay. Latency is often defined as time that is wasted in a system waiting for some activity to occur. A delay is the time between the initiation of a request or transaction by the user and the response received by that user. A number of different elements in the network, computer, and database can add delay into the process.

Database Management System

A **database management system** (DBMS) is a complex set of software programs that controls the organization, storage, management, and retrieval of data in a database. A DBMS consists of a collection of programs that provide users with the ability to store, modify, and extract information from a database. There are many different types of DBMSs, ranging from small systems that run on personal computers to large systems that run on mainframes. The following are examples of legacy database applications:

- Computerized library systems
- Automated teller machines
- Flight reservation systems
- Computerized parts inventory systems

The DBMS provides for a quick search and retrieval of information from a database. The DBMS also determines how data is stored and retrieved. It must address problems such as security, accuracy, consistency among different records, response time, and memory requirements. These issues are most significant for database systems on computer networks. Ever-higher processing speeds are required for efficient database management. Relational DBMSs, in which data is organized into a series of tables, and easily reorganized for accessing data in different ways, are the most widely used today. The internal organization can affect how quickly and flexibly users can extract information.

The set of rules for constructing queries is known as a query language. Different DBMSs support different query languages, although there is a semistandardized query language called a structured query language (SQL). Requests for information from a database are made in the form of a query, which is a stylized question. For example, the query **"SELECT ALL WHERE NAME = "JONES" AND AGE > 25"** requests all records in which the NAME field is JONES and the AGE field is greater than 25.

The information from a database can be presented in a variety of formats. Most DBMSs include a report writer program enabling users to output data in the form of a report. Many DBMSs also include a graphics component enabling users to output information in the form of graphs and charts.

A DBMS includes a software program that typically operates on a database server or mainframe system to manage structured data, accept queries from users, and respond to those queries. A typical DBMS has the following features:

- Provides a way to structure data as records, tables, or objects
- Accepts data input from operators and stores that data for later retrieval
- Provides query languages for searching, sorting, reporting, and other decision support activities that help users correlate and make sense of collected data
- Provides multiuser access to data, along with security features that prevent some users from viewing and/or changing certain types of information
- Provides data integrity features that prevent more than one user from accessing and changing the same information simultaneously
- Provides a data dictionary (metadata) that describes the structure of the database, related files, and record information

Most DBMS systems are client/server based and operate over networks. The DBMS is an engine that typically runs on a powerful server or cluster of servers, in a *storage area network (SAN)* environment or mainframe with a high-performance channel to a large data store. The DBMS accepts requests from clients that may require sorting and extracting data. Once the server has processed the request, it returns the information to the client.

Metadata is information that describes data in a database. A raw data file would appear as a mass of letters, numbers, and symbols, but by knowing the format and structure of how that data is stored, users can display the data as records, fields, attributes, and elements with specific rules and properties (i.e., some fields are locked or the data is displayed a certain way). A metadata file provides this knowledge.

Most databases are operational databases, meaning data going into the database is used in real time to support the ongoing activities of a business or operation. A point-of-sale business accounting system is an example. As items are sold, the inventory database is updated and the inventory information is made available to the sales staff. The invoicing, order entry, and related systems are also updated. Large merchandising companies would use this type of system for inventory control and management.

Data analysts use *online analytical processing (OLAP)* systems to analyze database information in order to find trends or make business decisions. A data warehouse is a large-scale OLAP and/or data mining system that is specifically designed to extract, summarize, combine, clean up, and process information from a number of data sources, such as the operational databases, legacy (historical) databases, and online subscription databases for the purpose of analysis. Metadata is especially important in this environment. **Data mining** is the process of searching for unknown information or relationships in large databases. Various software tools are used to accomplish this task.

Capabilities of a DBMS

One can characterize a DBMS as an "attribute management system" where attributes are small chunks of information that describe something. For example, "color" is an attribute of a car. The value of the attribute may be a color such as "red," "blue," or "silver."

Alternatively, and especially in connection with the relational model of database management, the relation between attributes drawn from a specified set of domains can be seen as primary. For instance, the database might indicate that a car that was originally "red" might fade to "pink" in time, provided it was of some particular "make" with an inferior paint job. Such higher similarity relationships provide information on all of the underlying domains at the same time, with none of them being privileged above the others.

Throughout recent history, specialized databases have existed for scientific, geospatial, imaging, document storage, and like uses. Functionality drawn from such applications has lately begun appearing in mainstream DBMSs as well. However, the main focus there, at least when aimed at the commercial data processing market, is still on descriptive attributes on repetitive record structures. Capabilities commonly offered by database management systems include:

- Query ability
- Backup and replication
- Rule enforcement
- Security
- Computation
- Change and access logging

- Automated optimization
- Metadata repository

Query Ability

Querying is the process of requesting attribute information from various perspectives and combinations of factors. Example: "How many trucks in Georgia are red?" A database query language and report writer allows users to interactively interrogate the database, analyze its data, and update it according to a user's privileges on data. It also controls the security of the database. Data security prevents unauthorized users from viewing or updating the database. Using passwords, users are allowed access to the entire database or subsets of it called subschemas. For example, an employee database can contain all the data about an individual employee, but one group of users may be authorized to view only payroll data, while others are allowed access to only work history and medical data. If the DBMS provides a way to interactively enter and update the database, as well as interrogate it, this capability allows for managing personal databases. However it may not leave an audit trail of actions or provide the kinds of controls necessary in a multiuser organization. These controls are only available when a set of application programs are customized for each data entry and updating function.

Backup and Replication

Copies of attributes must be made regularly in the event primary disks or other equipment fails. A periodic copy of attributes may also be created for a distant organization that cannot readily access the original. DBMS usually provide utilities to facilitate the process of extracting and disseminating attribute sets. When data is replicated between database servers, information remains consistent throughout the database system and users cannot tell or even know which server in the DBMS they are using, the system is said to exhibit replication transparency.

Rule Enforcement

Often users want to apply rules to attributes so the attributes are clean and reliable. For example, users may have a rule that says each car can have only one engine associated with it (identified by engine number). If somebody tries to associate a second engine with a given car, users want the DBMS to deny such a request and display an error message. However, with changes in the model specification such as—in this example—hybrid gas-electric cars, rules may need to change. Ideally such rules should be able to be added and removed as needed without significant data–layout redesign.

Security

Often it is desirable to limit who can see or change which attributes or groups of attributes. This may be managed directly by individuals, or by the assignment of individuals and privileges to groups, or (in the most elaborate models) through the assignment of individuals and groups to roles which are then granted entitlements.

Computation

There are common computations requested for attributes such as counting, summing, averaging, sorting, grouping, cross-referencing, etc. Rather than have each computer application implement these from scratch, they can rely on the DBMS to supply such calculations.

Change and Access Logging

Often users want to know who accessed what attributes, what was changed, and when it was changed. Logging services allow this by keeping a record of access occurrences and changes.

Automated Optimization

If there are frequently occurring usage patterns or requests, some DBMS can adjust themselves to improve the speed of those interactions. In some cases the DBMS will merely provide tools to monitor performance, allowing a human expert to make the necessary adjustments after reviewing the statistics collected.

Metadata Repository

Metadata is data describing data. For example, a listing that describes what attributes are allowed to be in data sets is called "metainformation."

Integrity

Database integrity is to ensure that data entered in the database is accurate and valid. This means that if there is any rule applied to any entity, the entered data must obey that rule.

- Entity integrity allows no two rows with the same identity in a table.
- Domain integrity allows only predefined values, e.g., dates.
- Referential integrity allows only the consistency of values across related tables, e.g., only IDs of registered customers.
- User-defined integrity allows only those predefined.

Major Features of a DBMS

DBMS software controls the organization, storage, retrieval, security, and integrity of data in a database. It accepts requests from the software application and instructs the operating system to transfer the appropriate data. The major DBMS vendors are Oracle, IBM, Microsoft, and Sybase. MySQL is a very popular open source product. DBMSs may work with traditional programming languages (COBOL, C, etc.) or they may include their own programming language for application development. [1]

DBMSs let information systems be changed more easily as the organization's requirements change. New categories of data can be added to the database without disruption to the existing system. Adding a field to a record does not require changing any of the programs that do not use the data in that new field. The major DBMS features include:

- Data security
- Data integrity
- Interactive query
- Interactive data entry and updating
- Data independence
- Data storage management

Data Security

The DBMS can prevent unauthorized users from viewing or updating the database. Using passwords, users are allowed access to the entire database or a subset of it known as a "subschema." For example, in an employee database, some users may be able to view salaries while others may view only work history and medical data.

Data Integrity

The DBMS can ensure that no more than one user can update the same record at the same time. It can keep duplicate records out of the database; for example, no two customers with the same customer number can be entered.

Interactive Query

A DBMS provides a query language and report writer that lets users interactively interrogate the database. These essential components give users access to all management information as needed.

Interactive Data Entry and Updating

A DBMS typically provides a way to interactively enter and edit data, allowing users to manage their own files and databases. However, interactive operation does not leave an audit trail and does not provide the controls necessary in a large organization. These controls must be programmed into the data entry and update programs of the application.

This is a common misconception about using a desktop computer DBMS. Creating lists of data for a user's own record keeping is one thing. However, although complete information systems can be developed with such software, it cannot be done without understanding how transactions and files relate to each other in a business system. In addition, some type of programming is required, whether at a graphical drag-and-drop level or by using traditional languages.

Data Independence

When a DBMS is used, the details of the data structure are not stated in each software application program. The program asks the DBMS for data by field name; for example, a coded equivalent of "give me customer name and balance due" would be sent to the DBMS. Without a DBMS, the programmer must reserve space for the full structure of the record in the program. Any change in data structure requires changing all application programs.

Data Storage Management

Today's customers, employees, suppliers, and business partners expect to be able to tap into information any time of day, from any location. At the same time, the organization must be increasingly sensitive to issues of customer privacy, data security, and regulatory requirements. To keep operations running and the organization secure, managers need a comprehensive strategy that addresses three primary aspects of business continuity: high availability, disaster recovery, and continuous operations.

Database and Repository Security

An *enterprise single sign-on (E-SSO)* is a system designed to minimize the number of times that a user must type an ID and password to sign into multiple applications. The E-SSO solution automatically logs users in and acts as a password filler where automatic login is not possible. Each client is typically given a token that handles the authentication; in other E-SSO solutions each client has E-SSO software stored on his computer to handle the authentication. An E-SSO authentication server is also typically implemented into the enterprise network. **Personally identifiable information (PII)** is any piece of information which can potentially be used to uniquely identify, locate, or contact a person or steal the identity of a person. [2]

A travel document that contains an integrated circuit chip based on international standard ISO/IEC 14443 and that can securely store and communicate the e-passport holder's personal information to authorized reading devices is called an **e-passport.** A **message authentication code (MAC)** is a short piece of information used to support authentication of a message. A MAC algorithm accepts as input a secret key and an arbitrary-length message to be authenticated, and outputs a MAC (sometimes known as a tag or checksum). The MAC value protects both a message's integrity as well as its authenticity by allowing verifiers (who also possess the secret key) to detect any changes to the message content. MACs are computed and verified with the same key, unlike digital signatures. The **detection and identification rate** is the rate at which individuals, who are in a database, are properly identified in an open-set identification (watch list) application. [3]

Biometric Data and Databases

Biometric data is a catch-all phrase for computer data created during a biometric process. It encompasses raw sensor observations, biometric samples, models, templates and/or similarity scores. Biometric data is used to describe the information collected during an enrollment, verification, or identification process, but does not apply to end-user information such as user name, demographic information, and authorizations.

The database is a collection of one or more computer files. For biometric systems, these files could consist of:

- Biometric sensor readings
- Templates
- Match results
- Related end user information

A **record** consists of the template and other information about the end user (e.g., name, access permissions). A **gallery** is the biometric system's database, or set of known individuals, for a specific implementation or evaluation experiment.

A **probe** is the biometric sample that is submitted to the biometric system to compare against one or more references in the gallery. A biometric sample consists of data representing a biometric characteristic of a user as captured by a biometric system. **Biometric information** consists of stored electronic information pertaining to a biometric. This information can be in terms of raw or compressed pixels or in terms of some characteristic (e.g., patterns). The formatted digital record used to store an individual's biometric attributes is called a biometric template. This record typically is a translation of the individual's biometric attributes and is created using a specific algorithm.

Important terms that have been introduced in previous chapters will be re-introduced in the following sections. An enrollee is a person who has a biometric reference template stored in a biometric package. Enrollment is the process of collecting biometric samples from a user and the subsequent preparation, encryption, and storage of biometric reference templates representing that person's identity.

Identification is a task where the biometric system searches a database for a reference matching a submitted biometric sample; and if found, returns a corresponding identity. A biometric is collected and compared to all the references in a database. Identification is "closed-set" if the person is known to exist in the database. In "open-set" identification, sometimes referred to as a "watch list," the person is not guaranteed to exist in the database. The system must determine whether the person is in the database, then return the identity. The identification rate is the rate at which an individual in a database is correctly identified. Verification is a task where the biometric system attempts to confirm an individual's claimed identity by comparing a submitted sample to one or more previously enrolled templates. The verification rate is a statistic used to measure biometric performance when operating in the verification task. It is also the rate at which legitimate end users are correctly verified.

Watch list is a term that describes one of the three tasks that biometric systems perform. It answers the questions: Is this person in the database? If so, who are they? The biometric system determines if the individual's biometric template matches a biometric template of someone on the watchlist, as illustrated in Figure 4-2.

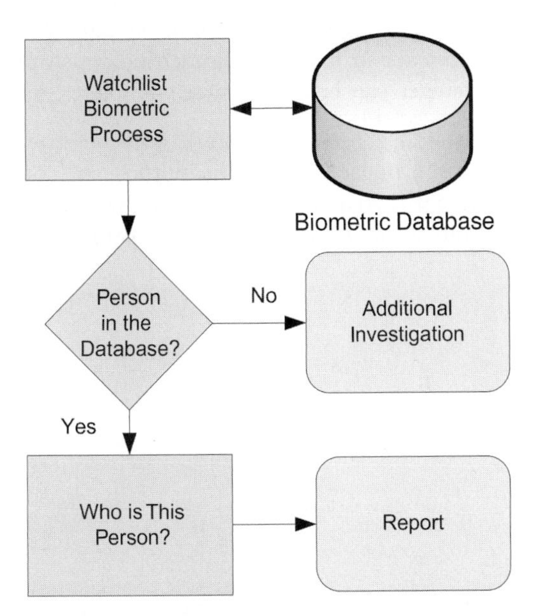

Figure 4-2 Watch List Open-Set Identification

Course Technology/Cengage Learning

The individual does not make an identity claim, and in some cases does not personally interact with the system whatsoever. Open-set identification is the biometric task that more closely follows operational biometric system conditions to (1) Determine if someone is in a database and (2) find the record of the individual in the database.

Closed-set identification is a biometric task where an unidentified individual is known to be in the database and the system attempts to determine his/her identity. Performance is measured by the frequency with which the individual appears in the system's top rank (or top 5, 10, etc.).

Comparison is the process of comparing a biometric reference with a previously stored reference or references in order to make an identification or verification decision. Match is a decision that a biometric sample and a stored template comes from the same human source, based on their high level of similarity (difference or hamming distance) such as a loop fingerprint pattern. Threshold is a user setting for biometric systems operating in the verification or open-set identification (watch list) tasks. The acceptance or rejection of biometric data is dependent on the match score falling above or below the threshold. The threshold is adjustable so the biometric system can be more or less strict, depending on the requirements of any given biometric application.

One-to-many is a phrase used in the biometrics community to describe a system that compares one reference to many enrolled references to make a decision. The phrase typically refers to the identification or watch list tasks. One-to-one is a phrase used in the biometrics community to describe a system that compares one reference to one enrolled reference to make a decision. The phrase typically refers to the verification task (though not all verification tasks are truly one-to-one) and the identification task that can be accomplished by a series of one-to-one comparisons. Figure 4-3 depicts the difference between a one-to-many and one-to-one comparison.

A **claim of identity** is a statement that a person is or is not the source of a reference in a database. Claims can be positive (I am in the database), negative (I am not in the database) or specific (I am end user 123 in the database). A *reference* includes the biometric data stored for an individual for use in future recognition. A reference can be one or more templates, models, or raw images.

Biometric reference data can include data stored on a card for the purpose of comparison with biometric verification data. Data that is not stored on an ID card or to a computation that is not performed by the integrated circuit on an ID card is deemed **off-card. On-card,**

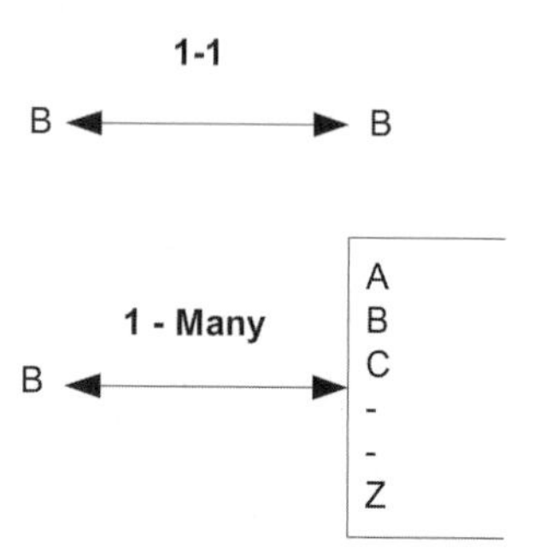

Figure 4-3 One-to-One and One-to-Many Comparisons

Course Technology/Cengage Learning

however, refers to data stored on an ID card or to a computation that is performed by the integrated circuit chip on an ID card.

Options for Biometric Database Storage

While each biometric device and system has its own operating methodology, there are some general rules of thumb that managers can expect to find in any system. The process for a given user will usually begin with an enrollment process. Here, the system captures one or more (typically three) samples of the biometric. These samples are stored in a biometric template and used for future comparison during authentication. Key elements in choosing a biometric system include ensuring that the enrollment process is relatively simple for the user, requires a short period of time, and provides for a high-quality template.

After it is generated, the template needs to be stored on some kind of media. Since templates range from 9 bytes to around 1.5K in size, storage space is not typically a major issue except in very large implementations. There are typically three options for template storage:

- Within the biometric reader itself: This provides for quick response during future authentication. However, it does not lend itself to situations where the user will need to authenticate at multiple locations. For example, a bank's ATM machines could not use this method since customers won't always use the same machine.
- Stored remotely in a central repository: This overcomes the problem of users authenticating from multiple locations. There is the potential for "sniffing" the biometric data off the network and replaying the authentication session unless encryption is used. In addition, some users are very privacy conscious and do not like the idea of information such as fingerprint data being stored centrally.
- On a portable token such as a smart card: This method addresses the drawbacks of both previous methods. The biometric data is not centrally stored, does not traverse the network, and the user carries the information from location to location. Users also have a feeling that they control their personal identification data. The one drawback is that the cost of the biometric implementation is higher. A device is required to read the smart card as well as the biometric data.

Once enrollment and storage are complete, users authenticate themselves by matching the template against current input, usually referred to as "live data." Most commonly, the user enters a username or PIN and then enters the live data (i.e., scans their fingerprint). Comparison of the live data and the template results in a simple binary yes/no match. Verification biometric systems tie the username or PIN to the template for a one-to-one match. While this is not the only method, it is the most common and reliable.

Biometric Database Issues

To be effective, a biometric system must compare captured biometric data to a biometric database. The national ID system page highlights issues surrounding database abuse, which has both static and dynamic dimensions. Details are available at http://www.cato.org/pubs/pas/pa237.html. As an example, privacy organizations oppose any such scheme because:

- No compelling case has been presented for its utility or effectiveness as a crime-fighting tool

- There are inevitable costs (in dollars, privacy, and liberty)
- There is a high potential for abuse by entities in both the public and private sectors

The static issues surrounding databases are mainly about safeguarding large and valuable collections of personally identifying information. If these databases are part of an important security system, then they, and the channels used to share PII, are natural targets for attack, theft, compromise, and malicious or fraudulent use.

The dynamic issues surrounding databases mainly concern the need to maintain reliable, up-to-date information. Databases seeking to maintain accurate residence information must be updated whenever someone moves. Databases used to establish eligibility for benefits must be updated to exclude persons no longer eligible. The broader the function of the system, the more frequent and broader the updating that is required, increasing the role of general social surveillance in the system.

Trust

It appears one of the issues plaguing token-based ID systems, like ID cards, does not apply to biometric systems, because "you are your ID." But the question of the reliability of the token is really a question about trust. In an ID card system, the question is whether the system can trust the card. In biometric systems, the question is whether the individual can trust the system. If someone else captures a personal signature, fingerprint, or voice, for instance, what prevents it from being used by others? Any use of biometrics with a scanner run by someone else involves trusting someone's claim about what the scanner does and how the captured information will be used.

Vendors and scanner operators may say they protect privacy in some way, perhaps by hashing the biometric data or designing the database to enforce a privacy policy. But the end user typically has no way to verify whether such technical protections are effective or implemented properly. End users should be able to verify any such claims, and to leave the system completely if not satisfied. Exiting the system, of course, should at least include expunging the end user's biometric data and records.

Linking

An oft-noted risk of biometric systems is the use of biometrics as a linking identifier. This risk, of course, depends to some extent on standardization. Consider, for instance, the use of a Social Security Number as a linker across disparate databases. While the private sector would not have been able to develop anything like the SSN on its own, once the government created this identifier, it became a standard way of identifying individuals. Standardization therefore creates new privacy risks because information gathered for one purpose can be used for completely unrelated purposes.

Currently, automated fingerprint ID systems (AFIS) are heavily used by the government in connection with law enforcement, but there is, at present, little standardization within the AFIS industry. If law enforcement and private industry were to unify their fingerprint databases to one common standard, such as under a national ID system, this would potentially put one's entire life history in interoperating databases, only a fingerprint away.

Tracking

By far the most significant negative aspect of biometric ID systems is their potential to locate and track people physically. While many surveillance systems seek to locate and track, biometric systems present the greatest danger precisely because they promise extremely high accuracy. Whether a specific biometric system actually poses a risk of such tracking depends on how it is designed.

Why should users care about perfect tracking? Detractors believe that perfect tracking is inimical to a free society. A society in which everyone's actions are tracked is not, in principle, free. It may be a livable society, but would not be our society.

Detractors also believe that perfect surveillance, even without any deliberate abuse, would have an extraordinary chilling effect on artistic and scientific inventiveness and on political expression. This concern underlies constitutional protection for anonymity, both as an aspect of First Amendment freedoms of speech and association, and as an aspect of Fourth Amendment privacy.

Implemented improperly, biometric systems could:

- Increase the visibility of individual behavior: This makes it easier for measures to be taken against individuals by agents of the government, by corporations, and by peers.
- Result in politically damaging and personally embarrassing disclosures, blackmail, and extortion: This hurts democracy, because it reduces the willingness of competent people to participate in public life.
- Increase the "circumstantial evidence" available for criminal prosecution: This might dramatically affect the existing balance of plausible-sounding evidence available to prosecutors, and hence increase the incidence of wrongful conviction. Many criminal cases are decided by plea bargaining, a process that is sensitive to the perceived quality of evidence. Even ambiguous or spurious evidence generated by complex technical systems may be difficult for overburdened public defenders to challenge.
- Enable the matching of people's behavior against predetermined patterns: This could be used by the government to generate suspicion, or by the private sector to classify individuals into micromarkets, the better to manipulate consumer behavior.
- Aid in repressing readily locatable and trackable individuals: While the public's concern is usually focused on the exercise of state power, these technologies may also greatly empower corporations. If proper privacy safeguards are not constructed into such systems, they would prove useful in dealing with such troublesome opponents as competitors, regulators, union organizers, whistleblowers, and lobbyists, as well as employees, consumer activists, customers, and suppliers.

As previous stated, biometrics includes automated methods of recognizing a person based on a physiological or behavioral characteristic. Biometric technologies are becoming the foundation of an extensive array of highly secure identification and personal verification solutions. As the level of security breaches and transaction fraud increases, the need for highly secure identification and personal verification technologies is becoming apparent.

Confidential Data

Biometric-based solutions are able to provide for confidential financial transactions and personal data privacy. Organization-wide network security infrastructures, government IDs, secure electronic banking, investing and other financial transactions, retail sales, law enforcement, and health and social services are already benefiting from these technologies.

Biometric-based authentication applications include workstation, network, and domain access, single sign-on, application logon, data protection, remote access to resources, transaction security, and Web security. Trust in these electronic transactions is essential for the healthy growth of the global economy. Used alone or integrated with other technologies such as smart cards, encryption keys and digital signatures, biometrics are set to pervade nearly all aspects of the economy and our daily lives. Utilizing biometrics for personal authentication is becoming convenient and considerably more accurate than current methods, such as the utilization of passwords or PINs. This is because biometrics includes the following benefits:

- It links the event to a particular individual: A password or token may be used by someone other than the authorized user
- It is convenient: There is nothing to carry or remember
- It is accurate: It provides for positive authentication
- It provides an audit trail
- It is becoming socially acceptable and cost effective

More information about biometrics, standards activities, government and industry organizations and research initiatives on biometrics will be presented in Chapter 12.

Database Authentication

As previously stated, biometrics includes the science and technology of measuring and analyzing biological data. In information technology, biometrics refers to technologies that measure and analyze human body characteristics, such as fingerprints, eye retinas and irises, voice patterns, facial patterns and hand measurements, for authentication purposes. Authentication by biometric verification is becoming increasingly common in corporate and public security systems, consumer electronics and point of sale (POS) applications. In addition to security, the driving force behind biometric verification has been convenience.

To prevent identity theft, biometric data is usually encrypted when it's gathered. Here's how biometric verification works on the back end: To convert the biometric input, a software application is used to identify specific points of data as match points. The match points in the database are processed using an algorithm that translates that information into a numeric value. The database value is compared with the biometric input the end user has entered into the scanner and authentication is either approved or denied.

Biometrics.gov is the central source of information on biometrics-related activities of the federal government. Two sister Web sites provide a repository of biometrics-related public information (www.biometricscatalog.org) and opportunities for discussion (www.biometrics.org). These sites, working together, were developed to encourage greater collaboration and sharing of information on biometric activities among government departments and agencies; commercial entities; state, regional, and international organizations; and the general public.

Positive Aspects Concerning Biometrics

Why be concerned about biometrics? Proponents argue that:

- Biometrics by themselves are not dangerous because all the real dangers are associated with the database behind the biometric information, which is little different from the problems of person-identifying information (PII) databases generally

- Biometrics actually promote privacy by enabling more reliable identification and thus frustrating identity fraud

But biometric systems have many components. Only by analyzing a system as a whole can one understand its costs and benefits. Moreover, users must understand the unspoken commitments any such system imposes.

Surveillance

The chronic, longitudinal capture of biometric data is useful for surveillance purposes. Biometric systems entail repeat surveillance, requiring an initial and subsequent captures. Another major issue relates to the "voluntariness" of capture. Some biometrics, like faces, voices, and fingerprints, are easily "grabbed." Other biometrics, at least under present technology, must be consciously "given." It is difficult, for instance, to capture a scan of a person's retina or to gather a hand geometry image without the subject's cooperation. Easily grabbed biometrics is a problem because people can't control when they're being put into the system or when they're being tracked. But even hard-to-grab biometrics involves a trust issue in the biometric capture device and the overall system architecture.

Biometric Databases

A number of biometric database and database systems are currently under development by both government and commercial entities. The most ambitious projects are those by the government, as they are the most expensive and time- and resource-consuming. Future requirements often call for a change in design and specifications, therefore the details provided in this section are subject to change. Several biometric databases will be described in this section, including:

- Department of Homeland Security (DHS): IDENT
- FBI: Next Generation Identification Database
- IEEE: MCYT bimodal database
- Biometric Storage System (BSS)
- Motorola Metro ID

DHS IDENT biometric database

The Homeland Security Department is expanding its automated biometric identification system (IDENT) to include biometric and limited personal information collected for immigration, intelligence, law enforcement, and national security programs. DHS uses that system and its Enforcement Operational Immigration Records (Enforce) to store biometric and case management information gathered from the Bureau of Immigrations and Customs Enforcement. Originally, the two systems operated as one under the name Enforce/IDENT. Under the revised regulation, the systems will operate independently.

The U.S. Visitor and Immigrant Status Indicator Technology program will manage IDENT. Biometric information collected for the system will apply to: [4]

- The implementation and enforcement of laws, regulations, treaties or orders related to DHS missions

- Background checks or security screening in connection hiring new employees or issuing licenses and credentials
- Data collected by federal, state, local, and foreign agencies pertaining to wanted individuals and subjects of interest

Biometric information in the system will be available to other organizations, including federal, state, local, tribal, foreign, and international agencies.

Next Generation Identification Database

The FBI has announced its plans to assemble the world's largest biometric database, nicknamed the Next Generation Identification system. Currently, the FBI stores fingerprints, facial features, and palm print characteristics at its facilities in Washington, D.C. This existing FBI database is called the National Crime Information Center (NCIC). The agency's $1 billion dollar database, however, will hold far more information on any given person. [5]

The FBI expects to make this new comprehensive biometric database available to a wide variety of federal, state, and local agencies, all in the name of keeping Americans safe from terrorists and illegal immigration. The FBI also intends to retain, upon employer request, the fingerprints of any employee who has undergone a criminal background check, and will inform the employer if the employee is ever arrested or charged with a crime.

The biometric database the FBI envisions will rely heavily on real-time, or very nearly real-time, comparisons. This database could include general face recognition, specific feature comparison, walking stride, speech patterns, and iris comparisons. To date, facial-recognition technology has been less than stellar. A study showed some progress in the technology, where existing implementations proved more than 60 percent effective during the day, but accuracy fell to 10 to 20 percent at night. Law enforcement officials have stated they would accept a 0.1 percent error rate across a 24-hour period, which leaves current technology with quite a gap to close.

The FBI plans to work closely with the Center for Identification Technology Research (CITeR) to improve existing metrics and create new ones. CITeR is reportedly working on an iris scanner that can identify people at up to 15 feet as well as a facial recognition scanner capable of identifying faces accurately at a range of up to 200 yards.

The FBI's decision to implement this kind of tracking and identification system raises a number of concerns regarding citizen privacy, as well as serious questions about the accuracy of collected data. Any database not closely monitored and continuously updated will inevitably become obsolete. An open issue with this type of biometric system is that it might become confused simply by the natural aging process, weight loss, weight gain, injury, or permanent disability of individuals. While there are proven methods of identification that remain accurate even in the presence of such factors, none of them yield real-time results that can be immediately pegged as belonging to an individual even in a crowd of people.

Previously, the FBI exempted its National Crime Information Center (NCIC), the Central Records System, and the National Center for the Analysis of Violent Crime from subsection (e) (5) of the 1974 Privacy Act. This subsection mandates that each agency maintaining a system of records shall "maintain all records which are used by the agency in making any determination about any individual with such accuracy, relevance, timeliness, and completeness as is reasonably necessary to assure fairness to the individual in the determination."

According to the FBI, discharging this duty conflicts with the agency's primary purpose as a law enforcement organization, because it is "impossible to determine in advance what information is accurate, relevant, timely, and complete." Information once thought innocuous may also eventually shed critical details as an investigation continues; and the restrictions of (e) (5) "would limit the ability of trained investigators and intelligence analysts to exercise their judgment in reporting on investigations and impede the development of criminal intelligence necessary for effective law enforcement."

At this point, the proposed biometric identification system contains no recourse for citizens who are misidentified, no formal method for the update and correction of biometric information, and no indication that citizens would even be allowed to view their own biometric profiles. Even in the best of scenarios, it's unclear whether or not any national database of biometric information could be kept secure, updated, and available for citizen review.

IEEE Bimodal Database

The Institute of Electrical and Electronics Engineers (IEEE) is an international professional society for electrical and electronics engineers, computer scientists, and educators. The IEEE establishes standards for emerging electronic-based technologies. The Spanish Ministerio de Ciencia y Tecnología (MCYT) project has produced a large-scale biometric bimodal database involving fingerprint and signature traits.

The current need for large multimodal databases to evaluate automatic biometric recognition systems has motivated the development of the MCYT bimodal database. [6] The main purpose has been to consider a large-scale population, with statistical significance, in a real multimodal procedure, and to include several sources of variability that can be found in real environments. The acquisition process, contents, and availability of the single-session baseline corpus are fully described. Some experiments showing consistency of data through the different acquisition sites and assessing data quality are also presented. Keywords that are often applied to the MCYT bimodal database include:

- MCYT baseline corpus
- Bimodal biometric database
- Multimodal databases
- Automatic biometric recognition systems
- Large-scale population
- Statistical significance
- Variability
- Acquisition process
- Single-session baseline corpus
- Data consistency
- Data quality assessment

Biometric Storage System

The U.S. Citizenship and Immigration Services (USCIS) of the Department of Homeland Security (DHS) has been tasked by Congress with processing all immigration benefit applications and petitions. Many applications, petitions, and other benefits require that fingerprints and other biometrics be captured to conduct background checks, verify the identity of applicants, petitioners, or beneficiaries, and produce benefit cards with biometrics and documents. To fulfill its statutory mandate, USCIS is establishing a new system of records that will consolidate all biometrics collected by USCIS into one centralized system. This new system of records is called the *biometric storage system (BSS)*.

Implemented as a part of a USCIS enterprise-wide "Transformation Program," BSS will help transition the agency's data management practices to a paperless, more centralized, and unique identity driven methodology. BSS will become the centralized repository for all biometric data captured by USCIS from applicants filing immigration applications. This new system will eventually replace existing legacy systems. USCIS captures biometric data from applicants to facilitate three key operational functions:

- Conducting fingerprint-based background checks
- Verifying an applicant's identity
- Producing benefit cards/documents

Currently, USCIS does not have a centralized, long-term storage program for fingerprint biometrics. Accordingly, applicants are sometimes required to return to an USCIS Application Support Center (ASC) to provide fingerprints again during the case adjudication process. BSS will store the biometric information, thereby decreasing the burden on applicants by negating the need to provide multiple sets of biometric data.

Further, BSS will consolidate storage of information from multiple, separate systems into a centralized database, allowing for greater control, security, and management of the data. BSS also will provide increased functionality over current systems, and improved communication between government databases and personnel, facilitating more efficient processing of applications. This furthers USCIS's goals of reducing immigration benefit and petition case backlog, and improving the process for vetting and resolving applications for immigration benefits. [7]

Motorola Metro ID Storage System

A key crime fighting tool for large law enforcement agencies is the automated fingerprint identification system (AFIS). Already in use for many years, AFIS systems allow officers to take fingerprints from an individual or crime scene and electronically compare them to a database of known prints. This comparison allows police to identify people and connect them to any criminal record or unsolved crimes. Today, most agencies require their AFIS systems to store and match more than just fingerprints; many departments have a need to store latents, palm prints and facial and iris images. These servers are a part of a larger network of technology that provides law enforcement officials with biometric solutions. These powerful technologies require large resources to obtain and maintain. High costs prevent law enforcement agencies of small cities from owning these proven crime-solving tools. This section addresses the need of small-city agencies and their need for AFIS capabilities and the roadblocks they face. Also presented is an alternative solution for a local biometric storage system with matching services. An alternative, Motorola Metro ID Biometric Storage Systems, provides matching capabilities for local agencies.

Motorola LiveScan Station captures fingerprints, palm prints, mug shots and other biometric data, supports data entry, and interfaces with Metro ID and includes a central site for other applications, such as a state AFIS. The system provides proven workflows to create primary booking profiles. Mobile ID allows the scanning of two fingers out in the field and wirelessly submits those prints to the department's Metro ID database. Mobile ID performs a 1:N search to connect individuals to their records. Mobile ID combines Motorola's strengths in radio communications, mobile applications and AFIS technologies into one powerful tool. The Mobile ID option of the solution Verify Station is designed to identify individuals, using a single finger scanner, at the time of processing when the subject is present. When used in conjunction with Metro ID, this solution performs 1:1 search via a badge or other secure document or a 1:N search against the Metro ID database.

A system to process fingerprint, palm print, and latent searches on local sites is the ideal solution to the local agency problem. The delay in processing prints or the time it takes to search against an entire central database increases the risks that a suspect will be released. Two of the most critical time frames for local law enforcement to resolve crimes are the first 24 hours after the crime has been committed, and immediately after the police stop a suspect. The majority of the local law enforcement agency requirements are summarized as:

- Fingerprint identification in the field
- Fast local identification for local crime resolution
- An efficient booking process
- Electronic archiving of local bookings and records, as well as efficient submission to central sites
- Expert tools for 10-print, palm print, and latent processing
- A safeguard against technology obsolescence to protect their investment
- A solution that fits local budgets and makes efficient use of limited resources

A biometric database and matching system designed to function as an AFIS solution with small agency budgets in mind would be an ideal solution. Motorola's Metro ID, with all its optional components, provides local agencies with comparable benefits to an AFIS solution in a package design specifically for local agencies. Furthermore, the Metro ID system can be easily upgraded to add modular components as future system demands necessitate.

This comprehensive solution allows local law enforcement officers to capture, store, encode, match, and search biometric data. Using advanced matching algorithms such as Minutiae Matching and Expert Matching, Metro ID can search in variety of ways using unsolved latents, latents, latent palm prints, and palm prints. The Metro ID system comes equipped with sufficient computational power to perform the most advanced matching algorithms in seconds. Metro ID is compliant with National Institute of Standards and Technology (NIST) file formats as well as National Fingerprint File (NFF) FBI certified.

NIST Database Tools

NIST provides numerous database tools at its official Web site. [8] This biometric data provides for the statistical analysis of biological observations and phenomena. The fingerprint and mug-shot databases facilitate analysis for researchers in the law enforcement field. The NIST Digital Video 1 is a public-domain collection of digital video created to encourage more researchers to address real-world problems and support the scientific comparison of solutions of digital video search, retrieval, and display. Most of the NIST products listed below are available for purchase, however some are free:

- NIST Digital Video of Live-Scan Fingerprint Database Disk 1
- NIST 8-Bit Gray Scale Images of Fingerprint Image Groups
- NIST Mated Fingerprint Card Pairs
- NIST Supplemental Fingerprint Card Data
- NIST Mated Fingerprint Card Pairs II
- NIST Fingerprint Minutiae from Latent Matching 10-Print Images
- NIST Plain and Rolled Images from Paired Fingerprint Cards
- NIST Dual Resolution Images from Paired Fingerprint Cards
- NIST Mug-Shot Identification Database
- Facial Recognition Technology Database (FERET)

Database Security and Integrity Issues

A rash of well-publicized database hacking incidents brings security to the forefront of the issues facing the database community. These links will help ensure that database systems are secure from unauthorized access. Three relevant Web sites include: [9]

- Data security management: www.sans.org/score/checklists/ISO_17799_checklist.pdf
- Database security: www.databasesecurity.com/
- Enterprise data security: www.wallstreetandtech.com/resourcecenters/datamanagement/

There are a number of database security and integrity issues that must be addressed by network administrators and managers. These include:

- PCI DSS
- Inference
- Aggregation
- SQL injection
- Database roles and rights
- SQL injections
- SQL access controls
- Database servers
- HIPAA compliance (privacy and security)
- Oracle
- Protecting the warehouse

Physical Security Overview

Physical security is the term used to describe protection provided outside the computer and networking system. Typical physical security components are guards, locks, fences, and cameras to deter direct attacks. Many physical security measures are the result of commonsense thinking and planning, and are often obvious solutions. Sources that can compromise an organization's ability to function include natural events, human vandals, power loss, fire, water, and heat and humidity levels. These physical vulnerabilities to security can and often occur simultaneously. Issues to be considered include the cost of replacing the resource, the speed with which equipment can be replaced, the need for available computing power, and the cost or difficulty of replacing operating system and application software.

The physical element of computer and network security involves the hardware, the facility, cable runs, network demarcation points, and the software and database systems that reside in the physical location. An initial step when evaluating physical security is to conduct an audit that looks at all the elements that comprise the enterprise environment that must be protected. Analyzing the results of this audit can reveal shortcomings in the organization's physical risk.

How easy is it for someone to access sensitive hardware components and software and database storage devices? It is essential to note that in today's distributed and e-commerce environment, a number of physical threats can be located at remote branches and locations. Devices that can be an access point for attackers include remote terminals, communications controllers, multiplexers, servers, printers, and PCs. A number of devices can be a source of eavesdropping and monitoring.

The organization cannot function without the physical assets that include computer systems and networking systems. It is important to note that a database might be the "company." A credit-reporting company database might be its major asset. Several good examples include the records that are located on the databases of the Social Security Administration, the Internal Revenue Service, stock exchanges, Internet Service Providers, and airlines. This section suggests precautions that can be taken to prevent or minimize damage from the disasters involving biometric databases.

Physical Security Categories

As mentioned previously, physical security issues cover a broad spectrum of possibilities, which include natural disasters, vandals and destructive individuals, fire and water damage, power loss, and heat and humidity. Each of these issues can negatively impact the operation and content of the biometric database and storage media assets. This section will provide an overview of human-initiated and power loss situations and discuss possible ways to mitigate the cost of such occurrences.

Human-Initiated Damage

As distributed computing increases, protecting a system from outside access becomes more difficult and important. A protection mechanism and policy are needed to prevent unauthorized people from obtaining access to the systems and to verify the identity of accepted users. (Physical access methods will be addressed in a subsequent section.)

Humans cause a high percentage of outages and damage, which can be accidental or deliberate. These incidents include work errors, sloppy work, or deliberate sabotage caused by disgruntled employees. Or attackers may include bored operators, ignorant employees, intruders, or thrill seekers.

Because computers and their database media are rather sensitive, a vandal could inflict a considerable amount of destruction quickly. An unskilled vandal may use a brute-force frontal attack with an ax or brick; however, these people will probably be noticed and stopped before any serious damage can occur. More skilled vandals can short-circuit a processor or disable or crash a disk drive with small tools, which would be difficult to detect. Competitors or real saboteurs can also cause destruction.

All telecommunications areas, computer rooms, and networking facilities must be kept secure from unauthorized access. The most effective access control method is a card-operated lock that allows access to authorized people and maintains a record of the event. With this method it is easy to control access and to determine who was in the area at the time of failure. Good records are essential for recovering from damage. Equipment location, inventory, and configuration records should be kept where they are readily available to anyone restoring service.

Power Loss

A constant source of predictable power is required to keep a computer system operating properly. For some time-critical applications, loss of service from a system is intolerable, and efforts must be made to provide some alternative power source if this occurs. Because of possible damage to media by sudden loss of power, many disk drives monitor the power level and quickly retract the recording head if power fails.

One protection against power loss is an **uninterruptible power supply** (UPS). This device stores energy during normal operation so it can provide backup energy if power fails. One form of UPS uses batteries that are continually charged when the normal power is on, which then provides power if the normal source fails (Figure 4-4). Several problems with batteries include heat, size, flammability, fumes, leakage, and insufficient output. Size and limited duration of energy

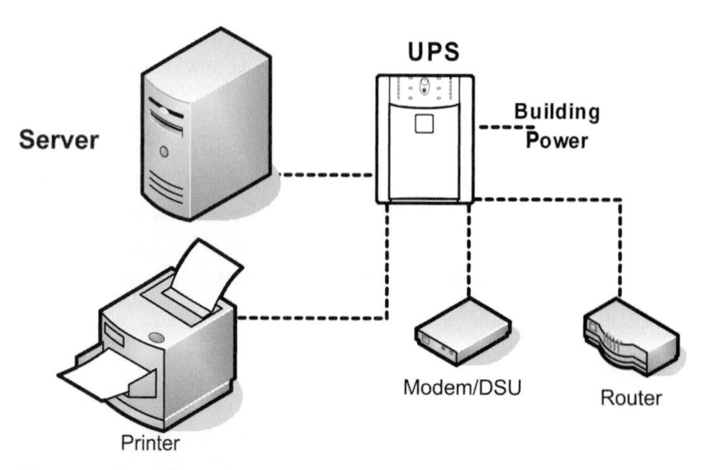

Figure 4-4 UPS for Server Configuration

Course Technology/Cengage Learning

output are limitations with the standard UPS system. Some UPS systems use outside generators and motor generators to provide support to the primary backup power supply; however, this is a very expensive alternative. If backup power is to be provided for an extended period of time, the outside generator will be a requirement. All telephone central offices use both UPS systems and outside generators because this service is expected to operate continuously and seamlessly.

Another issue with power is its quality. Most people are unaware that AC power fluctuates plus or minus 10 percent, which means that the supply could be between 108 volts to 132 volts. A voltmeter can be utilized to measure this variation. A disk drive or air conditioner power cycling can cause a temporary drain on the system and cause the lights to dim. When a motor stops, a temporary surge can be sent across the system. It is also possible for lightning to send a momentary large surge across the system. Instead of being constant, the power delivered along a power line shows many fluctuations, called drops, spikes, and surges. A drop is a momentary reduction in voltage, whereas a spike or surge is a rise. For computing equipment, a drop is less serious than a surge, however most electrical devices is tolerant of some fluctuations of current.

Voltage fluctuations can be destructive to sensitive electronic equipment. Simple devices such as surge suppressors filter spikes from a power source, clocking fluctuations that could affect the electronic devices. These surge suppressors are rated in joules—the higher the rating, the more protection afforded. The cost of these devices depends upon the joule rating. These protection devices should be installed in every computing and networking device in the installation. This includes PCs, printers, servers, routers, and communication equipment devices such as modems and DSUs.

Another surge that can damage electronic equipment is a lightning strike. To increase protection, personal computers can be unplugged when not in use and during electrical storms. Another source of destruction is when lightning strikes a telephone line and passes through the modem to the computer. Disconnecting a modem during a storm is also an effective preventive measure. A lightning strike on a telephone line can follow a path to the modem card or standalone modem and destroy the electronics. Additional lightning suppression devices may be available from the local telephone company.

Environmental Safeguards Recap

Adequate environmental safeguards must be installed and implemented to protect computer, database, and networking assets. The criticality and sensitivity of each asset will determine the level of security required. The more critical the asset is for accomplishing the mission of the organization, the higher the level of security required. The following safeguards are important elements that should be considered:

- Electrical power supply provisions
- Protection from electromagnetic magnetic interference (EMI) and radio frequency interference (RFI)

Summary

- Biometric data is used to describe the information collected during an enrollment, verification, or identification process, but does not apply to end user information such as user name, demographic information, and authorizations. It encompasses raw sensor observations, biometric samples, models, templates and/or similarity scores.

- Biometric information consists of stored electronic information pertaining to a biometric. This information can be in terms of raw or compressed pixels or in terms of some characteristic, such as patterns. To be effective, a biometric system must compare captured biometric data to a biometric database.
- Various government entities are involved in the deployment of biometric database systems. Several large projects are ongoing at the Department of Homeland Security, the FBI, and the IEEE.
- Typical physical security components include guards, locks, fences, and cameras to deter direct attacks. Sources that can compromise an organization's ability to function include natural events, human vandals, and power loss. These physical vulnerabilities to security can and often occur simultaneously. Their impact on the biometric database asset can be significant.

Key Terms

Biometric information Consists of stored electronic information pertaining to a biometric.

Claim of identity A statement that a person is or is not the source of a reference in a database.

Data Items representing facts, text, graphics, bit-mapped images, sound, and analog or digital live-video segments.

Database (DB) A collection of information organized in such a way that a computer program can quickly select desired pieces of data.

Database management system (DBMS) A complex set of software programs that controls the organization, storage, management, and retrieval of data in a database.

Data mining The process of searching for unknown information or relationships in large databases.

Data warehouse A database that consolidates an organization's data.

Data warehousing A software strategy in which data are extracted from large transactional databases and other sources and stored in smaller databases, which makes analysis of the data easier.

Detection and identification rate The rate at which individuals, who are in a database, are properly identified in an open-set identification (watch list) application.

e-passport A travel document that contains an integrated circuit chip based on international standard ISO/IEC 14443 and that can securely store and communicate the e-passport holder's personal information to authorized reading devices.

Gallery The biometric system's database, or set of known individuals, for a specific implementation or evaluation experiment.

Information Data that has been processed to add or create meaning and knowledge for the person receiving it.

Message authentication code (MAC) A short piece of information used to support authentication of a message.

Metadata Information that describes data in a database.

Off-card Data that is not stored on an ID card or to a computation that is not performed by the integrated circuit on an ID card.

On-card Refers to data stored on an ID card or to a computation that is performed by the integrated circuit chip on an ID card.

Personally identifiable information (PII) Any piece of information which can potentially be used to uniquely identify, locate, or contact a person or steal the identity of a person.

Probe The biometric sample that is submitted to the biometric system to compare against one or more references in the gallery.

Record Consists of the template and other information about the end user (e.g. name, access permissions).

Schema Data is stored in a database according to a predefined format and an established set of rules.

Structured Query Language (SQL) A standard language for making interactive queries from and updating a database.

Uninterruptible power supply (UPS) This device stores energy during normal operation so it can provide backup energy if power fails.

Review Questions

1. A ________________ is a collection of information organized in such a way that a computer program can quickly select desired pieces of data.
2. A ________________ is a complex set of software programs that controls the organization, storage, management, and retrieval of data in a database.
3. ________________ is information that describes data in a database.
4. ________________ is the process of searching for unknown information or relationships in large databases.
5. ________________ is to ensure data entered in the database are accurate and valid.
6. List the major features of a DBMS.
7. ________________ is any piece of information which can potentially be used to uniquely identify, locate, or contact a person or steal the identity of a person.
8. Detection and identification rate is the rate at which individuals in a database are properly identified in a closed-set identification application. (True or False?)
9. Biometric data is a catch-all phrase for computer data created during a biometric process. (True or False?)
10. The biometric database is a collection of one or more computer files. Identify these files: ________________.
11. The formatted digital record used to store an individual's biometric attributes is called a ________________.

12. A device that provides protection against power loss for computers and other electronic devices is an ________________.
13. Heat and humidity has no negative impact of a database system. (True or false?)
14. The Department of Homeland Security and the U.S. Citizenship and Immigration Services have been tasked with processing ________________.
15. The FBI has announced its plans to assemble the world's largest biometric database, nicknamed the ________________.
16. The identification rate is a statistic used to measure biometric performance. (True or False?)
17. The watch list is a term used with closed-set identifications. (True or False?)
18. Enrollment is the process of collecting biometric samples from a user representing that person's identity. (True or False?)
19. Comparison is the process of comparing a biometric reference with a previously stored reference to make an identification or verification decision. (True or False?)
20. Data that is stored on a card for the purpose of identification is called on-card. (True or False?)

Discussion Exercises

1. Identify major biometric database and DBMS systems.
2. Identify and describe the various software tools that can be used to access and update biometric databases.
3. Provide an overview of the various safeguards that should be considered for biometric systems and databases.
4. Describe the capabilities commonly offered by database management systems.
5. Research the topic of uninterruptible power supplies. Provide an overview of the capabilities and features for various sizes of these devices.
6. Identify physical data elements that could be stored in a biometric database.
7. Search for organizations that have experienced some computer facility disaster that impacted the database. Describe the recovery process and issues that surfaced during this situation.
8. Describe the online analytical processing (OLAP) system. Provide a list of capabilities.
9. Develop a list of PII (i.e., eye color).
10. Provide metadata for some database file.

Hands-On Projects

4-1: Biometric Database Solution Awareness

Time required—30 minutes

Objective—Identify biometric database solutions that have been implemented in some organization.

Description—Use various Web tools, Web sites, and other tools to identify biometric database solutions that are in development or production for both commercial and government organizations.

1. Identify organizations whose only or major asset is a database. Identify the vendor of the database product. Also describe the capabilities to provide for a biometric system access.
2. Research the SQL programming language and construct a biometric query for some modality feature.
3. Identify three vendors that provide biometric database and DBMS systems. Develop a comparison of features and cost.
4. Identify an organization that added a biometric database application to a legacy computer system. Provide an overview of the issues and solutions for this conversion project.

4-2: Evaluate the Database Requirements for a Biometric Solution

Time required—45 minutes

Objective—Identify those requirements for a biometric database that can provide some level of access control and security for OnLine Access assets and resources.

Description—The owners and managers of OnLine Access have identified several applications and systems that have the potential to satisfy the organization's access control and security requirements. Note that data has been compromised and/or stolen in the past. Security and access control of the database system is a priority.

1. Research the topic of database access control and security. Identify those initiatives that can improve the database security issue.
2. Produce a matrix describing each option's requirements and potential cost.
3. Rank order those solutions that would best provide the desired results and provide a rationale for each.

chapter 5

Legacy and Biometric Systems

After completing this chapter, you should be able to do the following:

- Understand the basic concepts and terms that apply to computer and network systems
- Understand the issues of security in the context of legacy systems
- Identify computer and electronic components that can benefit from the various biometric technologies
- Learn the terms and concepts used in biometric information systems design and implementation
- Become familiar with various types of applications that can be incorporated into biometric systems
- Look at sample biometric systems that have been developed for government applications

Legacy computer systems include all of the hardware and software applications and solutions that have been in a production status for many years. There is a significant capital investment in these systems and they include policies and procedures for the ongoing successful operations of the organizations. Security and access control systems are antiquated or nonexistent in many of these organizations and management may not be receptive in making major changes to implement any system, especially a new technology.

Many of the biometric applications and systems rely on some computer and/or network infrastructure. Because many legacy applications use the Internet and Web resources, the expectation is that a biometric solution will also use these same resources. The explosion of e-commerce transactions will necessitate additional security and access control for the protection of an organization's resources. Biometric solutions may well be the solution to these requirements.

Introduction

Security systems design incorporates a number of technologies and techniques to provide for a secure computing technology and database environment. Biometrics is in the forefront of technology in providing access control techniques for these computer networks. Biometrics, therefore, is the study and application of biological data and biometric-based authentication for access control.

All biometric technology systems share certain aspects: they are dependent upon an accurate reference or "registration" sample; if a biometric system is to identify a person, it first must have this sample, positively linked to the subject, to compare against; modern biometric identification systems, based on digital technology, analyze personal physical attributes at the time of registration and distill them into a series of numbers.

Biometric systems rely on the use of physical or behavioral traits—such as fingerprints and face, voice, and hand geometry—to establish the identity of an individual. The deployment of large-scale biometric systems in commercial, government, and public service applications increases the public's awareness of this technology. This rapid growth also highlights the challenges associated with designing and deploying biometric systems. The recent years have seen a significant growth in biometric research, resulting in the development of innovative sensors, robust and efficient algorithms for feature extraction and matching, enhanced test methodologies, and novel applications. These advances have resulted in robust, accurate, secure, and cost-effective biometric systems.

Many of the early biometric devices were rather cumbersome in use and priced at a point which prohibited their implementation in all but a few very high-security applications where they were considered viable. This situation is changing: not only has considerable technical progress been made, providing more accurate, more refined products, but unit cost has dropped to a point which makes them suitable for broader scale deployment where appropriate. In addition, the knowledge base concerning their use and integration into other processes has increased dramatically.

This chapter discusses biometric systems, secure protocols, and secure electronic transactions. Biometric recognition (or simply "biometrics") is a rapidly evolving field with applications ranging from accessing one's computer or physical assets to gaining entry into a country.

With the advent of e-commerce systems, a secure process was required to handle credit card, debit card, and other financial transactions over the network. Major specifications that accomplish this are called the Secure Electronic Transaction and the Payment Card Industry Data Security Standard.

Computer and Information System Basics

A brief overview of basic information concerning computer and information systems will provide the reader some basis for understanding the issues relating to biometric-based information systems. This section includes information that identifies the computer system infrastructure components that are required for biometric systems to operate successfully. Computer systems, computer programs, and the all-important communication network components will be discussed.

Computer Systems and Computer Programs

A *computer system* includes not only the computer, but also any software and peripheral devices that are necessary to make the computer function. Every computer system, for example, requires an operating system. Peripherals include disk drives, tape drives, printers, and communication devices. Figure 5-1 depicts a common configuration for a mainframe computer system.

A *computer program* (also called software program or program) provides the instructions for a computer. A computer requires programs to function, and a computer program does nothing unless its instructions are executed by a central processor. Computer programs refer to either an executable program or the source code from which an executable program is derived.

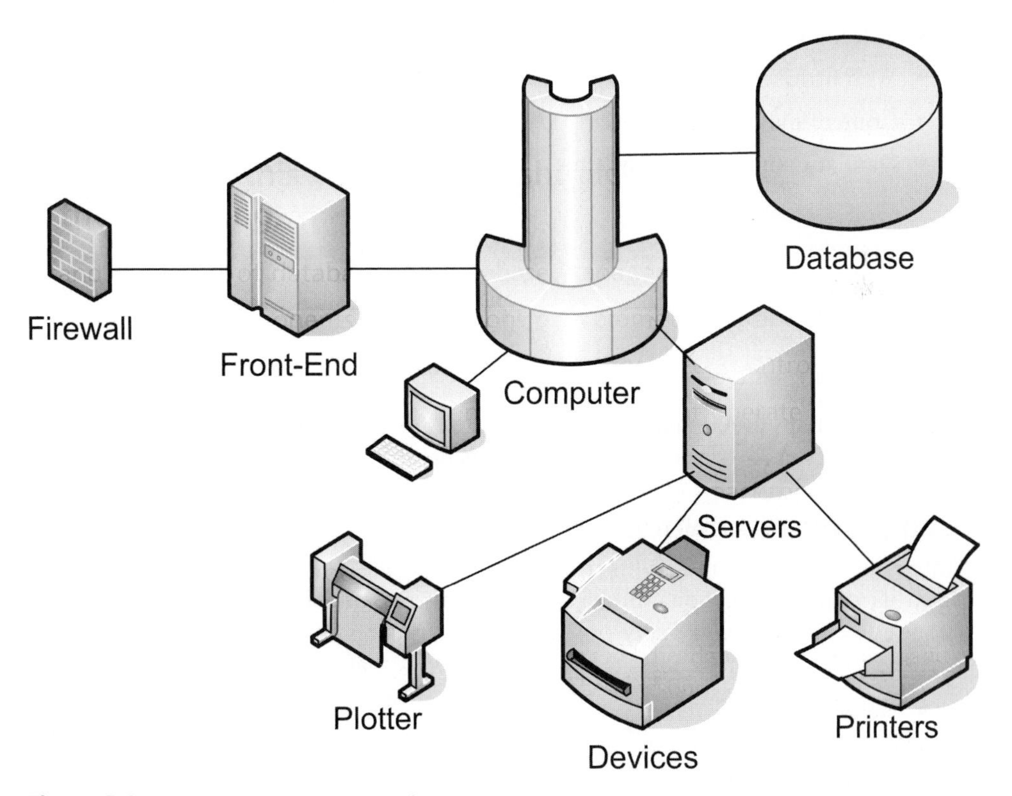

Figure 5-1 Mainframe Computer Configuration

Course Technology/Cengage Learning

Data and Information

Data refers to a collection of organized information, usually the results of experience, observation, or experiment, or a set of premises. This may consist of numbers, words, or images, particularly as measurements or observations of a set of variables. *Information* includes processed data that enhances the recipient's knowledge. Raw facts, called data, are transformed into something meaningful and useful.

Application Software Systems

A *software system* is a system based on software that forms part of a computer system (a combination of hardware and software). The term *software system* is often used as a synonym for a computer program or software and is related to the application of systems theory approaches in software engineering and in an architecture context. This approach is often used to study large and complex software, because it focuses on the major components of software and their interactions. Major categories of software systems include application software, programming software, and system software, although the distinction can sometimes be difficult. Examples of software systems include:

- Computer reservations system
- Air traffic control software
- Military command and control systems
- Telecommunication networks
- Web browsers
- Content management systems
- Database management systems
- Expert systems
- Spreadsheets
- Window systems
- Word processors
- Biometric systems

Software systems are an active area of research for groups interested in software engineering in particular, and in systems engineering in general.

Communication Network System

Numerous tasks and operations must occur in the telecommunications network for it to perform its required functions. Some of these activities are simple, but most are complex and require interactions with many other elements to successfully transport communications traffic. The following list includes the major categories that must be addressed.

- Network management
- Security
- Recovery
- Addressing

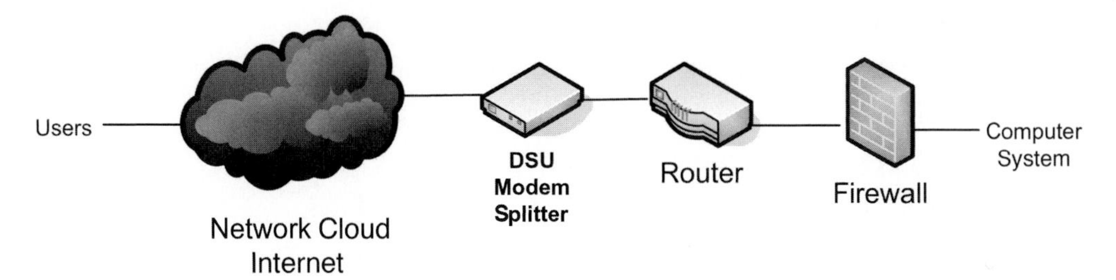

Figure 5-2 Network Configuration

Course Technology/Cengage Learning

- Synchronization
- Utilization
- Flow control
- Error detection/correction
- Interfacing
- Signal generation
- Handshaking
- Routing
- Formatting

Even though many of the activities and functions listed are automatic, managing the system is often time-intensive and requires a considerable amount of hands-on management from systems administrators and network technicians. The accessibility and availability of the communications network warrants a high priority. Figure 5-2 depicts a common network configuration associated with a mainframe computer system.

Security Systems Design

This section is concerned with the design of the legacy perimeter security and the security policies that must be included in this development. *Note that a legacy system is an older computer system that has become central to business operations.* It is essential that the administrator and security manager know how the components function and interact with both internal and external networks. A number of issues must be addressed to successfully protect the organization's computer and networking assets via access control methods and techniques. These include the following:

- Identify and understand the enemy
- Measure the cost or lack of security
- Identify security assumptions
- Look at the human factors
- Identify the organization's weaknesses

- Limit the scope of system access
- Understand the computer and network environment
- Limit trust in software
- Look at physical security
- Make security part of the normal operating procedures

An important document in systems design is the specification. The specification is documentation that reflects agreements on products, practices, or operations produced by one or more organizations (or groups of cooperating entities), some for internal usage only, others for use by groups of people, groups of companies, or an entire industry.

Trusted Systems

As previously stated, trust is the composite of availability, performance, and security, which includes the ability to execute processes with integrity, secrecy, and privacy. Technical solutions that provide legacy and biometric support to organizational trust systems will be presented in subsequent sections. Each of these systems provides a part of the total solution for protection of the organization's computer and network assets and resources. The task of management and systems administration is to identify, secure, and implement security systems that will protect the organization from the untrusted networks. E-commerce security and integrity are usually maintained by implementing a number of protocols, hardware, and software solutions.

E-Commerce Security and Secure Protocols

There is considerable apprehension concerning the lack of security across the various networks. To transact business, organizations must have a means to conduct secure, confidential, and reliable transactions. To achieve this goal, four cornerstones of secure electronic transactions must be present. These include the following attributes:

- Possess the ability to prevent unauthorized monitoring of traffic
- Prevent the content of messages from being altered after transmission and provide a method to prove they have or have not been altered
- Determine whether a transmission is from an authentic source or a masquerade
- Prevent a sender from denying the receipt or transmission of a message (nonrepudiation)

Cryptographic systems provide a method to transmit information across an untrusted communication system without disclosing the content of the information to a snoop. These systems also provide confidentiality and can provide proof that a transmission's integrity has not been compromised. Additional details will be presented in subsequent sections.

Securing Electronic Transactions

Commercial and governmental organizations utilize the Internet to transmit e-mail, purchase goods and services, and transact banking and other services that depend upon security transmissions. There are methods and protocols that can provide secure transmissions for these activities.

E-mail often contains confidential information such as credit card numbers or Social Security numbers. It is essential that the contents and the sender of a message be authenticated. An e-mail might be considered a legal document in a transaction such as responding to an official competitive bid; therefore, it is essential that the sender and recipient be assured of the integrity of any transmission.

Many network users purchase goods and services from Web sites. Buyers select the items and authorize payment using the Web. These items are then delivered through the normal shipment channel. Typical items include books, CDs, and auction items. Software and recordings can be delivered immediately by downloading from the Internet. All of these transactions must be validated and tracked.

Most financial transactions now occur over some communication link through electronic banking. E-commerce systems offer users most of the services and functions provided by conventional banks. Functions include checking of account balances and statements, transferring funds between accounts, and electronic payments. Authentication and authorization are two security functions that must occur with these transactions. This means the user is identifiable and has authority to access a particular account, with a particular transaction.

There are a number of Internet transactions that fall in the general financial category. The Web lends itself to the supply of small quantities of information and other services to many users. The use of the Internet for voice and videoconferencing is another service that is likely to be supplied when it is paid for by end users. Transactions such as these can be safely executed only when they are protected by appropriate security policies and mechanisms. A client must be protected against the disclosure of credit and debit card numbers during transmission and against a fraudulent vendor who obtains payment without delivering the goods or service. Conversely, vendors must obtain payment before releasing the items selected for purchase, whether online or through the normal delivery conduit. In addition, the required protection must be achieved at a reasonable cost in comparison with the value of the transaction.

Secure Protocols

After authentication and authorization has been established, users (clients and servers) can engage in secure sessions and transactions. A number of protocols have been developed to handle secure sessions and transactions. A brief description of the various protocols is presented in this section. *Note: Additional details will be presented in Chapter 12.*

- Transport Layer Security (TLS): A cryptographic protocol that provides secure communications on the Internet for such things as Web browsing, e-mail, Internet faxing, instant messaging, and other data transfers.
- Secure Sockets Layer (SSL): A protocol developed by Netscape for transmitting private documents via the Internet. SSL uses a cryptographic system that uses two keys to encrypt data—a public key known to everyone and a private or secret key known only to the recipient of the message.
- Pretty Good Privacy (PGP): A computer program that provides cryptographic privacy and authentication. PGP is often used for signing, encrypting, and decrypting e-mails to increase reliability for e-mail communications.

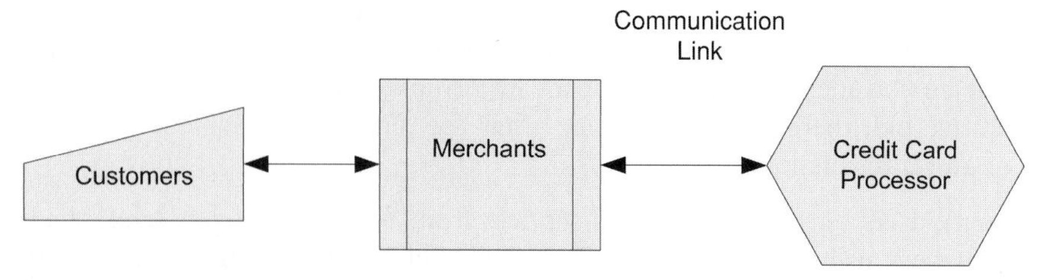

Figure 5-3 Generic Credit Card Flow

Course Technology/Cengage Learning

- Secure Hypertext Transfer Protocol (S-HTTP): An alternative mechanism to the https Uniform Resource Identifier scheme for encrypting Web communications carried over HTTP.
- Secure Electronic Transaction (SET): A system for ensuring the security of financial transactions on the Internet.

Credit Card Transactions

If e-commerce is to succeed, a method must exist for consumers to use credit cards over the Internet. Credit card and debit card usage on the Internet is growing at an astronomical rate. SET and SSL are utilized in the credit card transaction to provide security provisions. Figure 5-3 provides a generic example of a credit card transaction flow.

SSL encrypts a credit card number and other information using a 40-bit key. Due to its size, this key can be hacked; however, it may be adequate for some needs. Even though SSL keeps the credit card number and information private while it is being transmitted, it does not address the issue of whether the card is valid, stolen, or being used without permission. SET addresses these SSL limitations by using an "electronic wallet" that can identify the user and validate the transaction. An electronic wallet is a type of software application used by the consumer for securely storing purchasing information. Furthermore, SET-based systems have an advantage over other mechanisms, in that SET adds digital certificates that associate the cardholder and merchant with a particular financial institution and the Visa or MasterCard payment system.

Electronic Commerce

E-commerce involves Internet use for credit card purchases of such items as airline tickets, computer hardware and software, books, and miscellaneous products. It involves methods of securing transactions, authorizing payments, and moving money between accounts. One process used to accomplish these transactions is called **electronic data interchange (EDI)**. EDI is a process whereby standardized forms of e-commerce documents are transferred between diverse and remotely located computer systems. These forms include purchase orders and invoices. Figure 5-4 depicts the EDI environment.

Financial Transactions

Financial transactions are the lifeblood of e-commerce, which is why it is vitally important to keep these transactions secure. Unfortunately, unsecured and insufficiently secured financial

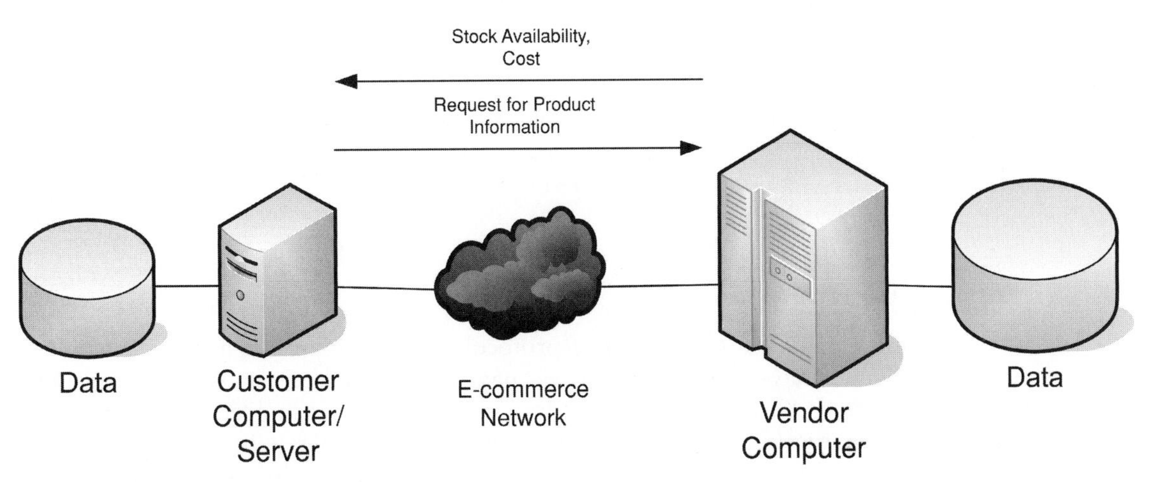

Figure 5-4 EDI Environment

Course Technology/Cengage Learning

transactions are more commonplace than most consumers realize. While industry statistics indicate improvements in security technologies have been implemented, many organizations are still not using them to their fullest advantage. Studies have indicated that financial losses can be attributed to computer-related breaches, stolen credit card numbers, and stolen login passwords. To help mitigate these security breaches, many organizations have implemented a security program that is supposed to protect them and their customers from data theft, fraud, and misuse. All is not well, however, because internal policies, monitoring of security measures, and follow-up of security breaches are often disregarded.

Technology alone will not protect an organization or a financial transaction. If there is a vulnerable spot within an enterprise's security framework, an intruder will find a way to access and exploit it. For this reason, multiple security technologies, combined with a well thought-out and monitored security plan must be implemented to ensure financial transactions are secured and protected. *Multibiometrics will be discussed in Chapter 6.*

It is expensive to maintain a comprehensive security system for financial transactions; however, it is much more expensive to maintain one that is inadequate. If an organization is going to participate in the e-commerce game, it must provide those security systems necessary to provide an adequate level of support. The components most often utilized to provide this support includes the following technologies:

- Digital signatures
- Public Key Infrastructure (PKI)
- Secure Socket Layer (SSL)
- Authentication certificates
- Firewalls
- Border managers (routers)

- Biometrics
- Passwords

If an organization does not have the available capital to improve or develop the infrastructure for electronic transaction security, outsourcing is an alternative. It should be noted that outsourcing might be considered as an interim solution, with a plan to provide these functions in-house eventually.

Payment Protocols

The Secure Electronic Transaction (SET) protocol was developed so merchants could—automatically and safely—collect and process payments from Internet (Web) clients. Basically, SET secures credit card transactions by authenticating cardholders, merchants, and banks. It also preserves the confidentiality of payment data. SET requires digital signatures to verify that the customer, merchant, and bank are legitimate. It also uses multiparty messages that prevent credit card numbers from ending up in the wrong hands. Electronic money options include electronic cash, wallets, and cybercash.

Electronic cash is a scheme that makes purchasing easier over the Web. One method is for electronic cash companies to sell "Web dollars" that purchasers can use when visiting Web sites. Web sites accept these Web dollars as if they are coupons.

A **wallet** is an electronic cash component that resides on the user's computer or another device such as a smart card. It stores personal security information such as private keys, credit card numbers, and certificates. It can be moved around so that users can work at different computers and have its access controlled by policies. The *Personal Information Exchange* is a protocol that enables users to transfer sensitive information from one environment or platform to another. With this protocol, a user can securely export personal information from one computer system to another.

Cybercash uses the wallet concept, where a consumer selects items to purchase and the merchant presents an invoice and a request for payment. The consumer can then have the cybercash wallet pay for the purchase electronically. Encrypted messages are exchanged between the merchant server, the consumer, the cybercash server, and the conventional credit card networks to transfer the appropriate funds.

Smart Card

A **smart card** is about the size of a credit card, which contains a microprocessor for performing a number of functions. A phone card is an example of a simple smart card. Unlike a stripe card, which has magnetic stripes on the outside, a smart card holds information internally and is more secure. It is often used as a token authentication device that generates access codes for secure systems. It has gained industry acceptance for a number of applications. Figure 5-5 shows a sample smart card.

Smart card manufacturers are hopeful that the SET standard will help build a market for the card in the United States. VISA has led the way; however, adoption of the SET standard has been slow. Adoption of the VISA offering within the U.S. could boost the SET standard, which would benefit e-commerce.

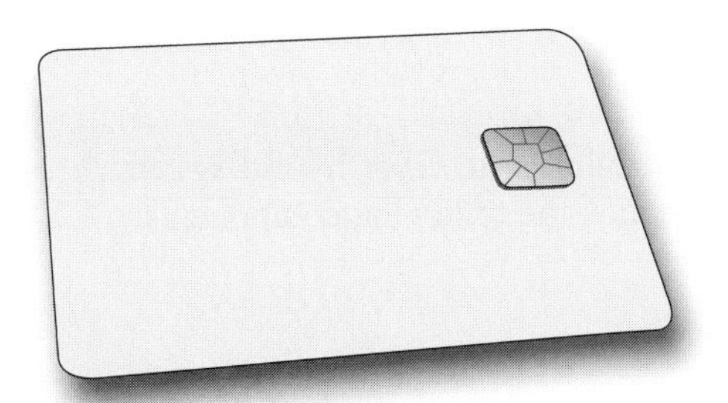

Figure 5-5 Sample Smart Card

Course Technology/Cengage Learning

Reactive and Proactive Security Systems

This section is concerned with those categories of security systems that operate in a reactive or proactive mode. These include intrusion detection and preventions systems and contingency planning and disaster recovery systems.

Intrusion detection is the art of detection and response to computer and network misuse. The benefits to the organization include the following security elements:

- Deterrence
- Detection
- Response
- Damage assessment
- Attack anticipation
- Prosecution support

Different intrusion detection techniques provide different types of benefits for a variety of environments, so it is essential to select and implement the most viable one. The key to selecting the correct detection system is to define the environment-specific requirements that best satisfy the organization's security needs.

System Configurations

The two categories of intrusion detection systems are network based and host based. Host-based technologies examine events relating to file and database access and applications being executed. Network-based technologies examine events such as packets of information exchanged between networked computers. A hybrid system that includes components from both host- and network-based detection systems offers the best solution.

The intrusion detection industry supplies tools with capabilities and features that go far beyond detecting intruders from outside the organization. Intrusion detection systems may provide the following capabilities:

- Event log analysis for insider threat detection
- Network traffic analysis for perimeter threat detection

- File and database integrity checking
- Security configuration management

Many products provide only one of these capabilities, however hybrid systems are available that are multifunctional. A hybrid system contains the following components:

- Alert notification
- Command console
- Database
- Network tap
- Network sensor
- Response subsystem

Network-Based Intrusion Detection

Network intrusion detection is effective at detecting outsiders attempting to penetrate the Enterprise network defenses. The two intrusion detection architectures include traditional sensor and distributed systems. The traditional sensor architecture is easy to deploy and operate, but is limited by high-speed, switched, or encrypted networks. The distributed architecture addresses these issues, but is significantly more difficult to deploy and manage.

Host-Based Intrusion Detection

Host-based intrusion detection systems are distributed systems that gather and process event logs and other data from computers in an organization. The data may be processed locally at the target location or transported to a centralized site for processing. Operationally, host-based systems can be effectively utilized for surveillance, damage assessment, intelligence gathering, and compliance.

Managing a host-based system is considerably more difficult than managing a network-based system. It is critical to have developed a good policy management and an efficient audit policy before implementing either system, which can significantly reduce the operational performance overhead.

An in-depth understanding of the issues regarding intrusion detection can be gained by reading *The Practical Intrusion Detection Handbook*. [1]

Intrusion Detection System Vendors

There are a number of providers of intrusion detection systems. These companies can provide different levels of support for acquisition, deployment, and maintenance. These system providers include:

- Axent Technologies, Inc.
- Cisco Systems, Inc.
- Computer Associates
- CyberSafe Corporation
- Internet Security Systems
- Network Associates

- Network Ice
- Network Security Wizards
- ODS Networks
- Pentasafe

Business continuity planning (BCP) is a term that has to do with failure of IT equipment, or the ability to employ it effectively. Items affecting business continuity range from a loss of power to floods, terrorist attacks, or anything else that causes a loss of business. Business continuity is a progression of disaster recovery aimed at allowing an organization to continue functioning after (and ideally, during) a disaster, rather than simply being able to recover after a disaster.

Disaster recovery is the process, policies, and procedures of restoring critical operations—including regaining access to data (records, hardware, software, etc.), communications (incoming, outgoing, toll-free, fax, etc.), workspace, and other business processes—after a natural or human-induced disaster. To increase the opportunity for a successful recovery of valuable records, a well-established and thoroughly tested data recovery plan must be developed. This task requires the cooperation of a well-organized committee led by an experienced chairperson. A **disaster recovery plan** (DRP) should also include plans for coping with the unexpected or sudden loss of communications and/or key personnel, focusing on data protection.

Biometric System Basics

A realistic threat model must be identified before a biometric access control system can be deployed. This includes identifying the categories of individuals the system is supposed to target, and the threat they pose in light of their abilities, resources, motivations, and goals. Any such system will also need to map out clearly in advance how the system will work, in both its successes and in its failures.

A **biometric system** is described as an automated system capable of:

- Capturing a biometric sample from a user
- Extracting biometric data from that sample
- Comparing the biometric data with that contained in one or more reference templates deciding how well they match
- Indicating whether or not an authentication of identity has been achieved

The system consists of integrated biometric hardware and software used to conduct biometric identification or verification. As stated in Chapter 3, biometric authentication requires comparing a registered or enrolled biometric sample (biometric template or identifier) against a newly captured biometric sample (for example, a fingerprint captured during a login). During enrollment a sample of the biometric trait is captured, processed by a computer, and stored for later comparison.

There is no "perfect" biometric that fits all needs. All biometric systems have their own advantages and disadvantages. There are, however, some common characteristics that are needed to make a biometric system usable. First, the biometric must be based upon a distinguishable trait. For example, for nearly a century, law enforcement has used fingerprints to identify people. There is a great deal of scientific data supporting the idea that "no two fingerprints are alike." Technologies such as hand geometry have been used for many years and technologies such as face or iris recognition have come into widespread use. Some newer biometric methods may be just as accurate, but may require more research to establish their uniqueness.

Another key aspect is the "user-friendliness" of a system. The process should be quick and easy, such as having a picture taken by a video camera, speaking into a microphone, or touching a fingerprint scanner. Low cost is important, but most implementers understand that it is not only the initial cost of the sensor or the matching software that is involved. Often, the life-cycle support cost of providing system administration and an enrollment operator can overtake the initial cost of the biometric hardware.

The Need for Biometrics

As discussed in previous chapters, modern rapid advancements in networking, communication, and mobility have increased the need for reliable ways to verify the identity of any person. Currently, identity verification is mainly performed by possession based and knowledge based systems and techniques.

Weaknesses of standard validation systems can be avoided if a body part provides the key. Particular characteristics of the body or habits are much more complicated to forge than a string, even if it is very long. It is evident that using biometrics adds a complexity to identification systems that would be hard to reach with a standard password-based approach. The main advantages of biometrics over compared to a standard system are:

- Biometric traits can not be lost or forgotten.
- Biometric traits are difficult to copy, share, and distribute.
- They require the person being authenticated to be present at the time and point of authentication.

The systems using biometrics have a high degree of robustness, speed, and accuracy in capturing the biometric sample. Automated biometric systems can be used in conjunction with passwords or tokens, improving the security of existing systems without replacing them. Among the different biometrics available to authenticate an individual, fingerprint recognition is the most widely implemented.

Biometric Systems Functions

Biometric systems are composed of multiple individual components (such as sensor, matching algorithm, and result display) that combine to make a fully operational system. A biometric system is an automated system capable of: [2]

- Capturing a biometric sample from an end user
- Extracting and processing the biometric data from that sample
- Storing the extracted information in a database
- Comparing the biometric data with data contained in one or more references
- Deciding how well they match and indicating whether or not an identification or verification of identity has been achieved

A biometric system includes biometric applications and may be a component of a larger system.

Attributes of Biometric Systems

Currently, other than personally recognizing someone or having a trusted third party personally swear to their identity, the only other technique for identifying a person is through the use of a "token." These tokens, which are in essence representations of the oath of a trusted third party, come in two basic forms:

- *Knowledge tokens*, such as passwords, secret PINs, or knowledge of personal data such as mother's maiden name.
- *Physical tokens* such as identification (ID) cards, passports, chip cards, or plain old keys.

Token IDs offer certain advantages over biometric identification. Security against "false acceptance" of impostors can be raised by increasing the complexity of the token used for identification. Also, in the event of loss or compromise, the token, be it a password, PIN, key, or ID card, can be revoked, changed or reissued, where a biometric measurement cannot.

The advantage of biometrics is that unlike tokens, biometrics cannot be lost, loaned, or forgotten. Token-based systems must verify that the presenter is the authorized user, not an unauthorized person who has come to possess the token. Used carefully, biometrics may be combined with token-based systems to mitigate the vulnerability of ID tokens to unauthorized use.

Functions of Biometric Systems

One useful way of thinking about biometrics is that they are used for one of two purposes:

- To prove that you are who you say you are (positive ID), or
- To prove that you are not who you say you are not (negative ID).

In a positive ID situation, the subject asserts that she is Jane Doe and submits a "live" sample (a fingerprint, for example) to the system. The system then checks its database of previously enrolled or registered samples to see if the live sample matches the reference sample. A positive ID system is designed to prevent more than one person from using a single identity. In a negative ID situation, John Roe claims *not* to be someone already known to the system. Here, the system checks its database to see that Roe is not on the watch list of suspected criminals and terrorists, whose biometrics are already in the system. A negative ID system is designed to prevent one person from using more than one identity.

When biometrics is employed to effect negative identification, one need not be enrolled. The only persons who must be "in" the database are those whom the operator is trying to keep out or catch. Biometrics alone cannot establish "true identity." A biometric system cannot prevent someone from furnishing fake credentials when they first enter the system. They can only prevent them from using another identity once enrolled.

Common Aspects of All Biometric Systems

All biometric technology systems have certain aspects in common. All are dependent upon an accurate reference or "registration" sample. If a biometric system is to identify a person, it first must have this sample, positively linked to the subject, to compare against. Modern biometric identification systems, based on digital technology, analyze personal physical attributes at the time of registration and distill them into a series of numbers. Once this reference sample is in the system, future attempts to identify a person are based on a comparison of a "live" sample and the reference sample or samples.

A perfect system would recognize a person 100 percent of the time, and reject an impostor 100 percent of the time. However, biometric samples are gathered from people in environmental conditions that are uncontrollable, over equipment that may slowly be wearing out, and using technologies and methods that vary in their level of precision. Consequently, the accuracy of biometric systems is assessed in light of these confounding variables via its tendency to experience either a "false match" (also called a "false accept") or a "false nonmatch" ("false reject"). The point at which these two rates intersect is called the equal error rate or crossover point.

Biometric systems may be "tuned" to diverge from the equal error rate to provide a match threshold that satisfies the designer's requirements. If a system compares a large number of persons against a small number of samples, and the consequence of a false match is low, a system biased towards a higher "false accept" or "false match" rate may be desirable. This might be the situation where officials at a border crossing or airport are looking for a short list of criminals. The advantage to biasing a system in this manner is that it is likely to err on the side of safety, and less likely to let a criminal slip through undetected. The disadvantage is the system will falsely associate innocent people with criminals or terrorists. If other safeguards are in place and the system operators understand the system's bias towards false match, the result can be a relatively trivial loss of convenience due to increased scrutiny, such as extra inspection of luggage, questioning, etc. Biasing such a system towards a high "false nonmatch" or "false reject" rate will result in fewer passengers slowed down at the gate, but at the cost of possibly losing the sought-after individuals. These measurements are described in a subsequent section.

When assessing the utility or the cost of a biometric system, it's important to understand the common features of all such systems. The following questions must be answered:

- How is the reference sample to be gathered and catalogued?
- How is the live sample going to be gathered?
- Can a live sample be captured without the subject's knowledge and cooperation?
- What is the value of a successful system, and what is the cost, to all parties, should it fail?

The implications of all four possible outcomes include:

- true match
- true nonmatch
- false match
- false nonmatch

Further, users should not assess failure simply from the perspective of the core biometric technology itself. Even an ideal system can be defeated easily if it is incorporated into an insecure or poorly designed overall system architecture. Any biometric system, especially one that involves a component of telecommunication, **must** be very carefully designed to prevent the loss or interception of user biometrics. Any deployed system must incorporate safeguards to prevent the interception of biometric data while it is being communicated. If a user's biometric is intercepted, criminals or hackers may be able to replicate either the sample itself or the string of binary data produced by a successfully matched sample. Armed with such intercepted biometric data, a criminal would be able to carry out a potentially very damaging identity theft.

Characterizing Different Biometrics

Different biometric features have characteristics that make them more or less useful for particular applications. Dr. James Wayman, director of the National Biometric Test Center at San Jose State University, categorizes biometric features in terms of five qualities: [3]

- Robustness: Repeatable, not subject to large changes
- Distinctiveness: There are wide differences in the pattern among the population
- Accessibility: Easily presented to an imaging sensor
- Acceptability: Perceived as nonintrusive by the user
- Availability: A user may present a number of independent measurable features

These qualities are explained by comparing fingerprinting to hand geometry.

Fingerprints are extremely distinctive, but not very robust. Fingerprints can be damaged in less than a minute of exposure to household cleaning chemicals. Many people have chronically dry skin and cannot present clear prints. Now, hands are very robust, but not very distinctive. Changing someone's hand geometry requires damaging the structure of the hand. And many people (somewhat less than 1 in 100) have similar hands, so hand geometry is not very distinctive. Hands are easily presented without much training required, but most people initially misjudge the location of their fingerprints, assuming them to be on the tips of the fingers. Both methods require some "real-time" feedback to the user regarding proper presentation. Both fingerprints and the hand are accessible, being easily presented. Summarizing the results of a lengthy survey, the study rated the public acceptance of electronic fingerprinting at 96 percent. With regard to availability, studies have shown that a person can present at least six nearly independent fingerprints, but only one hand geometry (a left hand may be a near mirror image of the right).

Issues with Biometric Systems

The issues of using biometric systems that must be addressed by managers and administrators are not many, but they are large in impact and ramifications. The following issues must be addressed for a successful implementation of a biometric access control system:

- Identification vs. verification problems
- Error rate impact by demographics, weather, illumination, and user behavior
- Determining which biometrics works with each application in the system
- Guidelines when selecting a biometric solution
- Pros and cons of each type of biometric system
- Techniques used to secure end-to-end biometric systems
- Components required for the system to operate effectively and efficiently
- Developing privacy-sympathetic biometric systems

Recognition Errors

There are two basic types of recognition errors: the false accept rate (FAR) and the false reject rate (FRR). A false accept is when a nonmatching pair of biometric data is wrongly accepted as a match by the system. A false reject is when a matching pair of biometric data is wrongly rejected by the system. The two errors are complementary: When administrators attempt to lower one of the errors by varying the threshold, the other error rate automatically increases. There is, therefore, a balance to be found, with a decision threshold that can be specified to either reduce the risk of FAR, or to reduce the risk of FRR.

In a biometric authentication system, the relative false accept and false reject rates can be set by choosing a particular operating point for a detection threshold. Very low error rates for both error types at the same time are not possible. By setting a high threshold, the FAR error can be close to zero; similarly, by setting a significantly low threshold, the FRR rate can be close to zero. A meaningful operating point for the threshold is decided based on the application requirements, and the FAR versus FRR error rates at that operating point may be quite different. To provide high security, biometric systems operate at a low FAR instead of the commonly recommended *equal error rate (EER)* operating point where FAR = FRR.

Compromised Biometric Data

Paradoxically, the greatest strength of biometrics is at the same time its greatest liability. An individual's biometric data does not usually change over time as the pattern in a user's iris, retina, or palm vein remains the same throughout life. Unfortunately, this means if a set of biometric data is compromised, it is compromised forever. The user only has a limited number of biometric features (one face, two hands, ten fingers, two eyes). For authentication systems based on physical tokens such as keys and badges, a compromised token can be easily canceled and the user can be assigned a new token. Similarly, user IDs and passwords can be changed as often as required. But if the biometric data is compromised, the user may quickly run out of biometric features to be used for authentication.

Vulnerable Points of a Biometric System

The first stage of a biometric system, called enrollment, involves scanning the user to acquire unique biometric data. During enrollment, an invariant template is stored in a database representing the particular individual. To authenticate the user against a given ID, this template is retrieved from a database and matched against the new template derived from a newly acquired input signal. This is similar to a password. Managers must first create a password for a new user; then when the user tries to access the system, she will be prompted to enter a password. If the password entered via the keyboard matches the password previously stored, access will be granted.

Attacks

There are seven main areas where attacks may occur in a biometric system:

- Presenting fake biometrics or a copy at the sensor
- Producing feature sets preselected by the intruder by overriding the feature extraction process.
- Tampering with the biometric feature representation

- Tampering with locally or remotely stored templates
- Attacking the channel between the stored templates and the matcher
- Corrupting the matcher
- Overriding the match result

Understanding the interface between biometric systems and general security systems is critical for the successful deployment of biometric technologies. The sensitivity of data passed between these two systems means that due care and diligence must be taken to avoid vulnerabilities such as identity, replay, and hill-climbing attacks. This is especially true as interfaces become standardized, such as with BioAPI, because these standard interfaces are available to developers and attackers alike. With this in mind, protection methods for data storage and transmission should be used to safeguard the systems.

Biometric Method Comparisons

This section is devoted to describing the common biometric methods used in developing security systems, including:

- Hand geometry and handwriting
- Fingerprint identification systems
- Retina and iris identification
- Voice verification

Hand Geometry and Handwriting

Handwriting is unique to every individual. Biometric devices which test such uniqueness are gaining popularity in many technologies which regulate sensitive materials or access controls. Most systems are based on a 3D analysis of a writing sample which takes into account pressure and form of script. [4]

Dynamic signature recognition is a behavioral authentication method used to recognize an individual's handwritten signature. This technology measures how a signature is formed by treating the signature as a series of movements that contain unique biometric data, such as rhythm, acceleration, pressure, and flow. The signature is captured when a person signs their name on a digitized graphics tablet, which can be attached to a computer or part of a PDA. The signature dynamics information is encrypted and compressed into a template.

Dynamic signature recognition technology can also track a person's natural signature fluctuations over time. Dynamic signature recognition systems are different from electronic signature capture systems, which treat the signature as a graphic image. Electronic signature capture systems are commonly used by merchants to capture electronic signatures in the authorization of credit card transactions.

Keystroke recognition requires no additional hardware with which to read, scan, view, record, or otherwise interrogate the requesting user because every computer is equipped with a keyboard. To authenticate an individual, keystroke recognition relies solely on software, which can reside on the client or host system. To create an enrollment template, the individual must type a user name and password multiple times. Best results are obtained if enrollment occurs over a period of time rather than at one sitting, as individual characteristics are identified more accurately over a period of time. With keystroke recognition, a user must type without making any corrections. If keystroke errors are made, the system will prompt the user to start over. Some of the distinctive characteristics measured by keystroke recognition systems are:

- The length of time each key is held down
- The length of time between keystrokes
- Typing speed
- Tendencies to switch between a numeric keypad and keyboard numbers
- The keystroke sequences involved in capitalization.

Each individual characteristic is measured and stored as a unique template. Some systems authenticate only at sign-on, whereas others continue to monitor the user throughout the session. As with other biometrics, the user's keystroke sample is compared with the stored template, and access is granted if the submitted sample matches the template, according to preestablished probabilities.

If the keystroke recognition software is used as one factor in a two-factor authentication system, it can be an effective layer of security. Keystroke recognition is not considered an effective single-factor authentication technique because hand injuries, variations in temperature that affect physical actions, fatigue, arthritis, and other conditions can affect authentication effectiveness. Also, since keystroke recognition is a relatively new biometric technology, reliable information concerning its effectiveness is not as available (as it is with fingerprint recognition).

Keystroke recognition biometrics is generally considered to be the easiest biometric technology to implement and use. No hardware is involved and software may be installed on the client or host. Because authentication is based on normal keyboard entry, individuals need only type the prescribed text to be authenticated.

Hand geometry systems use an optical camera to capture two orthogonal two-dimensional images of the palm and sides of the hand, offering a balance of reliability and relative ease of use (see Figure 5-6). They typically collect more than 90 dimensional measurements, including finger width, height, and length, distances between joints, and knuckle shapes. These systems rely on geometry and do not read fingerprints or palm prints. Although the basic shape and size of an individual's hand remains relatively stable, the shape and size of human hands are not highly distinctive. The system is not well suited for performing one-to-many identification matches. Hand geometry readers, as opposed to fingerprint sensors, can function in extreme temperatures and are not impacted by dirty hands. Hand geometry devices are able to withstand wide changes in temperature and function in a dusty environment. They are commonly used for access control to facilities, time clocks, or controlled areas.

The large size of the actual hand geometry readers restricts their use in widespread applications such as those requiring a small user interface, such as home computer user and keyboard. Hand-geometry readers could be appropriate for multiple users or where users access the system infrequently and are perhaps less disciplined in their approach to the system. Today, organizations are using hand-geometry readers in various scenarios, primarily for physical access control and recording work time and attendance. They are also used for the

Figure 5-6 Hand Geometry Capture

Course Technology/Cengage Learning

known traveler programs, such as the Transportation Security Administration's Registered Passenger Program, for streamlining airport security procedures for certain frequent travelers.

Fingerprinting

Fingerprinting is a highly familiar and well-established biometric science. The traditional use of fingerprinting has been as a forensic criminological technique, to identify perpetrators by the fingerprints they leave behind at crime scenes. Scientists compare a latent sample left at a crime scene against a known sample taken from a suspect. This comparison uses the unique features of any given fingerprint, including its overall shape, and the pattern of ridges, valleys, and their bifurcations and terminations, to establish the identity of the perpetrator.

In the context of modern biometrics, these features, called fingerprint minutiae, can be captured, analyzed, and compared electronically, with correlations drawn between a live sample and a reference sample, as with other biometric technologies. Fingerprints offer tremendous invariability, changing only in size with age and are highly resistant to modification or injury, and very difficult to "forge" in any useful way. Although the development of some sort of surreptitious sensor is not inconceivable, the reality is sensors remain obtrusive, requiring a willful finger pressure to gather a useful sample. Unlike other systems, based on cameras and high-tech sensors, fingerprint sampling units are compact, rugged, and inexpensive, with commercially available systems from multiple vendors offering very good accuracy. Next-generation scanners can analyze below the surface of the skin, and can add pore pattern recognition in addition to the more obvious minutia of the fingerprint. [5]

Fingerprint Identification Systems

Fingerprint identification is the method of identification using the impressions made by the minute ridge formations or patterns found on the fingertips. No two persons have exactly the same arrangement of ridge patterns, and the patterns of any one individual remain unchanged throughout life. Fingerprints offer an infallible means of personal identification. Other personal characteristics may change, but fingerprints do not.

Fingerprints can be recorded on a standard fingerprint card or can be recorded digitally and transmitted electronically to the FBI for comparison. By comparing fingerprints at the scene of a crime with the fingerprint record of suspected persons, officials can establish absolute proof of the presence or identity of a person.

At the forefront of fingerprint biometric technology is the integrated automated fingerprint identification system (IAFIS). This system is a national fingerprint and criminal history system maintained by the Federal Bureau of Investigation (FBI), Criminal Justice Information Services (CJIS) Division. The IAFIS provides automated fingerprint search capabilities, latent searching capability, electronic image storage, and electronic exchange of fingerprints and responses, 24 hours a day, 365 days a year. As a result of submitting fingerprints electronically, agencies receive electronic responses to criminal 10-print fingerprint submissions within two hours and within 24 hours for civil fingerprint submissions.

The IAFIS maintains the largest biometric database in the world, containing the fingerprints and corresponding criminal history information for more than 47 million subjects in the *Criminal Master File*. The fingerprints and corresponding criminal history information are submitted voluntarily by state, local, and federal law enforcement agencies.

Just a few years ago, substantial delays were a normal part of the fingerprint identification process, because fingerprint cards had to be physically transported and processed. A fingerprint check could often take three months to complete. The FBI formed a partnership with the law enforcement community to revitalize the fingerprint identification process, leading to the development of the IAFIS.

Facial Recognition

Facial recognition emerged into the national spotlight during the 2001 Super Bowl, when Tampa police scanned the faces of game fans without their knowledge for the purpose of spotting terrorists in the crowd. While this proved a public relations nightmare in January 2001, the use of this technology in New Orleans at the post-9/11 Super Bowl of 2002 generated little controversy. Facial recognition remains one of the more controversial biometric technologies because of its very *un*obtrusiveness. With good cameras and good lighting, a facial recognition system can sample faces from tremendous distances without the subject's knowledge or consent.

As previously stated, most facial recognition technology works by one of two methods: facial geometry or eigenface comparison. Facial geometry analysis works by taking a known reference point such as the distance from eye to eye, and measuring the various features of the face in their distance and angles from this reference point. Eigenface comparison uses a palette of about 150 facial abstractions, and compares the captured face with these archetypal abstract faces. In laboratory settings, facial recognition results are excellent and the accuracy of facial recognition has been good enough for casinos to have put the technology to use since the late 1990s as a means to spot banned players. Facial recognition technology proponents claim good performance even against disguises, weight changes, aging, or changes in hairstyle or facial hair.

Eye Biometrics: Iris/Retina

As stated previously, the human eye offers two features with excellent properties for identification. Both the iris and the veins of the retina provide patterns that can uniquely identify an individual. Retinal scanning is the older technology, and requires the subject to look into a reticule and focus on a visible target while the scan is completed. It is one of the more intrusive biometric technologies, with some subjects reporting discomfort at the scanning method.

Iris recognition has an advantage in ease of use, in that it merely requires the subject to look at a camera from a distance of three to ten inches. The pattern of lines and colors on the eye are, as with other biometrics, analyzed, digitized, and compared against a reference sample for verification.

Iridian Technologies, who hold the patents on iris recognition, claim the iris is the most accurate and invariable of biometrics, and their system is the most accurate form of biometric technology. Iridian's system also has the benefit of extremely swift comparisons. The company claims it can match an iris against a database of 100,000 reference samples in 2–3 seconds, whereas a fingerprint search against a comparable database might take 15 minutes.

Retina and Iris Identification

Iris recognition today combines technologies from several fields, including computer vision (CV), pattern recognition, statistical interference, and optics. The goal of the technology is near-instant, highly accurate recognition of a person's identity based on a digitally represented image of the scanned eye. The technology is based upon the fact that no two iris patterns are alike, thus the probability is higher than that of fingerprints. The iris is a protected organ, and can, therefore, serve as a lifelong password which the person must never remember. Development of recognition and identification techniques facilitates exhaustive searches through nation-sized databases. [6]

Iris recognition technology looks at the unique characteristics of the iris, the colored area surrounding the pupil. While most biometrics has 13 to 60 distinct characteristics, the iris is said to have 266 unique spots. Each eye is believed to be unique and to remain stable over time and across environments, such as weather, climate, and occupational differences. The iris is differentiated by several characteristics including ligaments, furrows, ridges, crypts, rings, corona, freckles, and a zigzag collarette.

Iris recognition systems use small, high-quality cameras to capture a black and white high-resolution photograph of the iris. Once the image is captured, the iris' elastic connective tissue, called the trabecular meshwork, is analyzed, processed into an optical "fingerprint," and translated into a digital form. Figure 5-7 depicts the process of generating an iris biometric. Given the stable physical traits of the iris, this technology is considered to be one of the safest, fastest, and most accurate, noninvasive biometric technologies. This type of biometric

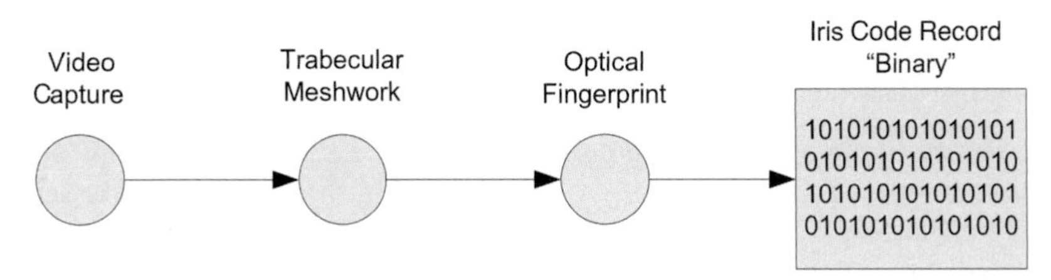

Figure 5-7 The Process of Generating an Iris Biometric

Course Technology/Cengage Learning

scanning works with glasses and contact lenses in place. Therefore, iris scan biometrics may be more useful for higher risk interactions, such as building access. Improvements in ease of use and system integration are expected as new products are brought to market.

Iris recognition technologies are now seen in a wide array of identification systems. They are used in passports, aviation security, physical and electronic access security, hospitals, and national watch lists. Iris recognition algorithms can be seen in more and more identification systems relating to customs and immigration. Future applications will include e-commerce, information security (InfoSec), authorization, building entry, automobile ignition, forensic applications, computer network access, PINs, and personal passwords.

Advantages and Disadvantages of the Iris for Identification

There are numerous advantages and disadvantages for employing iris recognition technologies. Advantages include:

- Highly protected, internal organ of the eye
- Externally visible; patterns imaged from a distance
- Iris patterns possess a high degree of randomness
- Uniqueness: set by combinatorial complexity
- Changing pupil size confirms natural physiology
- Prenatal morphogenesis (7th month of gestation)
- Limited genetic penetrance of iris patterns
- Patterns apparently stable throughout life
- Encoding and decision-making are tractable
- Image analysis and encoding time: 1 second
- Search speed: 100,000 IrisCodes per second on 300 MHz CPU

Disadvantages include:

- Small target (1 cm) to acquire from a distance (1 m)
- Moving target
- Located behind a curved, wet, reflecting surface
- Obscured by eyelashes, lenses, reflections
- Partially occluded by eyelids, often drooping
- Deforms nonelastically as pupil changes size
- Illumination should not be visible or bright
- Some negative (Orwellian) connotations

Retina recognition technology captures and analyzes the patterns of blood vessels on the thin nerve on the back of the eyeball that processes light entering through the pupil. Retinal patterns are highly distinctive traits. Every eye has its own totally unique pattern of blood vessels; even the eyes of identical twins are distinct. Although each pattern normally remains stable over a person's lifetime, it can be affected by disease such as glaucoma, diabetes, high blood pressure, and autoimmune deficiency syndrome.

The fact that the retina is small, internal, and difficult to measure makes the techniques for capturing its image more difficult than the processes performed in most biometric technologies. An individual must position the eye very close to the lens of the retina-scan device, gaze directly into the lens, and remain perfectly still while focusing on a revolving light while a small camera scans the retina through the pupil. Any movement can interfere with the process and can require restarting. Enrollment can easily take more than a minute. The generated template is only 96 bytes, one of the smallest of the biometric technologies.

One of the most accurate and most reliable of the biometric technologies, retina recognition is used for access control in government and military environments that require very high security, such as nuclear weapons and research sites. However, the great degree of effort and cooperation required of users has made it one of the least deployed of all the biometric technologies. Newer, faster, better retina recognition technologies are being developed.

Voice Verification

It is not remarkable when someone recognizes a voice on the telephone. However, this process is still a very difficult problem for computers, especially when the task is to identify someone positively. The prospect of accurate voice verification offers one great advantage, which allows a remote identification using the phone system, an infrastructure that's already been built and thus has zero client-side cost as no special reader needs to be installed in a home or workspace. Even without the phone system, the sampling apparatus, a microphone, remains far cheaper than competing, largely optically-based biometric technologies.

Voice recognition technology is still not ready to be used as a front-line biometric technology. Voice verification systems have to account for many more variables than do other systems, starting with the inevitable compression of a voice captured by cheap microphones, discriminating a voice from background noise and other sonic artifacts, and the human voice's tremendous variability, due to colds, aging, and simple tiredness. It is also possible for a person's voice to be captured surreptitiously by a third party and replayed, or a person's voice might be biometrically sampled remotely without consent during a fake door-to-door or telephone sales call. Because of these difficulties, commercial deployments of voice verification have been limited to "backup" status, systems in which there are other token-based methods of identification, with voice verification providing an added layer of protection. [7]

Signature Verification Systems

Signature verification systems are evaluated by analyzing their accuracy to accept genuine signatures and to reject forgeries. When considering forgeries, four categories can be defined from the lowest level of attack to the highest.

- *Random forgeries*: These forgeries are simulated by using signature samples from other users as input to a specific user model. This category actually does not denote intentional forgeries, but rather accidental accesses by nonmalicious users.
- *Blind forgeries*: These forgeries are signature samples generated by intentional impostors having access to a descriptive or textual knowledge of the original signature.
- *Low-force forgeries*: The impostor has access to a visual static image of the original signature. There are two ways to generate the forgeries. In the first way, the forger can use a blueprint to copy the signature, leading to low-force *blueprint forgeries*. In the second way, the forger can train to imitate the signature, with or without a blueprint,

for a limited or unlimited amount of time. The forger then generates the imitated signature, without the help of the blueprint and potentially after some time after training, leading to low-force *trained forgeries*. The so-called skilled forgeries provided with the MCYT-100 database correspond here to low-force trained forgeries.

- *Brute-force forgeries*: The forger has access to a visual static image and to the whole writing process, therefore including the handwriting dynamics. The forger can analyze the writing process in the presence of the original writer or through a video-recording or also through a captured on-line version of the genuine signature.

This last situation is realized when genuine signature data can be intercepted, for example when the user is accessing a networked system. In a similar way as in the previous category, the forger can then generate two types of forgeries. Brute-force blueprint forgeries are generated by projecting on the acquisition area a real-time pointer that the forger then needs to follow. Brute-force trained forgeries are produced by the forger after a training period where they can use dedicated tools to analyze and train to reproduce the genuine signature.

Card Systems

There are a number of terms and definitions that are used in card systems. *Cards* are defined as:

- A type of physical form factor designed to carry electronic information and/or human readable data
- a dual interface smart card–based ID badge for both physical and logical access that contains within it an integrated circuit chip (under FIPS 201)

A **card holder** is an individual to whom an ID card is issued or assigned. A **card serial number** is an identifier which is guaranteed to be unique among all identifiers used for a specific purpose. *Capture* is the process of taking a biometric sample from the user. In capture, a sample of the user's biometric is acquired using the required sensor, such as a camera, microphone, or fingerprint scanner. A **card reader** is any device that reads encoded information from a card, token, or other identity device and communicates to a host such as a control panel/processor or database for further action. A *card management system (CMS)* includes a smart card/token and digital credential management solution that is used to issue, manage, personalize, and support cryptographic smart cards and PKI certificates for identity-based applications throughout an organization.

The **Payment Card Industry Data Security Standard (PCI DSS)** was developed by the major credit card companies as a guideline to help organizations that process card payments prevent credit card fraud, hacking, and various other security issues. A company processing, storing, or transmitting credit card numbers must be PCI DSS compliant or it risks losing the ability to process credit card payments.

A smart card is a device that includes an embedded integrated circuit that can be either a secure microcontroller or equivalent intelligence with internal memory or a memory chip alone. The card connects to a reader with direct physical contact or with a remote contactless radio frequency interface. With an embedded microcontroller, smart cards have the unique ability to store large amounts of data, carry out their own on-card functions, such as encryption and mutual authentication, and interact intelligently with a smart card reader. A contact

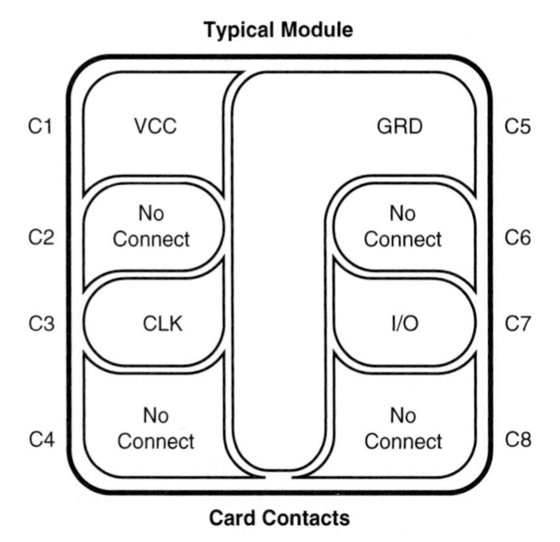

Figure 5-8 Typical Smart Card Module

Course Technology/Cengage Learning

card is the most common type of smart card. Electrical contacts located on the outside of the card connect to a card reader when the card is inserted. Figure 5-8 shows a typical smart card module indicating the card contacts.

Biometric System Components and Processes

A **Web access management** (**WAM**) system replaces the sign-on process on various Web applications, typically using a plug-in on a front-end Web server. The systems authenticate users once, and maintain that user's authentication state even as the user navigates between applications. These systems normally also define user groups and attach users to privileges on the managed systems. These systems provide effective access management and single sign-on to Web applications. They do not usually support any management of legacy systems such as network operating systems, mainframes, client/server applications, and e-mail systems.

An **identity management system** (**IDMS**) is composed of one or more computer systems or applications that manage the identity registration, verification, validation, and issuance process, as well as the provisioning and de-provisioning of identity credentials. Processes include the following activities:

- Identity proofing: The process of providing sufficient information, such as breeder documents, identity history, credentials, and documents, to establish an identity to an organization that can issue identity credentials
- Identity registration: The process of making a person's identity known to a system, associating a unique identifier with that identity, and collecting and recording the person's relevant attributes into the system
- Identity verification: The process of confirming or denying that a claimed identity is correct by comparing the credentials (something known, something possessed, something biological) of a person requesting access with those previously proven and stored in an ID card or system and associated with the identity being claimed

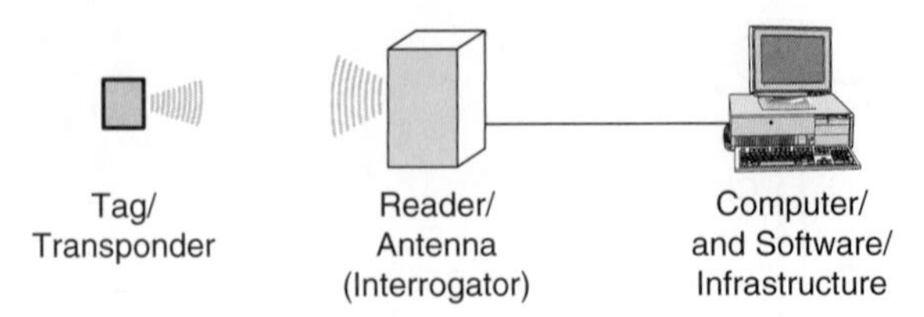

Figure 5-9 RFID Components

Course Technology/Cengage Learning

A **reader** includes any device that communicates information or assists in communications from a card, token or other identity document and transmits the information to a host system, such as a control panel/processor or database for further action.

Radio frequency (RF) consists of any frequency within the electromagnetic spectrum associated with radio wave propagation. Many wireless communications technologies are based on RF, including radio, television, mobile phones, wireless networks and contactless payment cards and devices. **Radio frequency identification** (**RFID**) is a technology used to transmit information about objects wirelessly, using radio waves. RFID technology is composed of two main pieces: the device that contains the data and the reader that captures such data. The device has a silicon chip and an antenna and the reader also has an antenna. The device is activated when put within range of the reader. The term RFID has been most commonly associated with tags used in supply chain applications in the manufacturing and retail industries. Figure 5-9 depicts the components of an RFID system.

Biometric System Design

Interoperability is the ability of two or more systems or components to exchange information and to use the information that has been exchanged. For the purposes of FIPS 201, the ability for any government facility or information system, regardless of the personal identity verification (PIV) issuer, is to verify a cardholder's identity using the credentials on the PIV card.

Robustness is a characterization of the strength of a security function, mechanism, service, or solution, and the assurance or confidence that it is implemented and functioning correctly. The Department of Defense (DoD) has three levels for robustness:

- Basic: Security services and mechanisms that equate to good commercial results
- Medium: Security services and mechanisms that provide for layering of additional safeguards above good commercial practices
- High: Security services and mechanisms that provide the most stringent protection and rigorous security countermeasures

Biometric security must be considered in the initial design of any system. Biometric technology can be used successfully to identify users; however it is necessary that the application be specifically designed, taking in consideration the ease of use, deployment, accuracy, and cost. Finger scanning appears to be the technology of choice. These systems are moderately accurate and easy to use. Acceptance by users is good and the technique is relatively inexpensive.

Biometric Identification Systems

Biometric identification systems can be grouped based on the main physical characteristic that lends itself to biometric identification. These systems are designed around various body components and attributes. *Note: Details concerning the various biometric modalities were presented in Chapter 2.* The development of biometric systems is being considered for a number of these modalities. A brief description of the major candidates are identified and described for each of the biometric methods. [8]

Fingerprint Identification

Fingerprint ridges are formed in the womb; humans have fingerprints by the fourth month of fetal development. Once formed, fingerprint ridges are like a picture on the surface of a balloon. As the person ages, the fingers do get larger. However, the relationship between the ridges stays the same, just like the picture on a balloon is still recognizable as the balloon is inflated.

Hand Geometry

Hand geometry is the measurement and comparison of the different physical characteristics of the hand. Although hand geometry does not have the same degree of permanence or individuality as some other characteristics, it is still a popular means of biometric authentication. Various projects were initiated to look at the potential of biometrics and one of these eventually led to a large and rather ungainly hand geometry reader being produced. It was very basic, but it worked and motivated its designers to further refine the concept. Eventually, a small specialist company was formed and a much smaller and enhanced hand geometry reader became one of the cornerstones of the early biometric industry. This device worked well and found favor in numerous biometric projects around the world.

Palm Vein Authentication

This system uses an infrared beam to penetrate the user's hand as it is waved over the system; the veins within the palm of the user are returned as black lines. Palm vein authentication has a high level of authentication accuracy due to the complexity of vein patterns of the palm. Because the palm vein patterns are internal to the body, this would be a difficult system to counterfeit. Also, the system is contactless and therefore hygienic for use in public areas.

Retina Scan

A retina scan provides an analysis of the capillary blood vessels located in the back of the eye; the pattern remains the same throughout life. A scan uses a low-intensity light to take an image of the pattern formed by the blood vessels.

Iris Scan

An iris scan provides an analysis of the rings, furrows, and freckles in the colored ring that surrounds the pupil of the eye. More than 200 points are used for comparison. Iris scans were proposed in 1936, but it was not until the early 1990s that algorithms for iris recognition were created and patented. All current iris recognition systems use these basic patents, held by Iridian Technologies.

Identity Automation

It is hardly surprising that for many years individuals and organizations both in the military and commercial sectors have been fascinated by the possibility of using electronics and the power of microprocessors to automate identity verification. At the same time, other biometric methodologies such as fingerprint verification were being steadily improved and refined to the point where they would become reliable, easily deployed devices. In recent years, security management has expressed an interest in iris scanning and facial recognition techniques which offer the potential of a noncontact technology, although there are additional issues involved. The last decade has seen the biometric industry mature from a handful of specialist manufacturers struggling for sales to a global industry shipping respectable numbers of devices and poised for significant growth as large-scale applications start to unfold.

Biometric ID System Components

Biometric ID systems could be viewed in terms of a generic biometric system made up of five basic components or subsystems, depending on the application:

- Data collection
- Transmission
- Signal processing
- Storage
- Decision

Biometric systems involve at least two discrete data collection steps.

- First, any biometric system must contain a biometric characteristic deemed "true" or canonical from the system's viewpoint. The term "enrollment" or "registration" refers to the first entry of biometric data into the database.
- Second, the system must compare a later-submitted sample ("live sample") to the sample in the database. Scale is crucial to the enrollment step because of the additional space required to handle the additional equipment and physical traffic needed for data collection.

Many biometric systems collect data at one location, but store and/or process it at another, requiring data transmission. Once a biometric is acquired, it must be prepared for comparison. There are four basic required tasks:

- Feature extraction
- Quality control
- Pattern matching
- Pattern classification

Feature extraction involves finding the true biometric pattern amid noise and signal degradation, preserving the critical information, and discarding redundant or unnecessary data. An example would be a text-independent speaker-recognition system. A properly implemented system isolates features that depend only on the speaker and not on the words being spoken.

At the same time, the system focuses on features that do not change even if the speaker has a cold or is not speaking directly into the microphone.

Quality control involves checking the quality of the signal. Ideally, it should be possible to make a quick determination so another measure can be taken if the signal is inadequate. **Noise** is unwanted components in a signal that degrade the quality of data or interfere with the desired signals processed by a system. **Failure to acquire** (FTA) results from the failure of a biometric system to capture and/or extract usable information from a biometric sample. **Failure to enroll** (FTE) occurs when there is a failure of a biometric system to form a proper enrollment reference for an end user. Common failures include end users who are not properly trained to provide their biometrics, the sensor not capturing information correctly, or captured sensor data of insufficient quality to develop a template.

Pattern matching involves comparing the live sample to the reference sample in the database. If the user claims to be Jane Doe, the pattern-matching process may only need to compare the sample to Jane Doe's stored template. In other situations, the sample must be compared to multiple templates. The pattern-matching process generates a quantitative "distance" measure of the comparison—how close are they? Even for the same person, the distance is rarely, if ever, zero.

Pattern classification is a technique aimed at reducing the computational overhead of pattern matching. In large-scale systems, it can be computationally taxing to match each sample against all stored templates in the database. If biometric patterns can be categorized, then it may be possible to perform the match against only the stored templates in that category. This is sometimes referred to as "binning." **Binning** is the process of parsing (examining) or classifying data to accelerate and/or improve biometric matching. A different technique with the same goal is **filtering**, which involves partitioning the database based on information not contained in the biometric itself. If the person is a man, it is not necessary to check against women's biometrics. Both of these techniques introduce additional error possibilities; if binning or filtering is erroneous, then the true template is not used and a false nonmatch results.

This subsystem implements the biometric ID system's actual policy with regard to matching. In general, lowering the number of false "non-matches" raises the number of false matches, and vice versa. The signal processing subsystem yields a quantitative "distance" measure, but "how close or far is enough?" is a matter of policy. In a high-security application where the cost of a false acceptance could be high, system policy might prefer very few false acceptances and many more false rejections. In a commercial setting, where the cost of a false acceptance could be small and treated as a cost of doing business, system policy might favor false acceptances to not falsely reject and thereby inconvenience large numbers of legitimate customers. The inevitable existence of these errors means any biometric ID system must also have well-designed policies for exception handling. False accepts and rejects will be described in subsequent sections.

Biometric reference samples must be stored somewhere for matching purposes. For systems only performing "one-to-one" matching, the database may be distributed on cards carried by each enrolled user. The user simply presents a biometric live input and the system checks to see if it matches the template stored on the card. Depending upon system policy, no central database need exist, although in this application a centralized database can be used to detect counterfeit cards or to reissue lost cards without re-collecting the biometric pattern.

In other cases, centralized storage is necessary because the system must match the live sample to multiple templates. As the number of templates grows, speed becomes an increasingly significant issue. One technique is to partition the database via binning or filtering so any sample need only be matched to the templates in one partition. This increases system speed and decreases false matches at the expense of increasing the false non-match rate owing to partitioning errors. System error rates thus change with increasing database size and ID systems do not linearly scale.

Full biometric patterns cannot be reconstructed from the stored reference samples if these are stored as templates, which reduce data richness dramatically. Templates themselves are often created using the system vendor's proprietary feature extraction algorithms. Whether stored templates themselves can be used to "spoof" the system internally is entirely dependent on the security of the system architecture.

Biometric ID systems may store the templates and also raw data. One reason to do so would be to allow changes to the system or to change system vendors without having to re-collect data from all enrolled users. Full raw-data storage is a riskier practice because new templates may be extracted from the data or the raw data itself may be used against the system. Subsequent sections provide specifics for at a number of biometric systems and vendors.

Biometric Access Control

Biometric access control systems consist of a reader or scanning device, software that converts the gathered information into digital form, and a database that stores the information for comparison with previous records. These readers, or scanning devices, can scan for a fingerprint, hand geometry, signature, iris/retina, facial recognition, voice print, and vascular pattern. This technology can be used for a number of applications including:

- Time and attendance reporting
- Building access control
- Verification of signatures
- Point-of-sale identity verification
- Process control security
- Cellular phone security

Although this can be considerably more convenient than current access methods, such as passwords and cards, many think of the technology as confined to heightened security applications. It is true that biometrics is used to check employees coming into almost every airport and to guard almost every nuclear plant. These access control systems are also the mode of entry at embassies around the world. However, the majority of implementations are used in common, everyday locales, including hair salons and restaurants.

Embedded Biometric Applications

Biometric sensors are now being embedded into systems such as cell phones or PDAs to customize features or settings and provide increased comfort and convenience. The standalone biometric reader, although seldom thought of as an embedded system, has been on the market for some time. A newer alternative features template management performed by a smart card.

Breaches of security can be minimized in many ways. One method limits the opportunity for infringement. For instance, if the database of those authorized for access to a facility does not rely on networked hardware systems, the chance of someone infiltrating the system is reduced, as are the possibilities of downtime.

Although new methods and attributes—such as vascular recognition and DNA matching, are promising—they are not yet mature enough for mainstream usage. Alternatives further along in development already exist and a variety of common methods are commercially available today. Selection of the appropriate biometric access control technology depends on a number of application-specific factors, including the environment in which the identity verification process is carried out, the user profile, accuracy requirements, and overall system cost.

Success Factors and Design Challenges

The main goal of any access control system is to deny entrance to some people and allow entrance to others. Although this sounds simple, some key factors must be considered early on when designing a biometric application. These include:

- User acceptance
- Throughput
- Accuracy
- Encryption
- Identity theft aversion

User acceptance of the access control device is one of the most critical factors in the success of a biometric-based implementation. To prevent improper use, which can cause access errors, the device should not cause discomfort or concern and must be easy to use.

Throughput, which is application-dependent, is the total time required to use the device. The elapsed time from presentation to identity verification is known as verification time. Most readers can verify identity within one second. However, when considering the use of biometrics for access control, the total time it takes a person to use the reader must be considered. This includes the time it takes to enter the ID number and the time required to get into the right position for scanning. The total time required for each person varies.

Accuracy is vital to the acceptance of the biometric type chosen. If it does not accurately read the person's biometric input, the system will no longer be used for access control because of its inaccuracies. Letting the wrong people in or denying access to the correct people poses serious problems. The two errors a unit can make are false acceptance and false rejection. When the live biometric template is compared to the stored template, a matching score confirms or denies the identity of the user. System designers set this numeric score to accommodate the system's desired level of accuracy, which is measured by the false accept rate (FAR) and false rejection rate (FRR).

The FAR is the probability that an unauthorized user will be allowed to pass for someone else. This error rate must be low enough to present a real deterrent for a given application. In today's biometric access control systems, FAR ranges from .0001 to 0.1 percent. In comparison, the biometric hand geometry reader used on the front entry area of 60 percent of U.S. nuclear power plants has a FAR of 0.1 percent. False rejection rate is just as crucial as FAR. The FRRs quoted for currently available systems vary from .00066 to 1.0 percent. A low FRR is important because this type of error can occur with almost every use of the access control device.

Error curves give a graphical representation of the device's performance. The point where the false accept and false reject curves intersect and the error rates equal one another is called the equal error rate (EER). It is desirable to have a smaller EER sensitivity setting. Setting the system to produce a very low rate of false accepts usually causes a higher rate of false rejects. Conversely, if the system is set to produce a low rate of false rejects, the rate of false accepts is higher. The application determines the acceptable levels of false accepts and false rejects.

Encryption keys and digital signatures enhance secure authentication in networked applications. For example, biometric information can be transferred using public key infrastructure systems that incorporate encryption to help make a networked system more tamper-resistant. The use of smart cards can help avoid identity theft. Since the card stores the user's ID and biometric template, there is no database that can be hacked. When the smart card is presented to the smart card reader embedded in a biometric reader, the user's hand is placed on the unit. The reader compares length, width, thickness, and surface area measurements against the template stored in the smart card to verify identity. This process takes approximately one second and is virtually foolproof.

Networked vs. Stand-Alone Biometric Systems

Biometric readers can either operate as stand-alone systems or be integrated into a network environment. Stand-alone biometric systems have been employed since the late 1980s. These are not only biometric readers, but also provide complete door control, including locking and unlocking a door based on who is authorized to enter and when. For example, the Schlage Recognition Systems HandKey II reader provides access for a very small group of people to a unique private library and museum. The HandKey II access control device uses field-proven hand geometry technology that maps and verifies the size and shape of a person's hand in less than one second. However, the majority of access control applications must manage more than one door. A network-enabled system of biometric readers can provide a central location where alarm conditions and door activities can be stored, viewed, and controlled via browser on a PC located virtually anywhere.

Networked systems also provide convenient template management and allow user profiles to be deleted or changed on the PC. This process also allows users to enroll at a single location while their templates can be made available at multiple locations.

All major biometric readers can be integrated into conventional access control systems in a variety of ways. The most common way is card-reader emulation, where the biometric device works with the access control panel in the same way a card reader does. The biometric reader's card reader output port is connected to the panel's card reader port. This method can be very effective for integration into existing card-based systems.

Integrating biometrics with cards or smart cards is becoming common and provides dual authentication. By adding a biometric reader to the access control system, a badge alone cannot be used to gain access. Both the badge and the person's biometric template are required, providing a higher level of security. Since the template only resides on the card, the solution also eases individual privacy concerns.

Requirements for a Network-Enabled Biometric System

A number of requirements must be met to connect the biometric reader to a networked environment. These include an ultra-high level of reliability, an operating system capable of working with the latest LANs (local area networks), WANs (wide area networks) and other networks, network standards, support for security, and support for biometric hardware/ software.

Building Ethernet connectivity into a product requires a significant investment in hardware and software integration. Device servers provide an easy and economical way of connecting the serial device to the network. They allow independence from proprietary protocols and the ability to fulfill a number of different functions. When the access control system is networked, the biometric template can be checked against a centralized database and records are made of all accesses made within the organization. [9]

Common System Requirements

While all biometric systems have their own advantages and disadvantages, there are some common characteristics needed to make a biometric system usable. First, the biometric must be based upon a distinguishable trait. Law enforcement has used fingerprints to identify people for a century. There is a great deal of scientific data supporting the idea that "no two fingerprints are alike." Newer biometric identification methods sometimes do not hold up in court.

Another key aspect is how user-friendly the system is. Most people find it acceptable to have their pictures taken by video cameras or speak into a microphone. In the United States, using a fingerprint sensor does not seem to be a problem. In some other countries, however, there is strong cultural opposition to touching something that has been touched by many other people.

While cost is always a concern, most implementers today are sophisticated enough to understand that it is not only the initial cost of the sensor or the matching software that is involved. Often, the life-cycle support cost of providing system administration support and an enrollment operator can overtake the initial cost of the hardware. Accuracy is also of key importance. Terms previously introduced that are used to describe the accuracy of biometric systems include false-acceptance rate (percentage of impostors accepted) and false-rejection rate (percentage of authorized users rejected).

When discussing the accuracy of a biometric system, it is often beneficial to talk about the equal-error rate or at least to consider the false-acceptance rate and false-rejection rate together. For many systems, the threshold can be adjusted to ensure that virtually no impostors will be accepted. Unfortunately, this often means an unreasonably high number of authorized users will be rejected. To summarize, a good biometric system is one that is low cost, fast, accurate, and easy to use.

Commercial and Government Biometric Systems

The Biometric Consortium has developed a list of biometric systems examples. [10] Because of the number of vendors identified, the list is included in the Appendix. The categories included in this list are systems that provide support for the following biometric methods:

- Face
- Fingerprint
- Palm print

- Hand and finger geometry
- Handwriting
- Iris
- Multimodal
- Retina
- Vein
- Voice/speaker
- Various others

The following sections will introduce both production and operational systems and provide capabilities and features for each:

- Project Semaphore
- US-VISIT
- IAFIS
- AFIS
- ABIS
- FaceIt®
- BioTime®
- BioWeb™
- IrisCode™
- TactileSense™
- PORTPASS
- TWIC
- IDENT-IAFIS
- Secure Flight

Note that these systems' features and functionalities could be in a development state; thus, the information provided in these sections could very well be outdated at the time of publication. Additional research must be conducted by any organization that is considering the implementation of a biometric solution.

Project Semaphore

An integrated approach to border management began in the UK in 2000. It included the concept of e-borders, which stressed using technology to extend the border to the point of departure rather than the point of entry. A number of projects were launched, including *Project Semaphore*. Its objectives are to provide the means for risk management of people across the entire travel process, from the moment a ticket is bought until the person arrives at their final destination. [11]

Project Semaphore was announced in September 2004 and IBM was selected as the preferred supplier for the pilot. Semaphore tested the technical and business process design for a major component of the UK's e-borders program. It was intended that Project Semaphore would lead to a process of shared information that would allow customs, police, and border management agencies to make decisions about the level of risk associated with individuals, and therefore indicate the level of scrutiny a person receives as they travel. Project Semaphore aims to process advance passenger information and passenger name records provided by the airlines, enabling airlines to record simply and effectively details of passengers intending to enter or leave the UK before they begin their journey.

e-Borders is a cross-cutting initiative coordinated by the Home Office in partnership with key border control, law enforcement, and intelligence agencies. e-Borders will use cutting-edge technology to play an important part in safeguarding against terrorism, serious and organized crime, and illegal immigration. At the same time, it will allow legitimate passengers to travel more efficiently and securely both into and out of the UK. The e-borders system provides the capability to:

- Identify people who have boarded transport destined for the UK
- Check them automatically against databases of individuals who pose a security risk
- Keep a simple electronic record of entry into the country
- Enable authorities to record people leaving the UK, helping to identify those who overstay

Routes have been chosen on the basis of risk assessments by the border agencies, including the Immigration Service, Customs and Excise, and the police. Project Semaphore will test and confirm the technical and business process design for the main e-borders program as well deliver immediate operational improvements across participating agencies.

US-VISIT

US-VISIT is part of a continuum of security measures that begins overseas and continues through a visitor's arrival in and departure from the United States. [12] It incorporates eligibility determinations made by both the Department of Homeland Security and the State Department. US-VISIT is helping demonstrate that the U.S. remains a welcoming nation, and that it can keep America's doors open, and the nation secure. US-VISIT is a top priority for the U.S. Department of Homeland Security because it:

- Enhances the security of U.S. citizens and visitors
- Facilitates legitimate travel and trade
- Ensures the integrity of the U.S. immigration system
- Protects the privacy of U.S. visitors

The U.S. Department of Homeland Security (DHS) intends to integrate biometric exit procedures into the existing international visitor departure process. The change will make the process of departing the United States more convenient and accessible for international visitors. DHS will take a number of steps toward full implementation of biometric exit procedures at airports. The first step will be the completion of a pilot program requiring international visitors to biometrically check out at select airports and seaports.

While the program demonstrated the technology works, it also revealed low traveler compliance. DHS has determined US-VISIT air exit procedures should be incorporated into the existing international visitor departure process to minimize the effect on visitors and to ensure seamless biometric collection regardless of the visitor's departure point. DHS, Congress, and the 9/11 Commission have consistently recognized biometric exit control as a priority in order to fully secure the nation's borders. Development of an automated exit capability is one of the department's congressional mandates. DHS has systematically tackled technical and operational challenges and deployed a biometrics-based entry process through US-VISIT.

DHS is now prepared to begin implementing exit procedures in the commercial air environment, where the significant majority of those subject to US-VISIT depart the United States. The department recently began discussing the air exit strategy with the airline industry and will be working with air carriers to implement it. DHS is scheduled to publish a regulation in the future outlining its plans for implementing an integrated air exit strategy. The biometric exit procedures have been tested in a number of airports.

The US-VISIT program provides biometrics-based identity management services to entities throughout the U.S. government. Most international visitors applying to enter the United States experience US-VISIT procedures, which includes a digital fingerprint collection and a photograph. Department of State consular officers and U.S. CBP officers use US-VISIT biometric identity management services to establish and verify international visitors' identities in order to make visa-issuance and admission decisions.

The process of obtaining the prints by way of laser scanning is called live scan, and the machine which scans live fingerprints into AFIS is called the *live scan device*. Live scan technology allows digitally scanned fingerprints to be submitted electronically to the Department of Justice within a matter of minutes and allows criminal background checks to be processed, usually within a specific time-frame.

The process of obtaining prints by putting a 10-print card (prints taken using ink) is occasionally called a 10-print or card scan. There are other devices which capture latent prints from crime scenes, as well as both wired and wireless devices to capture one or two live fingers. The most common method of acquiring fingerprint images remains the inexpensive ink pad and paper form. Scanning fingerprint cards in forensic AFIS complies with standards established by the FBI and NIST.

To match a print, a fingerprint technician scans in the print in question, and the computer marks all minutiae points according to an algorithm. In some systems, the technician then goes over the points the computer has marked, and submits the minutiae to a one-to-many (1:N) search. Increasingly, there is no human editing of features necessary in the better commercial systems. The fingerprint image processor generally will assign a "quality measure" that indicates if the print is useful for searching.

Fingerprint matching algorithms vary greatly in terms of Type I (false positive) and Type II (false negative) errors. They also vary in terms of features such as image rotation invariance and independence from a core reference point or center of the fingerprint pattern. The accuracy of the algorithm, the robust-to-poor image quality, and the characteristics noted above are critical elements of system performance.

Fingerprint matching has an enormous computational burden. Some larger AFIS vendors deploy custom hardware while others use highly optimized software to attain matching speed and throughput. In general, it is desirable to have at least a two-stage search. The first stage generally improves access precision by use of global fingerprint characteristics such as "pattern type combinations," while the second stage is the minutiae matcher.

In any case, the search systems return results with some numerical measure of the probability of a match score. In 10-print searching using a "search threshold" parameter to increase accuracy, there should seldom be more than a single candidate unless there are multiple records from the same candidate in the database. Many systems use a broader search in order to reduce the number of missed identifications, however, and these searches can return from one to 10 possible matches. Latent to 10-print searching will frequently return many candidates because of limited and poor quality input data. The validation of computer-suggested candidates is usually done by a technician in forensic systems. In recent years, though, "lights-out" or "auto-confirm" algorithms produce "identified" responses without a human operator looking at the prints, provided the matching score is high enough. "Lights-out" or "auto-confirm" is often used in civil identification systems, and is increasingly used in criminal identification systems as well.

IAFIS

The **Integrated Automated Fingerprint Identification System** (IAFIS) is a national fingerprint and criminal history system maintained by the Federal Bureau of Investigation (FBI). The IAFIS provides automated fingerprint search capabilities, latent searching capability, electronic image storage, and electronic exchange of fingerprints and responses. The IAFIS maintains the largest biometric database in the world, containing the fingerprints and corresponding criminal history information for millions of subjects.

Over the last few decades, a number of other biometric traits have been studied, tested, and have been successfully deployed in niche markets. Thanks to biometric systems portrayed in Hollywood films, the popularity of IAFIS, and the intuitive appeal of biometrics as a crime deterring security tool, completely automatic biometric systems give the appearance of being widespread and mature technologies. Not surprisingly, there is an overall misperception in the pattern recognition community that the important research problems have been largely solved but for the clever bells and whistles needed for making this technology work in the real world.

AFIS

The **Automated Fingerprint Identification System** (AFIS) is a system for automatically matching one or many unknown fingerprints against a database of known prints. It is a highly specialized biometric system that compares a submitted fingerprint record (usually of multiple fingers) to a database of records to determine the identity of an individual. AFIS is predominantly used for law enforcement, but is also being used for civil applications, such as background checks for soccer coaches, etc. Many nations, as well as many states and local administrative regions, have their own AFIS, which are used for a variety of purposes, including criminal identification, applicant background checks, receipt of benefits, and receipt of credentials, such as passports.

ABIS

This Department of Defense (DOD) system was implemented to improve the U.S. government's ability to track and identify national security threats. The **Automated Biometric Identification System (ABIS)** system includes mandatory collection of 10 rolled fingerprints, a minimum of mug shots from varying angles, and an oral swab to collect DNA.

As the amount of sensitive information stored in databases increases due to the current trend to automate command, control and communication (C3) systems; the impact of unauthorized access could be very detrimental to our nation's security. Access control hardware that uses retinal blood vessel pattern recognition may be the solution to the problem.

FaceIt®

FaceIt® is a facial recognition software engine that allows computers to rapidly and accurately detect and recognize faces. FaceIt® enables the broadest range of applications in the marketplace; from ID Solutions to information security to banking and e-commerce. Capabilities of FaceIt® includes:

- Face detection: Detects single or multiple faces—even in complex scenes
- Face recognition: Functions in either authentication (one-to-one matching) or identification (one-to-many matching)
- Image quality: Evaluates the quality of an image for face recognition and, if needed, prompts for an improved image
- Segmentation: Crops faces from background
- Faceprint: Generates digital code or internal template, unique to an individual
- Tracking: Follows faces over time
- Compression: Compresses facial images down to 84 bytes in size

BioTime®

BioTime® is fraud-free, easy-to-use computerized access control, attendance, and recording tool used by employers. All staff activity is recorded automatically at the points of entry and exit. Upon arrival at work, employees login quickly and easily. Upon exiting the premises at the end of the working day, this process is repeated. [13]

Because the system accurately and instantly identifies each employee, no unauthorized people can enter the facility, and all staff movement is fully recorded. This automatically eradicates time disputes, or "buddy clocking" by other staff members. These records can then be saved or printed, and analyzed at any time. Also, the figures can be integrated with most popular payroll packages, allowing for the complete automation of the entire payroll calculation process.

BioTime® is a complete system that includes software, clocking-in terminal, communications cables, a "Getting Started" guide, and a full installation manual. Because the software is completely menu-driven and user-friendly, it can be quickly installed and operated by someone with little computer experience.

The biometric access unit consists of a fingerprint reader and keypad enclosed in a powder-coated aluminum console with a backlit LCD display, internal buzzer, and red and green diodes. The unit is connected to the PC via both a USB cable (5m) and a serial RS232 (both cables are required together). It requires a power outlet. The unit can be fitted with an optional internal relay to power magnetic door locks or turnstiles.

The core of the BioTime® system is the 32-bit software, designed to run on Microsoft Windows 95B, 98, NT4, 2000, and XP systems.

BioWeb™

BioWeb™ is a biometric solution for Internet and intranet security. When examining the future of commerce and communication on the Internet, there remains one defining issue that could make or break any Web-based endeavor. This security issue must be dealt with before the Internet can truly fulfill its revolutionary potential. This system offers the ability to identify and authorize all communication over the Internet and networks, completely protecting customer's funds and information from fraud or hacking. Sequiam Biometrics BioWeb™ solution is an easily installed, ready-to-use software package that allows the instant biometric enabling of any Web site. It is ideal for secure business-to-business sites, payment authentication, and general Web site access security. [14]

BioWeb™ allows users to:

- Biometrically secure Web site and intranets quickly and easily
- Change the status and access levels on a person-by-person basis quickly and easily
- Use small record size to drastically reduce storage requirements
- Support multiple biometric devices and device types (finger, hand, iris, etc.) on one system
- Use a simple logon process and easy-to-use graphic and menu-driven administration system
- Instantaneously identify authorized users and prevent unauthorized access from anywhere in the world and record all logon activity for analysis
- Ensure this level of security and accountability at an extremely low cost

IrisCode™

The iris of each eye is protected from the external environment. It is clearly visible from a distance, making it ideal for a biometric solution. Image acquisition for enrollment and recognition is easily accomplished and most importantly is nonintrusive. No other biometric technology can rival the combined attributes of mathematical certainty, speed and noninvasive operation offered by iris recognition. [15]

This section explains the use of biometrics and the unique advantages that the IrisCode™ technology can provide. A high-resolution, noninvasive, noncontact, and extremely fast camera captures the image of an iris, translating it into an encrypted digital code. This technology is not retina scan technology and no laser is projected into the eye. The technology is widely acknowledged within the biometrics industry to be the most accurate, most stable, and scalable human authentication system in existence.

The iris recognition biometric technology used in IrisCode™:

- Is an access control system
- Has no requirement or costs for cards or PINs
- Is more accurate than DNA matching; there is no recorded instance of a false accept
- Has a very small record size (IrisCode 512 bytes)
- Uses identification, (one-to-many) not verification (one-to-one) matching
- Is noncontact; works with glasses, protective clothing, safety shields, and contact lenses
- Requires one enrollment only

- Is noninvasive and noncontact, using video-based technology
- Has extremely fast database matching (match rates in excess of 100,000 per second are achieved on a standard PC)

TactileSense™

Today, most security standards, often simple passwords, are easy to break. Since most text passwords are easy to remember or written down in obvious locations, it is relatively easy for gain unauthorized access to PCs, PDAs, electronic systems, Internet appliances, and set-top boxes.

TactileSense™ allows companies to integrate fingerprint-based security into existing products while opening up a range of new alternatives for personal security products. Without the need for a light source, the TactileSense™ polymer generates an image of the fingerprint patterns, identifying the unique characteristics that link it to an individual. TactileSense™ transforms the ridges, loops, and whorls of a fingerprint into an optical image pattern. This pattern is captured as an image by a custom designed camera, then transformed from an optical image into digital code. TactileSense™ eliminates the need to remember passwords and offers reliable protection from unauthorized use and data theft. This fingerprint biometric technology only senses a live finger, making it impossible to use a replica of a finger or fingerprint to authenticate a user.

PORTPASS

The US PORTPASS program, deployed at remote locations along the U.S.–Canada border, recognizes voices of enrolled local residents speaking into a handset. This system enables enrollees to cross the border when the port is not staffed. [16]

Voice biometrics works by digitizing a profile of a person's speech to produce a stored model voice print, or template. Biometric technology reduces each spoken word to segments composed of several dominant frequencies called *formants*. Each segment has several tones that can be captured in a digital format. The tones collectively identify the speaker's unique voice print. Voice prints are stored in databases in a manner similar to the storing of fingerprints or other biometric data. [17]

To ensure a good-quality voice sample, a person usually recites some sort of text or pass phrase, which can be either a verbal phrase or a series of numbers. The phrase may be repeated several times before the sample is analyzed and accepted as a template in the database. When a person speaks the assigned pass phrase, certain words are extracted and compared with the stored template for that individual. When a user attempts to gain access to the system, the pass phrase is compared with the previously stored voice model. Some voice recognition systems do not rely on a fixed set of enrolled pass phrases to verify a person's identity. Instead, these systems are trained to recognize similarities between the voice patterns of individuals when the persons speak unfamiliar phrases and the stored templates.

A person's speech is subject to change, depending on health and emotional state. Matching a voice print requires that the person speak in the normal voice that was used when the template was created at enrollment. If the person suffers from a physical ailment, such as a cold, or is unusually excited or depressed, the voice sample submitted may be different from the template and will not match. Other factors also affect voice recognition results. Background noise and the quality of the input device, such as a microphone, can create additional challenges for voice recognition systems. If authentication is being attempted remotely over the telephone, the use of a cell phone (instead of a landline) can affect the accuracy of the results. Voice recognition systems may be vulnerable to replay attacks: if someone records the authorized user's phrase and replays it, that person may acquire the user's privileges. More sophisticated systems may use liveness testing to determine that a recording is not being used.

Consumer voice-recognition systems are typically inexpensive and user-friendly. Most computer systems are equipped to support a microphone used to develop a voice template and later to collect the authentication request. Voice recognition is more often used in an environment in which voice is the only available biometric identifier, such as in telephony and call-center applications. Voice recognition systems have a high user acceptance rate because they are perceived as less intrusive and are one of the easiest biometric systems to use.

Voice verification technology uses the different characteristics of a person's voice to discriminate between speakers. These characteristics are based on both physiological and behavioral components. The physical shape of the vocal tract is the primary physiological component. The vocal tract is made up the oral and nasal air passages that work with the movement of the mouth, jaw, tongue, pharynx, and larynx to articulate and control speech production. The physical characteristics of these airways impart measurable acoustic patters on the speech that is produced. The behavioral component is made up of movement, manner, and pronunciation.

The combination of the unique physiology and behavioral aspects of speaking enable verification of the identity of the person who is speaking. Voice verification technology works by converting a spoken phrase from analog to digital format and extracting the distinctive vocal characteristics, such as pitch, cadence, and tone, to establish a speaker model or voiceprint. A template is then generated and stored for future comparisons.

Voice verification systems can be text dependent, text independent, or a combination of the two. *Text-dependent* systems require a person to speak a predetermined word or phrase. This information, known as a "pass phrase," can be a piece of information such as a name, birth city, favorite color, or a sequence of numbers. The pass phrase is then compared to a sample captured during enrollment. *Text-independent* systems recognize a speaker without requiring a predefined pass phrase. It operates on speech inputs of longer duration so it has a greater opportunity to identify the distinctive vocal characteristics of pitch, cadence, and tone.

Voice verification systems can be used to verify a person's claimed identity or to identify a particular person. It is often used where voice is the only available biometric identifier, such as over the telephone. Voice verification systems may require minimal hardware investment as most personal computers already contain a microphone. The downside to the technology is that, although advances have been made in recognizing the human voice, ambient temperature, stress, disease, medications, and other physical changes can negatively impact automated recognition.

Voice verification systems are different from voice recognition systems although the two are often confused. *Voice recognition* is used to translate the spoken word into a specific response, while voice verification verifies the vocal characteristics against those associated with the enrolled user. The goal of voice recognition systems is simply to understand the spoken word, not to establish the identity of the speaker. A familiar example of voice recognition systems is that of an automated call center asking a user to "press the number one on his phone keypad or say the word "one." In this case, the system is not verifying the identity of the person who says the word "one"; it is merely checking that the word "one" was said instead of another option.

TWIC

The *Transportation Worker Identification Credential (TWIC)* was created as a system-wide common credential that could be used across all transportation modes. TWIC could be used for all personnel requiring unescorted physical and/or computer access to secure areas of the national transportation system. TWIC was developed in response to threats and vulnerabilities identified in the transportation system. TWIC was also developed in accordance with the legislative provisions of the Aviation and Transportation Security Act (ATSA) and the Maritime Transportation Security Act (MTSA).

The TWIC program was a Transportation Security Administration and U.S. Coast Guard initiative. The TWIC program was developed to provide a tamper-resistant biometric credential to maritime workers requiring unescorted access to secure areas of port facilities, outer continental shelf facilities, vessels regulated under the Maritime Transportation Security Act, or MTSA, and all U.S. Coast Guard credentialed merchant mariners. Enrollment and issuance would take place over an 18-month period. To obtain a TWIC, an individual must provide biographic and biometric information such as fingerprints, sit for a digital photograph, and successfully pass a security threat assessment conducted by TSA.

While TWIC may be implemented across other transportation modes in the future, the TWIC Final Rule, published in the Federal Register on January 25, 2007, set forth regulatory requirements to implement this program in the maritime mode first. The program's goals were to:

- Positively identify authorized individuals who require unescorted access to secure areas of the nation's maritime transportation system
- Determine the eligibility of an individual to be authorized unescorted access to secure areas of the maritime transportation system
- Enhance security by ensuring that unauthorized individuals are denied unescorted access to secure areas of the nation's maritime transportation system
- Identify individuals who fail to maintain their eligibility qualifications after being permitted unescorted access to secure areas of the nation's maritime transportation system and revoke the individual's permissions

During the initial rollout of TWIC, workers will present their cards to authorized personnel, who will compare the holder to a personal photo, inspect security features on the TWIC, and evaluate the card for signs of tampering. The Coast Guard will verify TWICs when conducting vessel and facility inspections and during spot checks using hand-held scanners, ensuring credentials are valid. A second rulemaking will propose enhanced access control requirements, including the use of electronic readers by certain vessel and facility owners and operators.

An estimated 750,000 individuals will be required to obtain a TWIC. This includes Coast Guard-credentialed merchant mariners, port facility employees, longshoremen, truck drivers, and others requiring unescorted access to secure areas of maritime facilities and vessels regulated by the Maritime Transportation Security Act. The enrollment process consists of the following components:

- Optional preenrollment
- In-person enrollment
- Fee collection
- Security threat assessment
- Notification of the results
- Issuance of the TWIC to the applicant

Applicants may preenroll online to enter all of the biographic information required for threat assessment and make an appointment at the enrollment center to complete the process. Then applicants must visit the enrollment center where they will review and sign a TWIC Application Disclosure Form, provide biographic information and a complete set of fingerprints, sit for a digital photograph, and pay the enrollment fee. The applicant must bring identity verification documents to enrollment and in the case of aliens, immigration documents that verify their immigration status, so the documents can be scanned into the electronic enrollment record. [18]

IDENT-IAFIS

In a joint effort by the U.S. Department of Homeland Security (DHS) and the Department of Justice (DOJ), integrated 10-print biometric identification technology is operating in every U.S. Customs and Border Protection (CBP) Border Patrol station throughout the U.S. The newly advanced capability allows CBP Border Patrol agents to automatically search the FBI's fingerprint database, the Integrated Automated Fingerprint Identification System (IAFIS) and DHS Automated Biometric Identification System (IDENT) [19] provides rapid identification of individuals with outstanding criminal warrants through electronic comparison of 10-print digital finger scans against a vast nationwide database of previously captured fingerprints (roughly 10 million prints are stored by the FBI).

This capability is a fast, effective weapon in the war on terror. It allows law enforcement personnel to thoroughly check immigration and criminal backgrounds of people who have entered the U.S. illegally. The IDENT/IAFIS program was fully operational for several years within all Border Patrol stations and is in the process of being deployed to all the ports of entry nationwide. As part of US-VISIT deployment, all air and seaports of entry, all Immigration and Customs Enforcement (ICE) field locations, and the busiest land border ports of entry will have this capability.

This new ability to ID quickly and accurately, those apprehended for illegal border crossings gives Border Patrol agents an additional tool in their efforts to stop potential terrorists, criminals, or anyone else trying to enter the country who should be arrested, deported, or denied entry.

When a Border Patrol agent detains someone, the agent can take a live-scanned fingerprint of the suspect on the spot, and with the new technology, compare that print instantly against a database of previously captured fingerprints. The agent would learn within minutes whether that person has a criminal record or is wanted anywhere in the country on an outstanding warrant.

Until IAFIS was deployed at Border Patrol stations, getting back a positive hit, or definitive identity information, could take from eight days to three months. That was clearly too long a wait for agents who needed to decide right away whether to deny entry and return or detain the individual. It took that long because agents usually had to compare the old-fashioned, preelectronic, highly inaccurate eyeball method-inked fingerprints on cards with possibly imperfect index-print scans. This method also assumed the inked prints were not smudged or incomplete and the agent was a fingerprint expert, a skill now acknowledged as a specialty unto itself. Agents also had to wait for fingerprint cards to be faxed, which could distort the image of prints even further, or sent by snail mail. This lack of cutting-edge automation meant agents spent more time poring over cards and papers with less time available to spend on patrol, inspection, interdictions, and apprehensions.

Further complications included the fact that people trying to enter the country illegally were not about to state their real name to a Border Patrol agent. Whatever name an illegal immigrant did give the agent may or may not have been immediately verifiable through fingerprints. Yet agents still had to decide whether to hold or not hold the person.

With IDENT/IAFIS, all a Border Patrol agent needs to do is take a fingerprint scan. That person's criminal history, if there is one, and outstanding warrants are immediately accessible.

The ability to cross-reference millions of criminal records in a matter of minutes is more important than ever. Terrorists know that thousands of illegal immigrants try to cross the southern border every day, and they are trying to blend in with the crowd. Its original purpose, to identify immigration violators and hard-core criminals, has been superseded by the need to identify terrorists.

Secure Flight

Shortly after the CAPPS II program was cancelled, TSA announced it planned to develop a new passenger prescreening program called Secure Flight. TSA plans to use this centralized vetting capability to identify terrorist threats in support of various DHS and TSA programs. Further, TSA plans to use the platform to ensure that persons working at sensitive locations; serving in trusted positions with respect to the transportation infrastructure; or traveling as cockpit and cabin crew into, within, and out of the United States are properly screened depending on their activity within the transportation system. In addition to supporting the Secure Flight and Crew Vetting programs, TSA expects to leverage the platform with other applications such as:

- TSA screeners and screener applicants
- Commercial truck drivers with Hazardous Materials Endorsements
- Aviation workers with access to secure areas of the airports
- Alien flight school candidates
- Applicants for TSA domestic Registered Traveler program

According to TSA, Secure Flight will leverage the system development efforts already accomplished for CAPPS II, but will have several fundamental differences. Specifically, TSA is designing Secure Flight to incorporate only some of the capabilities planned for CAPPS II such as the core capabilities of watch list matching and CAPPS I rules application. Secure Flight will also only prescreen passengers flying domestically within the United States, rather than passengers flying into and out of the United States. Table 5-1 provides a summary of the

Capability in Program	Current Prescreening	CAPS II	Secure Flight
Watch list matching	x	x	x
CAPPS rules applications	x	x	TBD
Identity authentication		x	TBD
Criminal checks		x	
Search for unknown terrorists		x	
Use of opt-in lists		x	TBD
Use of alert lists		x	

Table 5-1 Planned Capabilities of CAPS II

capabilities planned for CAPPS II, as compared with the capabilities currently provided by the current passenger prescreening program and those planned for the Secure Flight program. As shown in the table, TSA does not plan to add additional features beyond the current passenger prescreening program, with the exception of matching passenger name record (PNR) data against an expanded terrorist watch list, which will be provided by the TSC. TSA is also exploring the feasibility of using commercial data as part of Secure Flight if the data are shown, through testing, to increase the effectiveness of the watch list matching feature. TSA does not currently plan for Secure Flight to include checking for criminals, performing intelligence-based searches, or using alert lists. TSA has not yet determined whether Secure Flight will assume the application of CAPPS I rules from the air carriers, or if an opt-in list capability will be used as part of Secure Flight.

In an attempt to achieve greater synergy and avoid duplication of effort, DHS has proposed to create an Office of Screening Coordination and Operations within DHS's Border and Transportation Security Directorate. The purpose of this office will be to coordinate a comprehensive approach to several ongoing terrorist-related screening initiatives:

- Immigration
- Law enforcement
- Intelligence
- Counterintelligence
- Protection of the border
- Transportation systems
- Critical infrastructure

When a passenger makes flight arrangements, the air carrier or reservation company will complete the reservation by entering PNR data in its reservation system, as previously done. Once the reservation is completed, the PNR will be electronically stored by the air carriers. Approximately 72 hours prior to the flight, the PNR will be sent to Secure Flight through a network connection provided by the DHS. Reservations that are made less than 72 hours prior to flight time will be sent immediately to TSA. Upon receipt of the PNR, TSA plans to process the PNR data through the Transportation Vetting Platform. During this process, Secure Flight will determine if the data contained in the PNR match the data in the TSC terrorist screening database and potentially analyze the passenger's PNR data against the CAPPS I rules, should TSA decide to assume this responsibility from the air carriers.

In order to match PNR data to information contained in the terrorist screening database, TSC plans to provide TSA with a subset of the database for use in Secure Flight, and provide updates as they occur. All individuals listed in the TSC data subset are to be classified as either selectees; who will be required to undergo secondary screening before being permitted to board an aircraft or no-flys; who will be denied boarding unless they are cleared by law enforcement personnel. When Secure Flight completes its analysis, each passenger will be assigned one of three screening categories:

- Normal screening required: No match against the terrorist screening database or CAPPS I rules
- Selectee: A match against the selectee list or the CAPPS I rules, or random selection
- No-fly: A match against the no-fly list

The results will be stored within the Secure Flight system until 24 hours prior to departure, at which time they will be returned to the air carriers. [20]

Biometric Products

EyeDentify Retina Biometric Reader

A person's eye is as unique as a fingerprint and can be used to identify someone. The EyeDentify retinal readers (Figure 5-10) claim to be the most secure biometric identification method. There have been no false acceptances and fraud seems impossible. Unfortunately, the false rejection rate is high, and the method will not be easily accepted by the general

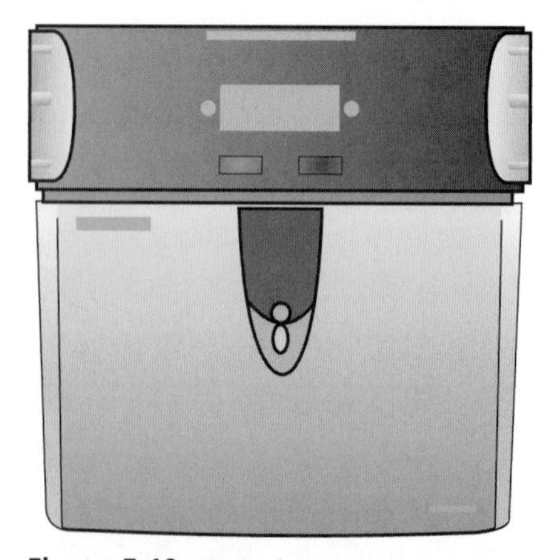

Figure 5-10 Retinal Scanner

Course Technology/Cengage Learning

public. Therefore, use of this method seems limited to high-security access control systems such as nuclear reactors or military installations. Features include:

- No false acceptances
- No counterfeits seem possible; false eyes, contact lenses and eye; transplants cannot breach the security of this device
- Difficult to use and socially difficult to accept because people do not like to have their eyes scanned
- Suitable for everyone with eyes
- Size for storage of identification tokens: 40 bytes
- Long-term stability: The retinal vascular pattern is very stable; only a small number of diseases or injuries will change this pattern
- Proven technology: Used in many systems

Many vendors who offer these types of devices, particularly fingerprint readers, are listed on the Web.

Operation and Performance

In a typical IT biometric system, a person registers with the system when one or more physical and behavioral characteristics is obtained. This information is then processed by a numerical algorithm, and entered into a database. The algorithm creates a digital representation of the obtained biometric. If the user is new to the system, enrollment occurs, which means the digital template of the biometric is entered into the database. Each subsequent attempt to use the system, or authenticate, requires the biometric of the user to be captured again, and processed into a digital template. That template is then compared to those existing in the database to determine a match. The process of converting the acquired biometric into a digital template for comparison is completed each time the user attempts to authenticate to the system.

The comparison process involves the use of a hamming distance, or the measurement of the similarity of two bit strings. For example, two identical bit strings have a hamming distance of zero, while two totally dissimilar ones have a hamming distance of one. Thus, the hamming distance measures the percentage of dissimilar bits out of the number of comparisons made. Ideally, when a user logs in, nearly most features match; then when someone else tries to log in, who does not fully match, and the system will not allow the new person to log in. Current technologies have widely varying equal error rates, varying from as low as 60 percent and as high as 99.9 percent. Figure 5-11 shows where the "errors to sensitivity" intersection lines occur.

Performance of a biometric measure is usually referred to in terms of the false accept rate (FAR), the false nonmatch or reject rate (FRR), and the failure to enroll rate (FTE or FER). The FAR measures the percent of invalid users who are incorrectly accepted as genuine users, while the FRR measures the percent of valid users who are rejected as impostors. One of the most common measures of real-world biometric systems is the rate at which both accept and reject errors are equal.

Stated error rates sometimes involve idiosyncratic or subjective elements. For example, one biometrics vendor set the acceptance threshold high, to minimize false accepts. In the trial, three attempts were allowed, and so a false reject was counted only if all three attempts failed. At the same time, when measuring performance biometrics, such as writing, speech,

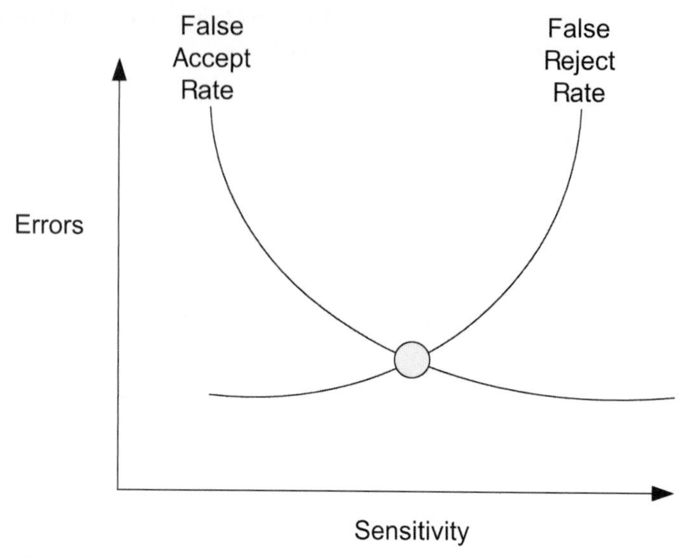

Figure 5-11 Biometric Sensitivity to Error Measure

Course Technology/Cengage Learning

etc., opinions may differ on what constitutes a false reject. If a signature verification system is trained with an initial and a surname, can a false reject be legitimately claimed when it then rejects the signature incorporating a full first name?

Despite these misgivings, biometric systems have the potential to identify individuals with a very high degree of certainty. Forensic DNA evidence enjoys a particularly high degree of public trust at present and substantial claims are being made in respect of iris recognition technology, which has the capacity to discriminate between individuals with identical DNA, such as monozygotic twins.

The State of Scientific Testing of Biometric ID Systems

Testing of biometric devices requires repeat visits with multiple human subjects. [21] Further, the generally low error rates mean many human subjects are required for statistical confidence. Consequently, biometric testing is extremely expensive and generally affordable only by government agencies. Few biometric technologies have undergone rigorous, developer/vendor-independent testing to establish robustness, distinctiveness, accessibility, acceptability, and availability in "real-world" (nonlaboratory) applications.

An in-depth discussion of the statistical methodology in testing biometric ID systems is beyond the scope of this discussion. Note, however, that it is very difficult to generalize from test results. At this time, scientists have no way of accurately estimating how large a test is needed to adequately characterize any biometric device in any application, even with advance knowledge of theoretical error rates.

A common authenticator for both physical and logical (information technology) authentication can provide a single credential that enables rights to corporate resources, timelier user life-cycle management, and improved physical access controls. The benefits are compelling;

however, implementing a common authenticator requires significant effort and can take several years. There are various dimensions to the problem, including:

- Smart card personalization and distribution
- Upgrade of physical access control systems
- Emergency access procedures

Discussion and Analysis

Any system assuring reliable person recognition must necessarily involve a biometric component. Because of the unique person identification potential provided by biometrics, they have and will continue to provide useful value by deterring crime, identifying criminals, and eliminating fraud. At the same time, managers are mindful of the need to provide controls to the problem of "function creep", creating systems that do not threaten basic rights to privacy and anonymity, and substantiate the business case for system deployment. Biometrics is an important and interesting pattern-recognition application with its associated unique legal, political, and business challenges.

While this chapter addresses some fundamental issues in biometrics, it should not be construed to imply that the existing biometric technology is not useful. In fact, there are a large number of biometric solutions that have been successfully deployed to provide useful value in practical applications. For example, the hand geometry system has served as a good access control solution in many deployments, such as university dorms, building entrance, and time/place applications. AFIS systems have been providing value to society by employing the integration of automatic and manual processes.

It should be stated that an emerging technology such as biometrics is typically confronted with unrealistic performance expectations and not fairly compared with existing alternatives, such as passwords, that are being tolerated. A successful biometric solution does not have to be 100 percent accurate or secure. A particular application demands a satisfactory performance justifying the additional investments needed for the biometric system. The system designer can exploit the application context to engineer the system to achieve the target performance levels.

This chapter explored several roadblocks for widespread adoption of biometrics as a means of automatic person identification: effective and efficient pattern recognition; and ensuring system integrity, system application integrity, and return on investment. From a pure pattern recognition perspective, large-scale identification and screening applications are the two most challenging problems. From a system perspective, both security and privacy are problems with no clear satisfactory solutions on the horizon, and cost savings need to be more thoroughly documented. It appears that surmounting these roadblocks will pave the way not only for the inclusion of biometrics into mainstream applications but also for other pattern recognition applications.

Recognition problems have historically been very elusive and underestimated in terms of the effort needed to arrive at a satisfactory solution. Additionally, since humans seem to recognize other people with high accuracy, biometrics has incorrectly been perceived to be an easy problem. There is no substitute to research, realistic performance evaluations, and standardization efforts that facilitate the cycle of build-test-share for transforming the technology into business solutions. Making the business case for biometrics has proved difficult for several reasons:

- The business value of security and deterrence is always difficult to quantify, regardless of technology
- Fraud rates and the costs of long-standing business systems (e.g., PINS and passwords) are not well understood
- Total costs for biometrics systems have not been well documented or reported

Many recent media reports have been critical of biometric systems on the issue of return on investment, but too little research has been done on this issue to reach any firm, general conclusions. Research funding in biometrics is negatively impacted by the lack of substantiated cost savings or increased productivity. It is difficult to justify funding for additional research in basic pattern matching algorithm development when the potential financial return is not immediately apparent. Biometrics is an ideal area for computer scientists to work closely with management scientists and business specialists to develop methods for assessing long-term financial returns attributable to deployed systems. The insistence on "return of investment" (ROI) issues is premature because there is no substitute to biometrics for effective positive identification; therefore, the development of reliable identity infrastructure is critical to effective functioning of society and this infrastructure will have to necessarily involve biometrics.

The community has a responsibility to initiate viable development of this emerging technology without encroaching on the fundamental rights of citizens. Considering the wide scope of the resulting impact on society, this responsibility needs to be substantially stimulated and shouldered by sustained and substantial research and development investment from government agencies worldwide. Considering the recent mandates of several governments for the nationwide use of biometrics in delivering crucial societal functions, there is a need to act with a sense of urgency. Pattern recognition systems have never been tried at such large scales, nor have they dealt with such a wide use of sensitive personal information. [22]

Summary

- Before implementing a biometric system, decision makers must identify the categories of individuals whom such systems are supposed to target, and the threat they pose in light of their abilities, resources, motivations, and goals. These requirement specifications will need to map out clearly, and in advance, the expectations of the biometric system. A biometric system is described as an automated system capable of a number of discrete functions. These include capturing a biometric sample from a user, extracting biometric data from that sample, comparing the biometric data with that contained in one or more reference templates, deciding how well they match, and then indicating whether or not an authentication of identity has been achieved.
- Commercial and government systems consist of integrated biometric hardware and software used to conduct biometric identification or verification, often for access control. The methods most often implemented include hand geometry and handwriting, retina and iris identification, voice verification, and fingerprint identification systems.

- A number of commercial and government biometric systems have been developed or are in the development process. These systems that provide support for fingerprint, face, palm print, retina and other biometric methods. Systems include Project Semaphore, US-VISIT, FACEIT®, IrisCode™, IDENT-IAFIS, Secure Flight, PORTPASS, and others.
- It is essential that the organization's administration personnel investigate thoroughly the products being considered due to the rapid change in biometric technologies. Vendors and products and services come and go, and biometric system solutions are expensive endeavors.

5

Key Terms

Automated Biometric Identification System (ABIS) System includes mandatory collection of 10 rolled fingerprints, a minimum of mug shots from varying angles, and an oral swab to collect DNA.

Automated Fingerprint Identification System (AFIS) A system for automatically matching one or many unknown fingerprints against a database of known prints.

Binning The process of parsing (examining) or classifying data to accelerate and/or improve biometric matching.

Biometric system An automated system capable of capturing a biometric sample from a user, extracting biometric data from that sample, comparing the biometric data with that contained in one or more reference templates, deciding how well they match, and indicating whether or not an authentication of identity has been achieved.

Business continuity planning (BCP) A term used involving concerns with failure of IT equipment, or the ability to employ it effectively.

Card holder An individual to whom an ID card is issued or assigned.

Card reader Any device that reads encoded information from a card, token, or other identity device and communicates to a host such as a control panel/processor or database for further action.

Card serial number An identifier which is guaranteed to be unique among all identifiers used for a specific purpose.

Cybercash Uses the wallet concept, where a consumer selects items to purchase and the merchant presents an invoice and a request for payment.

Disaster recovery The process, policies, and procedures of restoring operations critical to the resumption of operations, including regaining access to data (records, hardware, software, etc.), communications (incoming, outgoing, toll-free, fax, etc.), workspace, and other business processes after a natural or human-induced disaster.

Disaster recovery plan (DRP) Includes plans for coping with the unexpected or sudden loss of communications and/or key personnel, focusing on data protection.

Electronic data interchange (EDI) A process whereby standardized forms of e-commerce documents are transferred between diverse and remotely located computer systems.

Electronic cash A scheme that makes purchasing easier over the Web.

Failure to acquire (FTA) Results from the failure of a biometric system to capture and/or extract usable information from a biometric sample.

Failure to enroll (FTE) Occurs when there is a failure of a biometric system to form a proper enrollment reference for an end user.

Feature extraction Involves finding the true biometric pattern amid noise and signal degradation, preserving the critical information, and discarding redundant or unnecessary data.

Filtering Involves partitioning the database based on information not contained in the biometric itself.

Identity management system (IDMS) Composed of one or more computer systems or applications that manage the identity registration, verification, validation, and issuance process, as well as the provisioning and de-provisioning of identity credentials.

Integrated Automated Fingerprint Identification System (IAFIS) A national fingerprint and criminal history system maintained by the Federal Bureau of Investigation (FBI).

Interoperability The ability of two or more systems or components to exchange information and to use the information that has been exchanged.

Intrusion detection The art of detection and responding to computer and network misuse.

Noise Unwanted components in a signal that degrade the quality of data or interfere with the desired signals processed by a system.

Pattern classification A technique aimed at reducing the computational overhead of pattern matching.

Pattern matching Involves comparing the live sample to the reference sample in the database.

Payment Card Industry Data Security Standard (PCI DSS) Developed by the major credit card companies as a guideline to help organizations that process card payments prevent credit card fraud, hacking, and various other security issues.

Quality control Involves checking the quality of the signal. Ideally, it should be possible to make a quick determination so another measure can be taken if the signal is inadequate.

Radio frequency (RF) Any frequency within the electromagnetic spectrum associated with radio wave propagation. Many wireless communications technologies are based on RF, including radio, television, mobile phones, wireless networks and contactless payment cards and devices.

Radio frequency identification (RFID) Technology that is used to transmit information about objects wirelessly using radio waves.

Reader Any device that communicates information or assists in communications from a card, token or other identity document and transmits the information to a host system, such as a control panel/processor or database for further action.

Robustness A characterization of the strength of a security function, mechanism, service, or solution, and the assurance or confidence that it is implemented and functioning correctly.

Smart card A device that includes an embedded integrated circuit that can be either a secure microcontroller or equivalent intelligence with internal memory or a memory chip alone.

Wallet An electronic cash component that resides on the user's computer or another device such as a smartcard.

Web access management (WAM) A system replaces the sign-on process on various Web applications, typically using a plug-in on a front-end Web server.

Review Questions

1. ________________ is a process whereby standardized forms of e-commerce documents are transferred between diverse and remotely located computer systems.
2. ________________ is a scheme that makes purchasing easier over the Web.
3. A ________________ is about the size of a credit card, which contains a microprocessor for performing a number of functions.
4. Two techniques to verify the identity of someone are ________________ and ________________.
5. Tokens are available in two basic forms. These are ________________ and ________________.
6. ________________ is the process of taking a biometric sample from the user.
7. ________________ was developed by the major credit card companies as a guideline to help organizations that process card payments prevent credit card fraud, hacking, and various other security issues.
8. A ________________ system replaces the sign-on process on various Web applications, typically using a plug-in on a front-end Web server.
9. Name the five basic components or subsystems of biometric ID systems.
10. Once a biometric is acquired, it must be prepared for comparison. Identify the four basic required tasks.
11. An ________________ is composed of one or more computer systems or applications that manage the identity registration, verification, validation, and issuance process.
12. ________________ is unwanted components in a signal that degrade the quality of data or interfere with the desired signals processed by a system.
13. ________________ is the process of providing sufficient information, such as breeder documents, identity history, credentials, and documents, to establish an identity to an organization that can issue identity credentials.
14. ________________ is the process of making a person's identity known to a system, associating a unique identifier with that identity, and collecting and recording the person's relevant attributes into the system.
15. ________________ is the process of confirming or denying that a claimed identity is correct by comparing the credentials with references previously proven and stored.
16. ________________ is a technology used to transmit information about objects wirelessly, using radio waves.
17. ________________ is the ability of two or more systems or components to exchange information and to use the information that has been exchanged.
18. Binning and filtering are different names for the same biometric-data technique. (True or False?)

19. Pattern matching involves comparing the live sample to the reference sample in the database. (True or False?)

20. Text independent systems require a person to speak a predetermined word or phrase. (True or False?)

Discussion Exercises

1. Describe the four functions that an automated biometric system provides. Indicate the usefulness of each function.

2. Describe the five basic categories of biometric features. Why are these features relevant?

3. There are a number of issues that must be addressed by managers when planning to implement a biometric system. Provide an overview of each issue.

4. Provide an overview of the Project Semaphore system.

5. Provide an overview of the integrated automated fingerprint identification system (IAFIS).

6. Provide an overview of the common biometric methods. These include hand geometry and handwriting, retina and iris identification, voice verification, and fingerprint ID.

7. Research the Web to identify biometric systems. Provide an overview of the features and benefits of each system identified.

8. Provide an overview of the payment card industry data security standard. Why is this standard important to the industry?

9. Describe the capabilities of RFID. Provide a list of organizations that could benefit from the RFID technology.

10. Describe why a failure-to-acquire and failure-to-enroll are significant issues with quality control.

Hands-On Projects

5-1: Biometric System Solution Awareness

Time required—30 minutes

Objective—Identify biometric system solutions that have been implemented in some organization.

Description—Use various Web tools, Web sites, and other tools to identify biometric system solutions that are in development or production for both commercial and government organizations.

1. Research the topic on biometric-recognition errors. Provide an overview on the subject of rates and state the importance of them.

2. There are a number of vulnerable points in a biometric system. Describe these seven main areas where attacks could occur. What are the issues with these points?
3. Identify a biometric system that provides for fingerprint biometric access control. Describe the operation of this system.
4. Identify a biometric system that uses new categories (modalities) for access control. These could include face, voice, or signature methods.

5-2: Compare Legacy and Biometric System Requirements

Time required—45 minutes

Objective—Identify those requirements to migrate from a legacy system to a biometric solution that can provide some level of access control and security for OnLine Access assets and resources.

Description—The owners and managers of OnLine Access have identified the database requirements to implement a biometric access control solution. They must look at the current legacy computer system hardware and software and identify the various requirements to implement a biometric solution. The advantages and disadvantages of a biometric solution must be identified.

1. Research the topics of biometric hardware, software, and network components required for access control and security. Identify other organizations that have an initiative to migrate to a biometric solution.
2. Produce a matrix describing hardware, software, and network requirements and the potential cost.
3. Look at the various search engines for hardware, software, network, and other information that could impact an implementation of a biometric solution.

chapter 6

Biometric Multi-Factor System Design

After completing this chapter, you should be able to do the following:

- Understand the multimodal and multibiometrics environment
- Look at the issues of utilizing a multimodal access control system
- Identify those biometrics and legacy solutions that are integrated for access control and security solutions
- See how multibiometrics can enhance the capabilities of a security solution
- Understand how multibiometric systems are categorized into three system architectures according to the strategies used for information fusion
- Become familiar with the products, services, and providers of multimodal access control components, applications, and systems

Current security and access control often uses some physical device such as an ID card. This form of identification of "something that you possess" can be easily stolen, cloned, or lost. Logins and passwords that are "something that you know" have been implemented in most environments, but are easily forgotten and compromised. With the introduction of biometric technologies, the option of "something that you are" has been introduced. This multifactor authentication for identification purposes provides additional capabilities for security and access control in both physical and logical applications.

Introduction

A biometric system which relies only on a single biometric identifier in making a personal identification is often not able to meet desired performance requirements. Identification based on multiple biometrics represents an emerging trend. Multimodal biometric systems integrate face recognition, fingerprint verification, and speaker verification in making a personal identification. These systems take advantage of the capabilities of each individual biometric and can be used to overcome some of the limitations of a single biometrics. Preliminary experimental results demonstrate that the identity established by such integrated systems is more reliable than the identity established by a face recognition system, a fingerprint verification system, or a speaker verification system.

How can a production system be protected using biometric solutions? In many cases, multiple types of biometric solutions are required to provide a high level of security. This chapter will look at issues relating to this situation and will present various configurations that perform various access and physical security functions. These biometric methods might require integration into the currently operational legacy computer systems.

Reliable human authentication schemes are of paramount importance in a highly networked society. Advances in biometrics help address the myriad of problems associated with traditional human recognition methods. The performance and benefits of a biometric system can be significantly enhanced by consolidating the evidence presented by multiple biometric sources. Multibiometric systems are expected to meet the stringent performance requirements imposed by large-scale authentication systems.

Concepts covered in this chapter include fusion, multibiometrics, multiple sensors, and multisensory, multifactor and multimodal systems.

Legacy Computer Systems

Integrating biometric solutions into legacy infrastructures can be a daunting task. The following set of graphics provides a general overview of a legacy system, a stand-alone biometric system, and an integrated solution. Note that the major differences are the requirements of network access and database storage requirements at the central computer site. *Note: The computer hardware and software environments of legacy computer and network systems were presented in Chapter 5.*

Figure 6-1 depicts a common configuration for a host computer system located in some corporate headquarters facility. There would be a significant security and disaster recovery plan in force at this central site. Security devices such as routers and firewalls would be deployed to restrict access to the computer and database assets.

The self-contained biometric access control system depicted in Figure 6-2 could be used in a number of stand-alone application areas. This solution is only good for a local application, such as entrance control or verification of employees for time-card control.

Note: The issues of database and template storage were presented in Chapter 4. Accessibility and throughput are major issues when the network is used for authentication and identification. Another major concern (discussed in Chapter 3) involves the biometric applications that

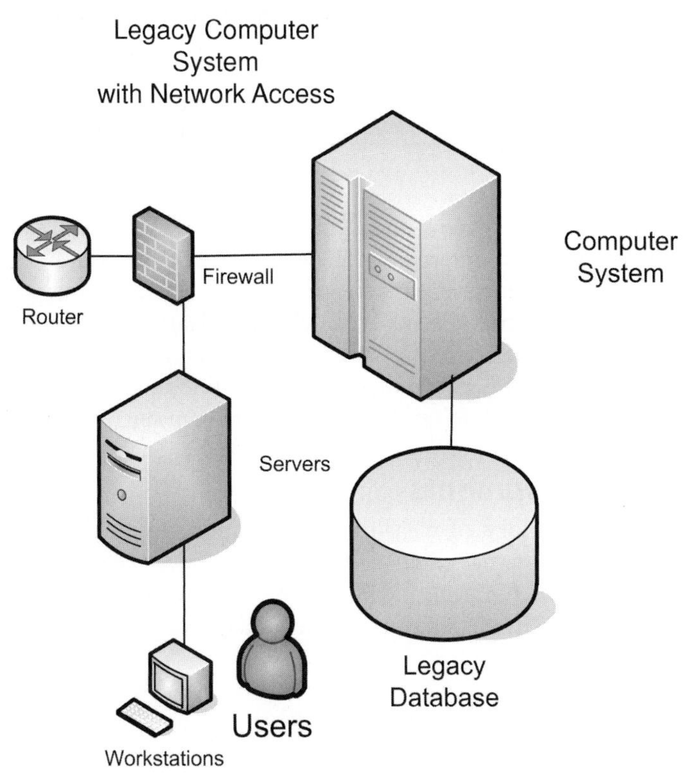

Figure 6-1 Legacy Computer System

Course Technology/Cengage Learning

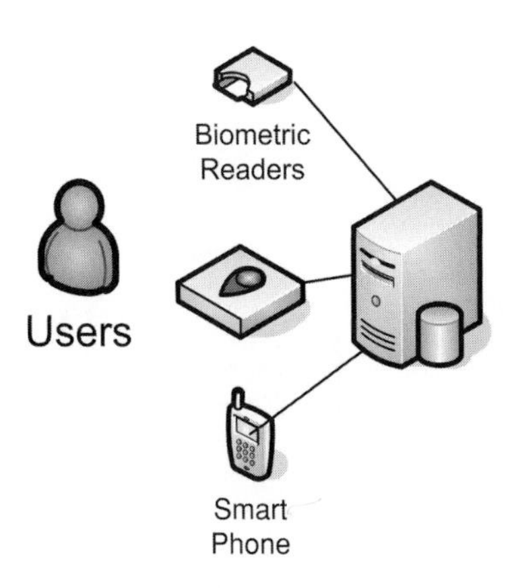

Figure 6-2 Self-Contained Biometric System

Course Technology/Cengage Learning

must reside on the computing devices and computer systems. The advantage of this distributed solution is that it can be used for access control anywhere there are network facilities.

Multibiometrics Defined

This section introduces multibiometric systems and demonstrates the noteworthy advantages of these systems over their unimodal counterparts. This chapter also describes in detail the various scenarios that are possible when fusing biometric evidence from multiple information sources. This overview on multibiometric systems concisely and clearly outlines the different fusion methodologies that have been proposed by researchers to integrate multiple biometric traits. **Fusion** is the union of different biometric modalities, resulting in an integrated system. Figure 6-3 depicts an integrated biometric solution with a remote computer system.

Fusion Technology Developments

Biometrics is the science and technology of measuring and statistically analyzing biological data. In information technology, biometrics usually refers to automated technologies for measuring and analyzing an individual's physical and behavioral characteristics, such as fingerprints, irises, voice patterns, facial patterns, and gait. The analysis of such data is then used

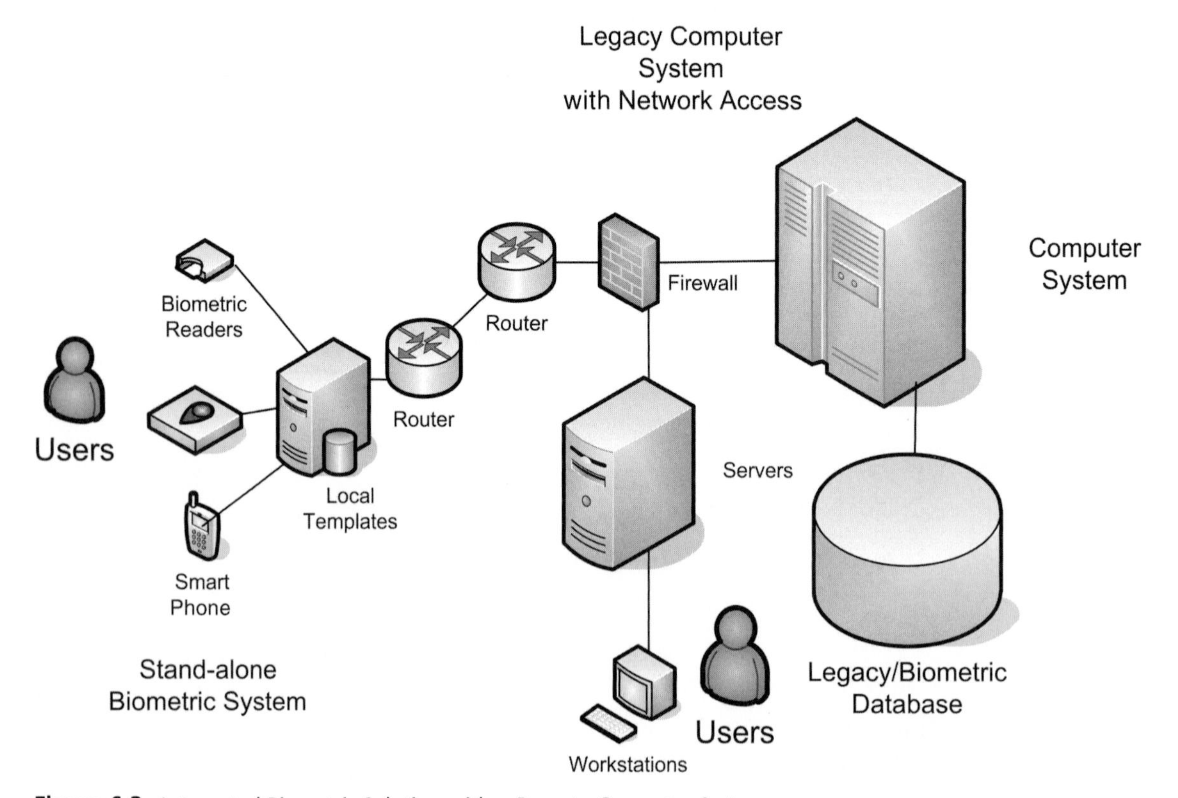

Figure 6-3 Integrated Biometric Solution with a Remote Computer System

Course Technology/Cengage Learning

for identification or verification purposes, depending on need. Multimodal biometrics, or **biometric fusion**, is the process of combining information from multiple biometric readings, either before, during, or after a decision has been made regarding identification or authentication from a single biometric.

This chapter presents theoretical and experimental state-of-the-art work as well as new trends and directions in the biometrics field. It offers security managers, administrators, and students a thorough understanding of how some building blocks of a multibiometric system are implemented. While this book covers a range of biometric traits, this chapter's main emphasis is on multisensory and multimodal face biometrics algorithms and systems. **Multisensory biometrics** refers to combining data from two or more biometric sensors, such as synchronized reflectance-based and temperature-based face images. **Multimodal biometrics** means fusing two or more biometric modalities, like face images and voice timber. A **multimodal biometric system** is, therefore, defined as a biometric system in which two or more of the modality components, such as biometric characteristic, sensor type, or feature extraction algorithm, occur in multiple components.

Multibiometrics encompasses a multimodal biometric system, which integrates face recognition, fingerprint verification, and speaker verification when making a personal identification. **Multifactor authentication** involves the use of multiple techniques to authenticate an individual's identity. This usually involves combining two or more of the following factors:

- Something the individual has: A card or token
- Something the individual knows: A password or personal identification number (PIN)
- Something the individual is: A fingerprint or other biometric measurement

The first part of this book addressed new and emerging biometrics, and emphasized biometric systems where single sensor and single modality are employed in challenging imaging conditions. This chapter provides information on the following multibiometric subjects:

- Multimodal face biometrics, which integrates voice and gait modalities with facial data
- Multibiometrics fusion methodologies and performance prediction techniques
- Multisensory face biometrics, which deals with the personal identification task in challenging variable illuminations and outdoor operating scenarios by employing visible and thermal sensors

Psychological research findings suggest that humans rely on the combined visual channels of the face and body more than any other channel when they make judgments about human communicative behavior. However, most of the existing systems attempting to analyze human nonverbal behavior are monomodal and focus only on the face. Research that aims to integrate gestures as a means of expression has only recently emerged. An approach to automatic visual recognition of expressive face and upper-body gestures from video sequences may be suitable for use in a vision-based, affective multimodal framework. Face and body movements are captured simultaneously using two separate cameras. For each video sequence, single expressive frames, both from face and body, are selected manually for analysis and recognition of emotions.

- First: Individual classifiers are trained from individual modalities.
- Second: Facial expression and affective body gesture information are fused at the feature and decision levels.

In the experiments performed, the emotion classification using the two modalities achieved better recognition accuracy, outperforming classification using the individual facial or bodily modality alone.

Sample Quality

The quality of biometric samples has a significant impact on the accuracy of a matcher. Poor quality biometric samples often lead to incorrect matching results, because the features extracted from these samples are not reliable. Therefore, dynamically assigning weights to the outputs of individual matchers based on the quality of the samples presented at the input of the matchers can improve the overall recognition performance of a multibiometric system. Proponents of these systems see the likelihood of a ratio-based fusion scheme, taking into account the quality of the biometric samples while combining the match scores provided by the matchers. Instead of estimating the quality of the template and query images individually, estimates include a single quality metric for each template-query pair based on the local image quality measures. Experiments on a database of numerous users with iris and fingerprint modalities have demonstrated the advantages of utilizing the quality information in multibiometric systems.

Neural Network

A **neural network** is a type of algorithm that learns from past experience to make decisions. The term neural network had been used to refer to a network or circuit of biological neurons; however, modern use of the term often refers to artificial neural networks. In the artificial intelligence field, artificial neural networks have been applied successfully to speech recognition, image analysis, and adaptive control in order to construct software agents, such as in computer and video games, or autonomous robots. Most of the currently employed artificial neural networks for artificial intelligence are based on statistical estimation, optimization, and control theory.

Multibiometric Components

A **multiapplication card** consists of a smart card that runs multiple applications using a single card. Applications could include physical access, logical access, data storage, and electronic commerce. A **multitechnology card** is an ID card that has two or more ID technologies that are independent and do not interact or interfere with one another. An example is a card that contains a smart card chip and a magnetic stripe.

A **multifactor reader** is a smart card reader that includes a PIN pad, biometric reader, or both to allow multifactor authentication. A **multitechnology reader** is a card reader/writer that can accommodate more than one card technology in the same reader. These could include both ISO/IEC 14443 and ISO/IEC 15693 contact-less smart card technologies or both 13.56 MHz and 125 kHz contact-less technology.

For applications requiring secure access, **mutual authentication** is the process used for the smart card-based device to verify the reader is authentic. It is also used to prove its own authenticity to the reader before starting a secure transaction.

Multibiometric System Issues

Consistent advances in biometrics help to address problems plaguing traditional human recognition methods and offer significant promise for applications in security as well as general convenience. In particular, newly evolving systems can measure multiple physiological or behavioral traits and thereby increase overall reliability. Multimodal biometrics provides an accessible, focused examination of the science and technology behind multimodal human recognition systems, as well as their ramifications for security systems and other areas of application. This chapter demonstrates the noteworthy advantages of these systems over their traditional legacy and unimodal counterparts. In addition, subsequent sections describe the various scenarios possible when consolidating evidence from multiple biometric systems and examine multimodal system design and methods for computing user-specific parameters. Topics and features:

- Introduce readers to the rapidly evolving field of multimodal biometrics and its exceptional utility for increasingly reliable human recognition systems
- Reveal the substantial advantages of multimodal systems over conventional identification methods
- Show how multimodal approaches address the problems of universality and spoofing; describe techniques to perform fusion, especially at the matching score level; examine existing multimodal databases and what they have to offer
- Use numerous examples and illustrations to support the explanations

This authoritative, comprehensive chapter on multimodal biometric systems concisely and clearly outlines their great promise for higher reliability than conventional verification systems.

System Design

As described previously, biometrics is the science of establishing human identity based on an individual's physical or behavioral traits. Research in this field of biometric security and access control continues to grow. Investigation into facial, voice, iris, and hand geometry traits has furthered the development of automated and semi-automated biometric identification systems.

Biometric systems are in place in venues as diverse as grocery stores, Disney World, and U.S. border crossings. As concerns about identity theft and identity fraud become paramount, the demand for these systems is sure to continue to rise. Most systems have relied on a single biometric trait, such as a fingerprint, a hand print, or an iris scan to establish identity. This reliance on a single trait poses certain limitations on the integrity and scalability of the systems.

As more and more users are enrolled into a system, it becomes difficult to rely on one trait to accurately identify those users, hence the need for multibiometrics. Multibiometrics involves consolidating the evidence presented by multiple biometric traits of an individual, such as face and fingerprint, face, and iris. Multibiometrics also helps deter "spoofing," or efforts to pass off artificial traits as real ones, since it is difficult for an impostor to spoof multiple biometric traits. It can therefore enhance the degree of confidence and decrease errors in an identification system.

As previously stated, traditional biometric systems rely on a single biometric identifier, such as fingerprint or face, each with its unique advantages A multibiometric system integrates two or more biometric identifiers and takes advantage of the capabilities of each biometric to provide even greater performance and higher reliability. Such advantage is especially important for large-scale biometric identification systems that require both access control and security functions.

Fused biometrics is a completely new breed of technology within the multibiometrics area. In **fused biometrics**, single or multiple sensors are used to collect different biometrical information, such as a face image and a fingerprint image, and a fused algorithm is used to create a single identification decision based on the results of those measurements. Fingerprint and face identification products are available for various systems, from large-scale cluster-based to embedded or mobile stand-alone devices.

Biometric System Uses

A biometric system which relies on a single biometric identifier to make a personal identification is not sufficient to meet desired performance requirements. Preliminary experimental results demonstrate that the identity established by such an integrated system is more reliable than the identity established by a face recognition system, a fingerprint verification system, and a speaker verification system.

It is important that managers understand the scientific nature of this new field, and, at the same time, the challenges it brings. The goal of this chapter is to consolidate all the work that has been done in the field so far. Multibiometrics is the combination of several different biometric technologies. Fingerprint, face, and iris biometrics are the most common form of data used for biometric applications. The demand for multibiometrics is increasing. Electronic passports are in the design stage in a large number of countries, and the EU has decided the electronic passports issued by member countries shall contain biometric information of facial features and, at a later stage, also fingerprints. Countries are working on including fingerprints and facial features within their initial rollout of electronic passports. The current implementation of multibiometrics includes:

- Multibiometric passports
- ID cards with multibiometrics
- Multibiometric verification systems: the technical infrastructure required for handling of biometric passports and travel documents

Integrating Faces and Fingerprints for Personal Identification

An automatic personal identification system based solely on fingerprints or faces is often not able to meet the system performance requirements. Face recognition is fast but not reliable, while fingerprint verification is reliable but inefficient in database retrieval. A prototype biometric system which integrates faces and fingerprints overcomes the limitations of face recognition systems as well as fingerprint verification systems. The integrated prototype system operates in the identification mode with an admissible response time. The identity established by the system is more reliable than the identity established by a face recognition system. In addition, the proposed decision fusion schema enables performance improvement by integrating multiple cues with different confidence measures.

Information Fusion in Biometrics

Reliable human authentication schemes are of paramount importance in a highly networked society. Advances in biometrics help address the myriad of problems associated with traditional human-recognition methods. The performance and benefits of a biometric system can be significantly enhanced by consolidating the evidence presented by multiple biometric sources. Multibiometric systems are expected to meet the stringent performance requirements imposed by large-scale authentication systems. Different fusion methodologies have been proposed by researchers to integrate multiple biometric traits.

With a very security-conscious society, biometrics-based authentication and identification have become the center of focus for many important applications, as it is believed that biometrics can provide accurate and reliable identification. Biometrics research and technology continue to mature rapidly, driven by pressing industrial and government needs and supported by industrial and government funding. As the number and types of biometrics architectures and sensors increase, the need to disseminate research results increases, as well.

Many of the applications require a higher level of accuracy performance not feasible with a single biometrics type today. It is also believed that fusing multiple biometrics will improve wider coverage of the population who may not be able to provide a single biometrics type and also improve security of the systems in terms of spoof attacks. Research issues include different modes and levels of fusion of biometrics samples, sensing modes, and modalities with a sole goal of improving performance of biometrics. Theoretical studies on sensor-fusion techniques applied to biometrics authentication, recognition, and performance are ongoing. Research areas and topics include:

- Sensing; intensity, depth, thermal, pressure, time-series
- Face, finger, ear, eye, iris, retina, vein pattern, palm, gait, foot
- Biometric template computation and feature extraction, matching
- Data and performance baselines
- Evolution of standards, competitions, and organized challenge problems
- Score-level, decision-level, and feature-level integration
- Architectures for integration, evidence integration
- Fusion-based identification techniques
- Normalization techniques involved in fusion techniques
- Machine-learning techniques in biometrics fusion
- Application dependence personalization of multibiometrics systems
- Theoretical studies in showing models for integration
- Performance modeling, prediction, and evaluation of multibiometrics systems
- Security improvement assessment for multibiometrics systems

Multibiometric Authentication

This section provides a comparative overview of the different architectures of multibiometric systems. It also looks at the effects of multibiometrics on the user.

In recent years, biometric authentication has seen considerable improvements in reliability and accuracy, with some biometrics offering reasonably good results overall. Some fairly technical information is presented here; however, it may challenge some readers to explore the technical aspects of biometric authentication.

To date, even the best biometrics faces numerous problems, some of them inherent to the technology itself. In particular, biometric authentication systems generally suffer from enrollment problems due to nonuniversal biometric traits, susceptibility to biometric spoofing, or insufficient accuracy caused by noisy data acquisition in certain environments. Multibiometrics are a relatively new approach to overcome those problems.

Driven by lower hardware costs, a multibiometric system uses multiple sensors for data acquisition. This allows it to capture multiple samples of a single biometric trait, called multisample biometrics and/or samples of multiple biometric traits, called multisource or multimodal biometrics. Multibiometric systems promise significant improvements over single biometric systems, such as higher accuracy and increased resistance to spoofing. They also claim to be more universal by enabling a user who does not possess a particular biometric identifier to still enroll and authenticate using other traits, thus eliminating enrollment problems. But can multibiometrics live up to the hype? At first glance, incorporating multiple biometrics into one system appears to be a very intuitive and obvious concept. There are very different ways to actually combine multiple sources of information to make a final authentication decision. Information fusion strategies range from simple Boolean conjunction to sophisticated statistical modeling. Without going into the mathematical details, this chapter reports on selected recent approaches. One goal is to analyze how well multibiometric systems are able to keep up with the vast promises made by advocates.

System Architectures Compared

Multibiometric systems are categorized into three system architectures according to the strategies used for information fusion. These include fusion at the:

- *Feature-extraction level*
- *Matching score level*
- *Decision level*

Systems are classified depending on how early in the authentication process the information from the different sensors is combined. Biometric authentication is a chain process.

Authentication Process Flow

Fusion at the feature-extraction level stands for immediate data integration at the beginning of the processing chain, while fusion at the decision level represents late integration at the end of the process. The following sections describe each of these architectures in detail and report on related research activities.

Fusion at the Feature-Extraction Level

In this architecture, the information extracted from the different sensors is encoded into a joint feature vector, which is then compared to an enrollment template, which itself is a joint feature vector stored in a database, and assigned a matching score as in a single biometric system.

The data captured from each sensor is used to create a feature vector. In the case of fingerprints, the vector might include three dimensions: the location, shape, and orientation of each minutia. In the ideal case, this vector uniquely identifies a given person in feature space; it is more likely that the vector identifies a subset of the enrolled population. Combining the feature vector from each biometric creates a vector that has a higher dimensionality and higher probability of uniquely identifying a person in feature space.

Fusion at the Matching-Score Level

Once the feature vectors from each biometric mode have been constructed, they are passed to their individual matching algorithms, which attempt to match them against previously captured templates. The individual matching probabilities are then combined to form a result from which a decision may be made. A variety of methods may be used, including:

- *Neural networks*
- *Discriminate analysis*
- *Weighted sums*

In a multibiometric system built on this architecture, feature vectors are created independently for each sensor and then compared to the enrollment templates, which are stored separately for each biometric trait. Based on the proximity of feature vector and template, each subsystem now computes its own matching score. These individual scores are finally combined into a total score, which is handed over to the decision module. The process flow inside a subsystem is the same as in a single biometric system, which allows the use of proven algorithms for feature extraction and matching. A very elegant example for this fusion strategy incorporates facial scan, fingerprint verification, and hand geometry scan into a common authentication system, using well-known methods for each identifier. This solution uses eigenfaces for the facial scan, minutiae patterns for the fingerprint system, and commonly used hand-geometry features. Matching scores for the three modalities are then normalized and combined using one of the following strategies:

- The *sum rule*: The sum rule is the probability that one or the other of two mutually exclusive events will occur is the sum of their individual probabilities. The rule that states the probability of the occurrence of mutually exclusive events is the sum of the probabilities of the individual events.
- *Decision tree determination*: A decision tree (or tree diagram) is a decision support tool that uses a graph or model of decisions and their possible consequences, including chance event outcomes, resource costs, and utility. A decision tree is used to identify the strategy most likely to reach a goal. Another use of trees is as a descriptive means for calculating conditional probabilities.
- *Linear discriminate analysis:* Discriminant analysis is a statistical technique used to classify objects into mutually exclusive groups based on a set of measurable object's features. It is often called pattern recognition.

The strategy of user-specific weights is the best solution seen so far to deal with nonuniversal biometric traits and enrollment problems. If a user does not possess a certain biometric identifier or shows only weak characteristics, the corresponding weight can be adjusted to a small value. The final question to be answered is whether this approach really leads to a higher accuracy. The experimental data suggests good performance for the combination of all three biometric identifiers. However, it is not significantly better than the best fingerprint systems tested in. This might be due to the fact that the individual subsystems used in this experimental system are rather weak, especially their hand-geometry verifier. Technicians can therefore hope to achieve even better performance when combining top-of-the-line verifiers for each biometric trait.

Fusion at the Decision Level

Strictly speaking, probabilistic-decision level fusion is not really a fusion, although it is generally termed as such. The principle behind this method is that each biometric system makes a match based solely upon the biometric it captures and then passes a binary "match/no-match" vector to a decision module, which forms a conclusion based on a majority vote scheme. Some systems also include a method to weight the decision toward more highly regarded biometrics, such as iris over retina scans.

Most observers accept that fusion produces better results when performed at the feature-extraction level rather than at the data-matching or decision levels, because the data is combined at its most information-rich stage and before external contamination from, for example, matching algorithms. However, feature level fusion is difficult, as the relationship between individual biometrics' feature spaces may not be straightforward (for example, how is fingerprint and iris data combined?). Additionally, the higher the dimensionality of the feature space, the longer the analysis computation time; therefore, there may need to be a trade-off between speed and accuracy at the application level. Modality independence is important when improving biometric systems through fusion; uncorrelated biometrics, such as fingerprint and iris, are expected to result in a higher improvement rate than correlated biometrics, such as two independent readings of the same fingerprint.

The problems associated with feature-level fusion are noted in the literature, with most fusion research being carried out in the data matching and probabilistic decision-level fusion areas. However, it should be noted that Daugman [1] presents arguments that dismiss decision-level fusion and conclude that a strong biometric is better alone than in combination with a weaker one at the decision level.

An improvement in the results of a biometrics system may be achieved through the use of a multimodal fusion biometric system. Successful systems used to produce positive results fused the data at the data matching stage. Improvement of using a fusion system over a single biometric, with the fused data improving the best single biometric, fingerprint in this case, increased by 20 percent.

In this fusion strategy, a separate authentication decision is made for each biometric trait.

Fusion at the decision level is a loosely coupled system architecture, with each subsystem performing like a single biometric system. This architecture has therefore become increasingly popular with biometric vendors, and is often advertised as "layered biometrics." The emergence of biometric standards like BioAPI (s 12) has further supported this concept. Many different strategies are available to combine distinct decisions into a final authentication decision, ranging from majority votes to sophisticated statistical methods. In practice, however, developers seem to prefer the easiest method using Boolean conjunctions. The BioNetrix Authentication Suite, for example, offers the following combination strategies:

The *AND rule* requires a positive decision from all verification modules. While this will certainly lead to low false authentication rates, it will also result in high false rejection rates.

The *OR rule* attempts to authenticate the user using one biometric trait. If this fails, the user is offered another attempt with another verification module. This policy trades a low false rejection rate for a high false authentication rate.

A very interesting rule is the *RANDOM rule*, where a biometric trait is randomly chosen. Although this is a very simplistic idea, it can definitely make it harder for intruders to spoof the system. But it comes without the inconvenience of a multilevel data acquisition for each authentication attempt.

Fusion at the decision level occurs at a very late stage of the authentication process. It can be assumed it does not show the same potential to improve the overall system performance as fusion at the matching-score level. Accuracy improvements can be guaranteed only under very specific conditions. If these conditions are violated by using biometric tests which differ significantly in their performance, their combination at the decision level can even lead to serious performance degradation.

Effects of Multibiometrics on the User

There are issues with the internal architecture of a multibiometric system; however, the effects of multibiometrics on the user have not been discussed in many references. Assumptions are that offering multiple biometric identifiers presents a negligible inconvenience to the user. Is this assumption justified? First and foremost, major privacy issues may be tied to multibiometrics. In a multibiometric system, the user has to reveal a whole spectrum of biometric identifiers, with all of them being stored in the template database after the initial enrollment. The user profiles stored in such a database are significantly more comprehensive than those stored in a single biometric system, and it is very attractive target for identity thieves. Biometrics vendors repeatedly claim the original data cannot be restored from the enrollment templates. However, users have no way of verifying this, since the feature extraction algorithms are always proprietary and never made available to the public. And in fact, there are commercially available systems for which the contrary has been shown.

In traditional password-based authentication, a user simply chooses a new password once the old one has been compromised. It is a major problem of biometrics that most biometric identifiers cannot be changed. If, for example, a fingerprint is compromised, you cannot just get a new one. You would need to switch to another biometric identifier, such as another finger. But in a multibiometric system, this one might be compromised as well, which makes the problem even more severe. The inconvenience of a multilevel data acquisition process to the user is another problem. Different biometric identifiers can either be obtained sequentially or simultaneously, but both ways have their disadvantages: if they are acquired one after another, it will take considerably more time for users to authenticate and thus reduce productivity. On the other hand, those who have already used biometric systems know that they rely on good data quality and are therefore sensitive to factors like positioning, clarity of voice, etc. It might be a challenge to provide good samples for multiple biometric identifiers at the same time (for example, if a user tried to position a thumb on the scanner, while at the same time rotating her head to pass the face recognition).

Not all multibiometric systems will be equally affected by the problems mentioned above. A good system design as well as a careful choice of the biometric traits to be used can certainly alleviate some concerns. Users should not forget about the obvious advantages which multi-biometric systems may offer to the user, such as lower failure-to-enroll rates and higher accuracy of authentication. It is still too early to predict whether these will be sufficient to make users accept the inconveniences. But in any case, all possible effects on the user should be discussed openly. This is still not happening today, with adverse effects being left out (as mentioned at the beginning of this chapter).

Multimodal Biometrics

Before considering multimodal biometrics, it is important to understand the core features of a conventional biometric system. Such a system can be decomposed into four components: [2]

- Biometric capture module: A fingerprint reader or iris scanner
- Feature extraction module: Software that selects the active matching data and produces a feature vector of the biometric, features extracted from the biometric reading will be used to create the feature vector
- Matching module: Compares the captured biometric with an existing database; measurements extracted from the active matching data describe the useful image features and thus are known as a feature vector
- Decision module: Provides a degree of confidence in any identity matched against the person providing the biometric sample

The ability of the system to perform well is based almost solely upon the quality of the biometric captured. A well-captured biometric is rich in distinguishing information, which in turn gives the feature extraction algorithms the best chance of finding a match with existing records. However, the ability of the system to capture high-quality biometric samples is reduced by many factors. Dirty fingerprint sensors, photographic light intensity, or a voice altered by a cold may all reduce the quality of a reading to the extent that multiple readings of the same biometric can produce a range of widely differing samples. Add to this the fact that a minority of individuals may not be able to provide a given biometric, and a conventional biometric system may soon become a burden to its operators.

As stated previously, the accuracy of a biometric system can be measured from two statistical measurements: FAR and FRR. The false match rate (FMR), also known as a Type I error or false acceptance rate (FAR) is a measure of the readings that the system incorrectly matches to its database. A low FAR metric increases a system's security performance. The false non-match rate (FnMR), also known as a Type II error or false rejection rate (FRR) is a measure of the readings that the system incorrectly fails to match to its database. The lower the FRR, the easier a system will be to use.

The FAR and the FRR are inversely proportional, forcing a trade-off between security and convenience when using biometric systems. A system that is easy to use, or convenient for both user and system administrator, may allow unauthorized access to a secure area, while a highly secure system may require continual human intervention to allow access to authorized users who are not recognized by the device. Multimodal biometric systems may be able to improve this tradeoff. Multimodal biometric systems capture two or more biometric samples and use fusion to combine their analyses to produce a better match decision by simultaneously decreasing the FAR and FRR.

Fusion Methodology

Multimodal biometrics systems can be designed to work in five ways:

- Multiple sensors may be used to capture the same biometric
- Multiple biometrics may be captured
- Multiple readings of the same biometric may be combined to achieve an optimal reading
- Readings of two or more units of the same biometric may be taken (such as two different fingerprints or both irises)
- Different matching and/or feature extraction algorithms may be used on the same biometric reading to give independent results

A combination of uncorrelated modalities, such as fingerprint and face or fingers, is usually expected to result in a better performance than a combination of correlated modalities. Fusion may take place at any point during the processing. However, while using multiple biometrics produces advantages, it also introduces disadvantages:

Advantages

- Improved population coverage is achieved by reducing the failure-to-enroll rate
- Lower FAR and FRR can be achieved
- Users could be challenged to enter a random and different sequence of biometrics each time
- The ability to spoof the system is reduced as replicating multiple biometrics traits is more difficult than replicating one
- A challenge–response authentication procedure may be facilitated that increases the probability that a live user is interfacing with the system

Disadvantages

- Multiple biometrics means multiple enrollment difficulties
- The process of determining the correct feature vector combinations required to fuse, such as iris and fingerprint data, is not simple and adds to a system's R&D overheads
- System processing times are increased due to the complex computations required

Biometric Fusion Weighting

A key feature of fusion technology is the ability to choose the degree of reliance on each of the biometric traits in order to maintain an acceptable balance between confidence and convenience for users and operators. An additional benefit is that the *weighting* (importance or influence) can be applied at population level, individual level, or indeed anywhere in between the weights. Weights could be used to ensure that a user with poor fingerprints had more importance placed on the other biometrics in a given system, increasing the likelihood of genuine acceptance. The biometric "weightings" may be hard-coded. An operator may more heavily weight the facial recognition component of a particular subset of workers who perform manual labor and therefore have poorly recognizable fingerprints; or organically learned over time by examining the stored template of the user. A system requiring consistent clarification over the identity of an individual, perhaps due to an eigenface that is easily matched to a large number of the enrolled population, may be set to learn the higher score achieved from the voice print which can be more heavily weighted, removing the burden of intervention from its operator.

Additional Multimodal Considerations

No biometric is truly universal. NIST reports fingerprints cannot be obtained from 2 percent of the U.S. population, while opponents show hand geometry and facial biometrics have a limited number of distinguishable patterns, 105 and 103 respectively, indicating they have a limited uniqueness factor. While multimodal biometrics systems overcome these problems, there are two overall limiting factors that affect them:

- The higher the dimensionality of the data vectors produced during fusion, the longer the computation time required to analyze them.
- The more biometrics used, the more expensive it becomes to set up and run the infrastructures.

The concept of "soft biometrics" has been introduced to overcome these factors. *Soft biometric traits* are those characteristics that provide some information about the individual, but lack the distinctiveness and permanence to sufficiently differentiate any two individuals. Examples include height, gender, ethnicity, and eye color. While these traits are easily identified by a human operator, research has been conducted to enable automated capture. When such automated systems become available, the integration of soft biometrics into a fusion system will diversify the differentiating traits that everyone exhibits, increasing the uniqueness factor that a biometric system so heavily relies upon.

Production Multibiometric Systems

Vendors have produced fully comprehensive solutions for biometric passports. A multibiometric *match-on-card* solution allows for the storage and matching of fingerprints, iris, and facial templates directly on a single smart card. The match-on-card technology and fingerprint technology have been combined with iris recognition technology from Iridian Technologies and facial recognition technology from OmniPerception.

Match-On-Card Process

1. *Enrollment*: The cardholder enrolls one or more fingerprints that are written to and stored on the tamper-proof smart card as a template using a fingerprint scanner. At this point, personal or confidential data, such as health care data or a certificate, is also stored on the smart card.
2. *Smart card reader*: This device must be available at the location where a person is being issued a card to establish identity, such as a driver's license, or for permission for access or transactions, such as a bank or office building. The smart card reader can be combined with the fingerprint reader.
3. *Presentation*: The individual's smart card is placed in a reader and prompts the person to present a previously enrolled finger. At this time the system can also show information about the person, depending on the application.
4. *Authentication*: The cardholder places one of the enrolled fingers on the fingerprint scanner. The live print is read and analyzed in a split second.

5. *Match*: If the fingerprint from the person and the template of the print on the card are compared, it is a match. If the two prints match, then the identity of the cardholder has been verified. Now the system can perform the requested actions such as uploading health care data or signing an e-mail using a certificate. If the prints do not match, the requested action is rejected, and the true cardholder's credentials are protected from fraud or misuse.

MegaMatcher 2.0

A provider of high-precision biometric identification technologies created MegaMatcher 2.0 SDK for the development of large-scale automated fingerprint identification systems (AFIS) and multibiometric face–fingerprint identification systems. MegaMatcher 2.0 includes both fingerprint and face identification engines with a fusion algorithm that allows the two technologies to work together to provide very fast one-to-many matching with even higher reliability than AFIS or facial recognition alone. MegaMatcher's powerful fused algorithm can produce up to 400,000 matches per second on a single processor PC; with MegaMatcher 2.0's fault-tolerant, scalable cluster software, this number can be multiplied across multiple PCs to perform extremely fast, parallel fingerprint and face matching using databases of practically unlimited size. [3]

With the increasing demand for automated biometrical identification systems for both forensic and civil applications, such as crime scene investigations, border crossings, passport and visa documentation, election control systems, and even credit card transactions, the need for both fast and accurate identification is ever more critical. While AFIS has been widely used for the past two decades, multibiometric systems incorporating both face and fingerprint identification offer a number of advantages. Using the multibiometric technology in MegaMatcher 2.0, developers and integrators can create systems where both face and fingerprint can be scanned at the same time using inexpensive hardware such as a fingerprint scanner and a simple Webcam or photo scanner (for example, to scan a passport photo). A multibiometric system based on MegaMatcher 2.0 provides both greater convenience and matching speed for the user and a much higher level of security, even when using very large databases, because the fusion of both face and fingerprint provides unambiguous identification.

The face and fingerprint algorithms in MegaMatcher 2.0 are the most powerful algorithms to date and include a number of enhanced features to ensure higher quality and accuracy in both enrollment and matching. New fingerprint measurement tools improve recognition by enabling poor fingerprints to be rejected at the enrollment stage. The new template generalization tool facilitates faster, more accurate matching by enabling a much higher quality template to be created from just a few fingerprints or faces. The MegaMatcher 2.0 fingerprint algorithm is tolerant to fingerprint translation, rotation, and deformation and has the ability to match between rolled and flat fingerprints.

Compared to the fingerprint-only algorithm in MegaMatcher 1.1, the MegaMatcher 2.0 algorithm offers considerably higher reliability in both fingerprint and multibiometric modes. When using fingerprints only, the FRR in MegaMatcher 2.0 is up to six times lower than in MegaMatcher 1.1 when set to the same FAR level. In the multibiometric mode, at the same FAR level, the FRR decreases by 10 to 100 times compared to the previous version.

Enhancements to the MegaMatcher 2.0 SDK make it easier to work with fingerprints both during and after enrollment. When multiple fingerprints have been scanned at the same time, the new fingerprint segmentation functionality in MegaMatcher 2.0 can automatically segment them into separate fingerprints. And for forensic applications, MegaMatcher 2.0 includes an API for the editing of latent fingerprint templates, such as those collected from crime scenes.

Instituto Federal Electoral

Sagem Défense Sécurité started deployment of a large-scale, multibiometric program in Mexico for the Instituto Federal Electoral (IFE). Comprising a turnkey system to check the identities of voters in electoral lists, this program features the largest multibiometric identification system in the world.

The multibiometric identification system integrates both digital fingerprints and facial recognition to ensure that each citizen is registered only once in the national voter rolls. It will eventually process some 72 million voters, with a capacity of 100,000 registrations and ID searches per day. The two-year program will be deployed in two phases.

The first involved reworking the existing electoral databases to set up the multibiometric system. Operated by the Mexican government, this system will search the 66 million digital fingerprints already recorded to detect any duplicates in the current voter rolls. During 2006, the 70 million ID photos were digitized to create a biometric portrait base.

The second phase was kicked off during 2006. The multibiometric system, integrating an AFIS and facial recognition, will start operation, handling the real-time processing of new requests. Using the multibiometric data, the IFE will check newly registered voters, to make sure that they only appear once on the rolls. By December 2007, the database was expanded to include some 72 million people.

Analysis of Fusion and Multibiometrics

Research and analysis of multibiometric systems includes multisensory and large-scale automatic biometric identification systems. These two efforts are described in this section. [4]

Multisensory Biometric Research

The field of biometrics can be divided into two main classes according to features that humans are born with, such as fingerprints or facial features, or their behavioral characteristics, such as a handwritten signature or the voice. The material presented in this section pursues the latter class, specifically how a person uses certain devices or tools.

Research results have suggested that biometrics can exploit people's habits to identity individuals by their handling of devices—the particular force a person applies to the keys in a keyboard, for example. There is also the time interval between each keypad when dialing a telephone number. Another example is the map described by the fingers as they navigate and solve a maze operation. Technology can extract these features through a haptic-based application and define the subsequent individual pattern. A framework that identifies behavioral patterns through physical parameters such as direction, force, pressure, and velocity has been built. The setup for the experimental work consisted of a multisensory tool, using the Reachin system. [5]

The advantages of such systems over traditional authentication methods such as passwords are well known, and so biometric systems are gradually gaining ground in terms of usage.

Current research explores the feasibility of automatically and continuously identifying participants in haptic systems. *Haptic* is a concept pertaining to the sense of touch, from the Greek word *haptein*, which means "to grasp." Such a biometric system could be used for authentication in any Haptic-based application, such as teleoperation or teletraining, not only at the beginning of the session, but continuously and throughout the session as it progresses. In order to test this possibility, a haptic system was designed in which position, velocity, force, and torque data from the tool was continuously measured and stored as users were performing a specific task.

Subsequently, several algorithms and methods were developed to extract biometric features from the measured data. Overall, the results suggest reasonable practicality of implementing haptic-based biometric systems, and that it is an avenue worth pursuing; however, it might be quite difficult to develop a highly accurate haptic ID algorithm.

Large-Scale, Automatic Biometric Identification System

Nowadays, the need for automated biometrical identification systems is increasing in civil and forensic fields of applications. Fast and accurate identification becomes particularly critical for large-scale applications, such as passport and visa documentation, border crossings, election control systems, credit card transactions control, and crime scene investigations. Many countries, including the U.S. and European nations, incorporate biometrical data into passports, ID cards, visas, and other documents for use in large, national scale automatic biometrical identification systems.

Automated fingerprint identification systems have been widely used in forensics for the past two decades, and recently they have become relevant for civil applications. Whereas large-scale biometrical applications require high identification speed and reliability, multibiometric systems that incorporate both face and fingerprint recognition offer a number of advantages for improving identification quality and usability. Large-scale automatic biometrical identification systems have a number of special requirements, which are different from those for small- or middle-scale biometrical systems:

- The system must perform reliable identification with large databases, as biometrical identification systems tend to accumulate false acceptance rates with an increase in database size, and using single fingerprint or face image for identification task becomes unreliable for large-scale application. Several biometrical samples should be used to increase identification reliability, and multibiometrical technologies, such as collecting fingerprint and face samples from the same person, are often employed for additional convenience.
- The system must show high productivity and efficiency, which correspond to its scale:
    - System scalability is important, as the system might be extended in the future, so a high productivity level should be kept by adding new units to the existing system.
    - Daily number of identification requests could be very high.
    - Identification request should be processed in a very short time, in real time, requiring high computational power.
    - Support for large databases, in tens or hundreds millions of fingerprints, is required.
    - The system must be tolerant to hardware failures, as even temporary pauses in its work may cause big problems (taking into account the application size).
- The system must support major biometrical standards. This should allow using system-generated templates or databases with systems from other vendors, and vice versa.

- The system must be able to match flat (plain) fingerprints with rolled fingerprints, as many institutions collect rolled fingerprint databases.
- The system must be able to work in the network, as in most cases client workstations are remote from the server with the central database.
- A forensic system must be able to edit latent fingerprint templates in order to submit latent fingerprints into AFIS for identification.

Despite all these requirements, the system price should be as low as possible. Many existing AFIS are specialized for law enforcement or other particular applications and are quite expensive. [6]

Information fusion at the matching score level involving user-specific weights and threshold levels could potentially reduce enrollment problems and at the same time improve accuracy of authentication. Furthermore, it is obvious that the simultaneous acquisition of multiple biometric identifiers makes it a lot harder for an impostor to spoof the system by presenting artificially created samples. However, these benefits are not free, as multibiometric systems are less cost-effective and they have significant effects on their users. Some of these could, in fact, lead to reduced user acceptance. Privacy issues and the inconvenience of multilevel data acquisition might cause acceptance problems. Many promising architectures for multibiometric systems are still at an experimental stage. Currently available multibiometrics are mainly layered, featuring only loose coupling between the different subsystems, sometimes even with different user interfaces.

Biometric systems using a single biometric trait have to contend with noisy data, restricted degrees of freedom, failure-to-enroll problems, spoof attacks, and unacceptable error rates. Multibiometric systems using multiple traits of an individual for authentication alleviate some of these problems while improving verification performance. Performance of multibiometric systems can be further improved by learning user-specific parameters.

The long-term vision for multimodal biometrics is to take capture units from different vendors and manufacturers and combine them in a simplistic manner. However, there are many barriers to this ideal. Primarily, the combination of individual biometric vectors into a coherent multimodal vector is not a simplistic task. It is dependent on proprietary, extraction algorithms. In addition, there is currently no standard biometric data exchange format that manufacturers can reliably assume data will be provided.

Analysis

Biometrics is in the news because of the national identity card debates occurring in some countries; biometrics are found in people's pockets because of the transition to electronic passports around the world. Biometrics has been a favorite in the movies for many years, and consumers can go to the Microsoft Web site and buy a keyboard with a built-in fingerprint reader. To understand how to develop strategies to make the most of the rapidly advancing technology of biometrics, biometrics must be viewed in the right context.

The biometrics industry is set to make major improvements in many areas requiring secure identification (fingerprint entry to homes and cars, travel and maternity units, for example). Biometrics is the single most convenient means of identification; everyone possesses this, and cannot forget it or leave it behind. However, biometric technology has a long and rough journey ahead of it to gain acceptance. It is not a panacea, as it is often portrayed by Hollywood. Any high-profile mistakes are likely to be quickly vilified by those who oppose its use and the propagation of what is seen as a "1984" Orwellian state.

Fused biometrics may make identification more secure and more accurate than single biometric systems. How can a compromised fingerprint be replaced or overcome? Biometric fusion could help by not relying on a single biometric, as could weighting or relying more heavily on those biometrics that are not compromised. Building such systems is not straightforward, however. How does one combine the multidimensional feature vectors of iris, fingerprint, and facial recognition systems?

It is more likely that a hybrid system of traditional digital security and biometrics—consisting of unimodal in the short term and progressing to multimodal in the future—will prevail. Digitally signed biometric tokens and/or databases, crossed-checked with the user, require a fraudster to not only spoof a biometric but also to break the encryption or tamper-resistant enclosure used to store the biometric template.

So is fusion required if current security techniques and unimodal biometrics can overcome security barriers? Even very good identification biometrics is imperfect on a large scale. A good system will deliver a "false reject" rate of 0.4 percent. This is enough to annoy a regular user, or more importantly, to encourage customers to flood call centers with complaints. Biometric fusion is slowly but surely showing that these FRR can be driven down; providing, in turn, a more user-friendly experience.

While biometrics will endure and improve verification procedures, a thought should be spared for a world where other areas of a security system, rather than the identification process, become the weak link and, therefore, are more likely to become the target of attack. One consequence to be considered is since biometrics is the one verification method that cannot be forgotten or left behind and multimodal biometric systems are decreasing the likelihood of spoofing, the individual becomes the weakest link. While biometrics are touted as being able to vastly increase authentication procedures, such as entry to secure areas, will they correspondingly increase attacks on one's person—such as kidnapping—to enable forceful use of one's iris pattern?

Biometrics will succeed only if it:

- can prove positive identification in a manner that the user finds nonintrusive, trustworthy, and secure
- is used in combination with other security methods and not seen as a universal solution

Summary

- Fusion is the union of different biometric modalities resulting in an integrated system. Multimodal biometrics, or biometric fusion, is the process of combining information from multiple biometric readings, like face images and voice timber, before, during, or after a decision has been made regarding identification or authentication from a single biometric.

- Multisensory refers to combining data from two or more biometric sensors, such as synchronized reflectance-based and temperature-based face images. A multimodal biometric system is defined as a biometric system in which two or more of the modality components—such as biometric characteristic, sensor type, or feature extraction algorithm—occur in multiple components.
- Multibiometrics encompasses a multimodal biometric system, which integrates face recognition, fingerprint verification, and speaker verification when making a personal identification. Multifactor authentication involves the use of multiple techniques to authenticate an individual's identity. This usually involves combining two or more of the following factors: something possessed, something known, or a personal physiological attribute.
- The number of multimodal biometric systems in the production or developmental phase is limited. The major systems include MegaMatcher and Instituto Federal Electoral.

Key Terms

Biometric fusion The process of combining information from multiple biometric readings, either before, during or after a decision has been made regarding identification or authentication from a single biometric.

Fused biometrics Single or multiple sensors are used to collect different biometrical information, such as a face image and a fingerprint image.

Fusion The union of different biometric modalities resulting in an integrated system.

Multiapplication card Consists of a smart card that runs multiple applications using a single card.

Multibiometrics Encompasses a multimodal biometric system, which integrates face recognition, fingerprint verification, and speaker verification when making a personal identification.

Multifactor authentication Involves the use of multiple techniques to authenticate an individual's identity.

Multifactor reader A smart card reader that includes a PIN pad, biometric reader, or both to allow multifactor authentication.

Multimodal biometrics Fusing two or more biometric modalities, like face images and voice timber.

Multimodal biometric system A biometric system in which two or more of the modality components, such as biometric characteristic, sensor type, or feature extraction algorithm, occur in multiple components.

Multisensory biometrics Refers to combining data from two or more biometric sensors, such as synchronized reflectance-based and temperature-based face images.

Multitechnology card An ID card possessing two or more ID technologies that are independent and that don't interact or interfere with one another.

Multitechnology reader A card reader/writer that can accommodate more than one card technology in the same reader.

Mutual authentication The process employed for the smart card–based device to verify the reader is authentic.

Neural network A type of algorithm that learns from past experience to make decisions.

Review Questions

1. ________________ is the union of difference biometric modalities resulting in an integrated system.
2. ________________ refers to combining data from two or more biometric sensors, such a synchronized reflective-based and temperature-based face images.
3. ________________ biometrics means fusing two or more biometric modalities, like face images and voice timber.
4. A ________________ is a type of algorithm that learns from past experience to make decisions.
5. A ________________ consists of a smart card that runs multiple applications using a single card.
6. A ________________ is an ID card that has two or more ID technologies that are independent and do not interact or interfere with one another.
7. A ________________ is a smart card reader that includes a PIN pad, biometric reader, or both to allow multifactor authentication.
8. A ________________ is a card reader/writer that can accommodate more than one card technology in the same reader.
9. ________________ is the process employed for the smart card–based device to verify the reader is authentic.
10. Multibiometric systems are categorized into three system architectures according to the strategies used for information fusion. These include fusion at the ________________, ________________, and ________________.
11. Fusion at the matching-score level includes a variety of methods that may be used, including ________________, ________________, and ________________.
12. Based on the proximity of feature vector and template, each subsystem computes its own matching score using ________________, ________________, and ________________ strategies.
13. ________________ are those characteristics that provide some information about the individual, but lack the distinctiveness and permanence to sufficiently differentiate any two individuals.
14. ________________ solution allows for storage and matching of fingerprints, iris, and facial templates directly on a single smart card.
15. The MegaMatcher system includes these two biometric modalities: ________________ and ________________.
16. The Mexican election system includes these two biometric modalities: ________________ and ________________.
17. ________________ is a concept pertaining to the sense of touch, from the Greek word *haptein*, which means "to grasp."

18. The accuracy of a biometric system can be measured from two statistical measurements, namely ________________ and ________________.

19. The relationship between FAR and FRR is ________________ between security and convenience.

20. The degree of reliance on each of the biometric traits in order to maintain an acceptable balance between confidence and convenience for its users and operators is called ________________.

Discussion Exercises

1. Before considering multimodal biometrics it is important to understand the core features of a conventional biometric system. Identify these four components.
2. Identify the five ways that multimodal biometric systems can be designed.
3. Describe advantages and disadvantages that can occur when implementing multiple biometric systems.
4. Identify the two overall limiting factors affecting multimodal biometric systems.
5. Describe the flow of the match-on-card process. Produce a flow graphic.
6. Identify unimodality biometrics that could be combined into multibiometric solutions.
7. Describe how a multifactor authentication can be used to authenticate someone's identity.
8. Provide an overview of a neural network and identify a production system.
9. Research the ISO/IEC 14443 and 15693 documents and provide an overview of these standards.
10. Identify multibiometric and multimodal systems that have been implemented successfully. Provide an overview of the features.

Hands-On Projects

6-1: Multibiometric System Solution Awareness

Time required—30 minutes

Objective—Identify multibiometric system solutions that have been implemented in any organization.

Description—Use various Web tools, Web sites, and other tools to identify multibiometric system solutions that are in development or production for both commercial and government organizations.

1. Research the Web and technical magazines for multibiometric systems. Provide an overview of the systems.
2. Research trade journals and technical magazines for biometric solutions that have been integrated into legacy computer systems. Describe the issues encountered in the integration.
3. Develop a comprehensive multibiometric or multimodal biometric solution using a stand-alone or host computer system. The solution can involve the identification or verification function.

6-2: Compare Multibiometric System Solution Requirements

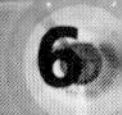

Time required—45 minutes

Objective—Research multimodal and multibiometrics solutions and identify the system that can provide some level of access control and security for OnLine Access assets and resources.

Description—The owners and managers of OnLine Access have identified hardware, software, and network components required to implement a biometric access control solution. They must look at the multimodal and multibiometrics solutions and identify the various requirements to implement a biometric solution. The advantages and disadvantages of these various biometric solutions must be identified.

1. Research the topics of unimodal, multimodal, and multibiometrics for access control and security. Identify other organizations that have an initiative to migrate to one of these biometric solutions.
2. Produce a matrix describing hardware, software, and network requirements and the potential cost for each option.
3. Look at the various search engines for hardware, software, network, and other information that could impact an implementation for each biometric solution.

chapter 7

Policy and Program Management

After completing this chapter, you should be able to do the following:

- Provide an overview of legacy corporate security policies, identity (ID) management, and system program management
- Become familiar with biometrics policies and procedures, methods assessment, and security and business ethics
- Understand the biometric information security management system operation
- Look at the features and functions of a biometric ID management system
- Understand the issue of governance in the IT environment
- Identify the necessary steps to improve cyber security

Organizations have a tendency to purchase automated systems without any idea of their daily operational requirements. Marketing representatives often do not possess any skills in the operations area; their major emphasis is on the sale. Buyers must beware if the product sales team consists of only the salesperson! Security and network administration managers must identify all the members who represent the product providers who can be expected to support the biometric solution implementation effort. Project management is mandatory in biometric solution procurements and implementations. Experience has shown projects tend to fail if project management is inadequate or non-existent.

Introduction

A **policy** is typically a document that outlines specific requirements or rules that must be met by all participants. In the information/network security realm, policies are usually point-specific and cover a single area. For example, an "acceptable use" policy would cover the rules and regulations for the appropriate use of computing facilities.

A **standard** is typically a collection of system-specific or procedural-specific requirements that must be met by everyone. For example, a computer center might have a standard that describes how to harden a Windows NT workstation for placement on an external (DMZ) network. Users would be required to follow this standard exactly if they needed to install a Windows NT workstation on an external network segment.

A **guideline** is typically a collection of system-specific or procedural-specific "suggestions" for best practice. These are not requirements that must be met, but are strongly recommended. Effective security policies make frequent references to standards and guidelines that exist within an organization. [1]

Biometric systems security and the management of biometric information security must become integrated into an organization's overall information security management program. This system should be based on policy defined to meet the business objectives of the organization, and a risk-based approach should be used to select and impose proper controls and monitor their effectiveness.

Biometric-based systems can range from moderately to highly complex. As with any complex software-based system, the assessment methodologies must be developed so they can address the nonfunctional requirements, such as availability, reliability, integrity and confidentiality, maintainability, etc. Even if a biometric-based system could provide perfect assurance of integrity and confidentiality, if it is rarely available, it is useless.

As organizations deploy a single ID management solution in an attempt to move away from retrofits, it is important that they look to a solution that integrates with existing system management technologies. This leads to reduced administrative costs and allows the effectiveness of the system to be constantly managed within the existing critical systems infrastructure. ID management is now entering mainstream utility within leading organizations. Biometric algorithms have reached the levels of accuracy that are required, and the cost-per-user has declined significantly in recent years, yielding a form of authentication that is more cost-effective and secure than in the past.

Misuse of personal information, including identity theft, has become more of a threat as information technology—including electronic commerce—has become ubiquitous. Where biometrics is authenticated remotely, for instance, by transmission of data from a sensor to a centralized data repository, a hacker could steal, copy, or reverse-engineer the biometric. This misappropriation could also come about through insider misuse. Without solid safeguards, the fear is that files could be misappropriated and transactions could be performed using other people's identities.

Boards and executive management have long known the need for enterprise and corporate governance. However, most are beginning to realize that there is a need to extend governance to information technology as well, and provide the leadership, organizational structures, and processes that ensure that the enterprise's IT sustains and extends the enterprise's strategies and objectives.

Corporate Security Policy

A corporate security policy is the gateway to a company's intellectual property. In today's world of information technology, the main threat to information security within a company is its employees. Employees are behind the firewall; furthermore, they have a user name and password on the network. A security policy should therefore be designed to explicitly list the dos and don'ts in the network. A security policy should serve as the company's constitution, governing how employees use the network and take care of both internal and external security issues. Policies should be well planned and periodically updated to reflect the organization's ever-changing challenges and the continuous evolution in the world of technology.

The overall program policy is usually a compilation of issue-specific policies. These policies are used to address specific parts of the network, such as the Internet and e-mail, antivirus programs, data backup and recovery, and software security. These issues could either be addressed in separate documents or embedded into the security policy. The team responsible for writing this policy should work in conjunction with upper management and legal counsel. Upper management's involvement in the design of the policy is important because this will also help in the acceptance and implementation of the policy. Although certain parts of the policy might be confidential, most of it should be made accessible to everyone within the organization, including business partners. In fact, some business partners now require a copy of the data-recovery policy to assure their data is safeguarded properly.

In developing a policy there are certain aspects of network security that should be taken into consideration. These tasks involve identification of assets, risk assessment, management of access to information, data backup and restoration, incident handling, and maintenance.

Identity Management

ID management is a broad administrative area that deals with identifying individuals in a system, such as a country, network, or enterprise, and controlling their access to resources within that system by associating user rights and restrictions with the established identity. The driver licensing system is a simple example of ID management. Drivers are identified by their license numbers and user specifications, such as restrictions, and are linked to the identifying number. Identities may manage themselves or other parties may manage them. These other parties may include private parties (e.g., employers or businesses) or public parties (e.g., personal record offices and immigration services).

Electronic Identity Management

Several interpretations of ID management have been developed in the IS/IT industry. Computer scientists now associate the phrase, quite restrictively, with the management of user credentials and the means by which users might log on to an online system. The focus on ID management reflects back to the development of directories, such as X.500, where a namespace serves to hold named objects that represent real-life "identified" entities, such as countries, organizations, applications, subscribers, or devices. The X.509 ITU-T standard defined certificates that carried ID attributes as two directory names: the certificate subject and the certificate issuer. X.509 certificates and public key infrastructure (PKI) systems operate to prove the online "identity" of a subject. Therefore, in IS/IT terms, one can consider ID management as the management of information held in a directory that represents items identified in real life, such as users, devices, services, etc. The design of such systems requires explicit information and ID engineering tasks.

The evolution of ID management follows the progression of Internet technology closely. In the environment of static Web pages and static portals of the early 1990s, corporations investigated the provision of informative Web content, such as the "white pages" of employees. Subsequently, as the information changed, due to employee turnover, provisioning, and deprovisioning, the ability to perform self-service and help-desk updates more efficiently morphed into what has become known as ID management today. Typical ID management functionality includes the following:

- User information self-service
- Password resetting
- Management of lost passwords
- Workflow
- Provisioning and deprovisioning of identities from resources

ID management also addresses the age-old problem: that every new application may require new data stores of users to be set up. The ability to centrally manage the provisioning and deprovisioning of identities and consolidate the proliferation of ID stores form part of the ID management process.

System Program Management

Organizational management is responsible for overseeing the design and implementation of department-wide management systems. The office's responsibilities are to:

- Oversee the design, development, and implementation of the system and the development of life-cycle and budgetary plans
- Monitor milestones, schedules, and budget expenditures
- Meditate and coordinate system activities throughout all levels of the unit
- Ensure that the system complies with applicable accounting concepts and standards, as well as all accounting policies and procedures
- Ensure that business requirements are met, the future direction of the initiative is consistent with unit planning, and the status of the project is appropriately communicated to internal and external organizations

- Oversee a comprehensive program of change management that includes addressing department communication, training plans, and human resource issues
- Coordinate with workgroups to maximize the input from the cross-functional areas of units into the implementation process
- Oversee all risk management plans to ensure that risks to the Program are identified and effective mitigation strategies developed

Administration and development responsibilities include:

- Developing uniform business rules, data standards, and accounting policy and procedures in support of financial systems implementations; ensuring the development of ongoing accounting policy that further supports the consistent development and implementation of these systems
- Developing department-wide policies and standards for financial and mixed financial systems
- Providing for the establishment of department-wide financial definitions and data structures
- Providing for the administration of a data integrity and quality control program to ensure compliance with applicable directives, departmental financial systems policy, and automated financial data exchange requirements
- Overseeing and monitoring existing department-wide and component accounting and financial management systems

Biometric systems security and its management must become integrated into the organization's overall information security management program. This system should be based on policy defined to meet the business objectives of the organization, and a risk-based approach should be used to select and impose proper controls and monitor their effectiveness.

Biometrics Policies and Procedures

Policy establishes the foundation, mandate, and authority for any organization and is an essential component to deploying biometrics effectively. The following policy documents outline the necessary roles and responsibilities of a policy for biometrics. Policy content will include privacy, planning guidance, acquisition, enterprise organization capabilities (collection, storage, use, access, and retrieval), and outreach and community education. Policies that impact the use of biometrics in an organization are listed below.

- Policy on biometrics
- Executive agent policies
- Information assurance policies
- Security/standards policies
- Acquisition policies
- Other general policies

The success of the new *Biometrics Task Force (BTF)* will directly correlate to the combination of effective technology and effective policy. The BTF is charting a course toward becoming an effective office that will direct and lead biometric efforts. The BTF will:

- Act as the DOD (Department of Defense) proponent for biometrics
- Lead in the development and implementation of biometric technologies for combatant commands, services, and agencies
- Deliver capabilities in order to contribute to the enhancement of the biometric community
- Increase joint service interoperability
- Empower the war fighter by improving operational effectiveness on the battlefield

Biometric Methods Assessment

Biometric-based systems can range from moderately complex to highly complex. As with any software-based complex system, the assessment methodologies must be developed so they can address the nonfunctional requirements, such as availability, reliability, integrity and confidentiality, maintainability, etc. Even if a biometric-based system could provide perfect assurance of integrity and confidentiality, if rarely available, it becomes useless. If improperly engineered, such that its maintenance and modification are difficult, the biometric system may prove to be a bad investment. In case of reliability assessment, false positives (verifying the signature of a wrong person) can have more serious consequences than false negatives. Nevertheless, false negatives occurring excessively could have a negative effect on system availability. These and similar problems have been addressed in the area of dependable computing, without focusing on the biometrics specifics. [2]

During the past several years, biometric devices and applications have experienced substantial growth in the United States and abroad. Total sales of biometric hardware, excluding sales to law enforcement and hardware integration revenue, have increased significantly. Propelling the expansion and use of applied biometrics is a combination of the falling cost of biometric devices, increasing sophistication of the technology, development of biometrics as a peripheral to common computer platforms, and efforts by the U.S. government. For instance, U.S. biometric identification and verification device sales are expected to increase in the near future due to various biometric initiatives. Furthermore, the average end-user price for a stand-alone physical security device utilized for ID verification, without installation, has increased. In addition, advancements in computer technology have meant that more personal computers have the processing power required to run biometric applications.

Biometric Security and Business Ethics

A variety of ethical concerns with biometric identification methods have been registered by users:

- Some biometric identification methods, like retinal scans, are relatively intrusive.
- The gathering of biometric information, like fingerprints, is associated with criminal behavior in the minds of many people.

- Traditionally, detailed biometric information has been gathered by large institutions, such as the military or police, and people may feel a loss of privacy or personal dignity.
- People can feel embarrassed when rejected by a public sensor.
- Automated face recognition in public places could be used to track everyone's movements without their knowledge or consent.

There are also many questions about how this data will be stored and used. This data consists of easily moved and duplicated electronic information, which requires security precautions and processes. A number of questions must be answered:

- How will masses of biometric data be stored and how will this information be safeguarded?
- Who will have access to this information?
- Will companies be allowed access to face biometrics, letting them use security cameras to positively identify customers on a routine basis?
- How would customers feel about walking into a store which they have never visited before, only to be greeted by name by a sales associate who has just read a summary of all of their recent purchases?

Biometric Information Security Management System

A **biometric information security management system** (ISMS) includes the biometric policy, security practices, operational security procedures, organizational structure, assigned responsibilities, and resources needed to protect biometric assets. Independent reviews of security practices should ensure that they are consistent with the biometric policy, and that adequate and effective controls are in place.

Formal mechanisms should be used to document and report biometric system events, including system malfunctions and other security incidents. Incident management results and metrics gathered from biometric event journals should be used in the review process to cause controls to be reevaluated over time. One goal of adopting an ongoing, systematic approach is for the continuous improvement of the biometric ISMS.

Biometric ISMS Standard

An information security management program to protect biometric assets is a prudent business practice that helps an organization identify and manage risk. A new biometric security standard, **ISO 19092**, provides a technology-specific extension to the ISO/IEC 17799/27001 Code of practice for information security management. Though developed in the financial services, ISO 19092 is a general-purpose standard that can be used by any industry that employs biometric technology as a policy-based authentication mechanism.

ISO 19092 defines core requirements for managing and securing biometric information for all applications and environments where biometric information is used. These requirements apply to the transmission and storage of biometric information. Validation of a biometric system relies on maintenance of a secure biometric event journal that can be used for legal and regulatory compliance and ISMS audit.

Core requirements can be met using physical protection when all biometric system components reside within the same tamper-resistant unit and there is no transmission of biometric information. Outside of this environment, requirements can be met using cryptographic mechanisms, such as a digital signature and encryption, to protect biometric information.

The standard defines control objectives and security controls that can be augmented or modified to meet the specific needs of an organization. Its protocols, security techniques, and message syntax can be applied by software applications to automate enforcement of biometric information security policy. By binding the biometric security policy and practices identifiers to the biometric reference template, applications using biometrics can transform biometric information into policy-based management action.

Biometric Policy

A **biometric policy** (**BP**) is a set of rules that indicate the applicability of a biometric reference template for use by some community or class of applications that have common security requirements. Each BP is a document identified by a globally unique name, an information object identifier assigned by the organization. BP identifiers can be bound cryptographically to a biometric reference template using a digital signature.

This cryptographic binding makes it possible for software applications to recognize and enforce the biometric security policy appropriate for a given template. A BP identifier may indicate that a template may be used to control access to customer or employee information and assets, or to authenticate a person acting in some role or engaging in some restricted activity. A BP identifier may indicate security requirements and conditions under which a biometric may be used, such as only with a valid pass phrase.

ISO 19092 defines a schema for representing BP information, and specifies cryptographic processing for binding this information to a biometric reference template. The BP information can be carried explicitly in a template by using a biometric policies extension, or implicitly associated with a template as an authenticated attribute that is bound to, but detached from, the template. In either case, a digital signature is used to bind BP information to a biometric reference template.

The BP schema defines a biometric security policy as a series of one or more biometric policy objects. Each object contains the biometric policy identifier assigned to a biometric security policy document by an organization. When more than one BP is carried in a biometric policies extension, it is appropriate to use the associated biometric template according to any of the policies included in the extension.

The biometric security policy object may optionally include the hash of the policy document and a *uniform resource identifier (URI)* that points to the document or to additional information. Including the hash and URI in the policy object provides using security applications with a network policy resource that allows policy information to be made available to relying parties. When bound to biometric reference templates, the BP will be included in biometric event journal entries and available for the incident management and Biometric ISMS reviews.

Biometric Practices

A **biometric practice statement** (**BPS**) describes the security practices that an organization follows during the biometric reference template life cycle, including business, legal, regulatory, and technical considerations. Each BPS document is identified by a globally unique name, an information object identifier assigned by the organization. Like BP identifiers, BPS identifiers can be cryptographically bound to a biometric reference template, making it possible for software applications to recognize and enforce the biometric security practices associated with a given template. While a BP is often a single, public document that consists of only a single page, BPS documents are not public and are less abstract. BPS documents are general purpose and more technical in nature. They are technology neutral, but may be applicable to specific biometric technologies or limited to well-bounded systems and applications within the organization.

Biometric ISMS Control Objectives

ISO 19092 defines biometric ISMS control objectives that can serve as the criteria for the evaluation and audit of a biometric security system. These criteria represent recommended practices for the operation and technical use of biometric technology within an organization. An existing ISMS may already specify some of the security policies and practices needed to meet these objectives.

Three types of control categories are identified in ISO 19092: biometric information life cycle, environmental, and key-management life cycle. The security of a biometric system implementation that resides within or relies upon an information and communication technology (ICT) infrastructure requires environmental controls. If the security of the implementation depends on cryptographic protection, key management life cycle controls are necessary.

The life cycle of biometric information in a biometric system is analogous to the life cycle for digital ID certificates. Individuals are enrolled in the system by the capture of their biometric data, and this data is used to generate a biometric reference template, which is then distributed, used, and eventually terminated and archived.

A biometric information security management program should establish control objectives for the biometric information life cycle. These control objectives should cover the biometric enrollment process, the biometric authentication process, the biometric reference template life cycle, and the biometric device life cycle.

Organizations should establish and maintain controls to ensure that the biometric enrollment process properly identifies enrollees and reliably authenticates their claimed identities. Controls should ensure that only authorized enrollments occur, and that the biometric data of the enrollee is accurate and complete.

The biometric authentication process includes identification of individuals and verification of a claimed identity. Both processes must be performed securely and in accordance with parameters specified by the biometric security policies and practices of the organization.

Throughout their life cycle, controls must be in place to ensure that biometric reference templates are properly handled. They must be securely and properly generated, then validated prior to their distribution and use. Reference templates must be securely transmitted and stored, and properly terminated. These security requirements must cover any period in which a template may be archived after termination and no longer in active use.

Controls must be in place to ensure that biometric devices are properly handled throughout their life cycle. Only authorized individuals may be allowed access to biometric devices, and operational security procedures must be in place to ensure that biometric devices function properly. The settings of biometric devices should be established by security practices and monitored using metrics from the biometric event journals.

When biometric technology is added to an ICT infrastructure managed by an existing ISMS, sufficient environmental controls to protect the security of biometric systems may already be in place. These can be applied to the protection of biometric assets. For environmental controls, a given implementation may require control objectives and control criteria for:

- The information security policy and infrastructure
- Biometric asset classification and management
- Personnel security management
- Physical security
- Operations and systems access management
- Systems development and maintenance
- Business continuity management
- Legal and regulatory monitoring and compliance
- Journaling of significant environment events

Some biometric technology must be protected by cryptography. If this technology is added to an infrastructure reliant on cryptographic techniques and managed by an existing ISMS, sufficient key management life cycle controls may already be in place, and these can be applied to the protection of biometric assets. For key management life cycle controls, a given implementation should consider control objectives and control criteria for: [3]

- Cryptographic key generation
- Storage
- Backup
- Recovery
- Distribution and usage
- Destruction and archival
- The life cycle of cryptographic devices

Biometric Identity Management

With rising security and fraud issues in day-to-day operations, every organization is looking to increase the levels of accountability and security among its employees, partners, and customers. This is true not only in the general enterprise, where employees need secure access to their desktops, but also in airports, law enforcement, and utilities, where reliable employee credentialing is critical in securing physical and logical access to resources and facilities. The most effective way to achieve this is to centralize an organization's ID management function in a single place so it can be effectively managed and the appropriate level of trust can be maintained in the authentication process.

A secure and effective method of authenticating an individual involves the verification of a unique and personal characteristic is a biometric. This is sometimes done in conjunction with a PIN or token, also known as multifactor authentication. Management of this biometric information, including its registration, storage, protection and verification, is known as *biometric identity management (BIM)*.

Often, BIM technology is evaluated from a narrow perspective: "Does it authenticate a small set of users in a controlled environment?" Although this approach may be sufficient to meet small, narrowly defined requirements, it is not sufficient where ID management needs to expand beyond the first implementation. This approach may lead either to poor implementations that do not move forward or to no implementation at all, where the solution does not address the need for future development. A properly implemented large-scale BIM system will allow an organization to enhance security, streamline processes, and significantly reduce its *total cost of ownership (TCO)* for IS/IT systems.

As previously described, ID management is the registration, storage, protection, issuance, and assurance of a user's personal identifier(s) and privilege(s) in an electronic environment and in a secure, efficient, and cost-effective manner. Previously, application providers have had to deal with the problem of identity and credential management as a result of insufficient ID management functionality within the operating system infrastructure or a complete lack of standards across various target platforms. This led most organizations to deploy various retrofits, where each application had its own ID management function built into the core system. However, supporting and maintaining the various retrofit systems creates a high cost of ownership for an organization. This is about to change, as more and more application providers move to a model in which they rely on ID management vendors to provide a single core application to credential and authorize users.

ID management is evolving in the same manner as data management did in the past. Today, few companies that are developing highly available, scalable systems are building their own database infrastructure. There are many reasons why ID management should be no different. ID management is not the core business of application providers. For example, SAP does not want to build authentication systems; it would rather focus on business process management workflows. In addition, having an external, trusted ID management provider allows organizations to provide the greatest flexibility and system security. Business systems should not be concerned whether the authentication of the individual is done through fingerprints, voice patterns, iris scans, or smart cards.

As complexity in IT infrastructure grows, every organization has difficulty implementing and managing various disparate ID management systems. This is a key problem for many organizations. To address this "organizational identity chaos" problem, organizations need a single authentication infrastructure that offers a centralized, policy-based approach to solve this problem and ensure that they not only have a single point of control, but also cost-savings gains.

BIM System Features and Functions

Using a biometric identifier to authenticate employees, partners, or customers/consumers allows an organization to have the highest level of assurance in the other participants' authenticity. There are many issues that should be understood by any organization deploying this kind of solution.

BIM is concerned with the large-scale, proper management of the biometric identities for an enrollment population. A narrow view of this is traditionally based on enrollment and authentication. Users are enrolled into a system and are authenticated, which satisfies the

ID registration and assurance functions. However, when the requirements for a large-scale BIM solution are segmented into a set of functional groups, it becomes apparent that a proper BIM system must be much broader in scope. The key is to deploy a solution that provides functionality to the organization in each of these seven areas:

- ID registration
- ID storage
- ID assurance
- ID protection
- ID issuance
- ID life cycle management
- System management

Identity Registration

ID capture consists of three main functions that facilitate the inclusion of users and their biometric data into the system: enrollment, quality capture of the biometric, and population coverage.

- **Enrollment:** A proper enrollment process is a key feature every system must provide to ensure a successful deployment. Organizations must understand the necessity for a good, managed enrollment process. A weak process will lead to system inaccuracies and an unreliable authentication infrastructure. Ideally, a good enrollment process is one in which the credentials of the user are properly ascertained at the time of enrollment. To drive quality enrollment, both the person enrolling in the system (enrollee) and the person asserting the validity of the credentials (enroller) should biometrically sign the enrollment record. Experience has shown that an enrollment process in which neither party can repudiate its participation in the transaction is the best way to drive quality of data and maintain a robust process.
- **Quality capture of the biometric:** An important consideration at enrollment time is the quality capture of the biometric. Every biometric capture activity has differing levels of quality. With fingerprints, for example, quality can vary depending on the condition of skin (oily or dry), dirt on the finger, dirt on the sensor, ambient temperature, etc. Enrolling a poor quality image yields a less accurate result. Efforts should be made to ensure that the highest quality enrollment images are captured. Generally, well-thought-out biometric systems employ a range of techniques to ensure that the best image is captured. This includes the enrollment of multiple fingers or iris images and the capture of multiple images (of each finger or iris) at enrollment. Ideally, for finger-based enrollment, the system should be capable of enrolling all 10 fingers.
- **Population coverage:** The ability to address all of the relevant population is also a key requirement. No single biometric can offer 100 percent coverage across all populations. The measure of a biometric implementation's ability to address the intended population is known as the failure to enroll (FTE) or failure to acquire (FTA).

Identity Storage

The ID storage function identifies some of the storage characteristics of a large-scale ID management system. Primarily, a large-scale ID system requires a data storage technology that can keep pace with the size of the enrollment population. This implies the use of a terabyte-capable relational database, such as IBM DB2 or Oracle.

Any BIM system must also support the backup, archiving, and restoration of information. The system should integrate into existing backup management solutions. This requirement is another reason to deploy a system only if it uses an industry-standard storage technology. The use of an industry-standard storage technology also lets the organization leverage the knowledge of its existing, highly trained database systems team.

Identity Assurance

This group of functions asserts an individual's identity to other applications within the organization.

Verification is the process of determining an individual's ID based on the presentation of a claim and a biometric in support of that claim. For a given claim, the system matches the presented biometric against the one stored at enrollment and returns the result (either a Boolean value or a score across some predefined range). It is very important that organizations choose a BIM solution that provides robust verification capabilities.

Some organizations have invested heavily in the deployment of PKI solutions. As this is an expensive investment, organizations should look to make a positive return on this investment. A BIM is a natural partner with an existing PKI system, as one key weakness of PKI solutions is they do not authenticate the individual, they simply prove that an individual's private key was used in a decryption or signing process. PKI implementations generally rely on passwords to form the link between the private key and the individual. The vulnerability of password systems is well known and documented and, therefore can be compromised.

Studies have shown employees commit 84 percent of fraud detected in commercial organizations. This includes the sharing or violation of the use of passwords in organizations. Biometric technologies provide the missing link between individuals and the cryptographic keys assigned to them.

The key benefit of biometrics is the ability to ensure the involvement of the individual in a transaction. This prevents individuals from repudiating their participation in a transaction later on. The most useful BIM deployments should provide adequate levels of security across the entire infrastructure and provide integration with PKI in such a way that it is capable of generating standards-compliant digital signatures. The use of a biometric to ascertain the identity of the individual is a strong form of a digital signature.

Authentication, which means validating a claimed ID through the comparison of a presented biometric with a biometric captured at enrollment, is only part of the functionality that a BIM solution should provide to the organization. Integrating authentication with authorization provides far more functionality than authentication alone. This authorization can be provided internally from the biometric ID management system or through integration with an external system.

Multifactor authentication is the combination of a biometric (something physiological) with some sort of token (something possessed). Two types of tokens are generally used: dumb and smart. Dumb tokens, such as magnetic stripe cards, are capable only of carrying a claim of ID or an encrypted biometric. There is no capability in a dumb token to carry out any processing. Furthermore, dumb tokens can be easily copied or duplicated. Smart tokens, on the other hand, are devices that have the ability to carry out independent processing within the token (i.e., smart cards). Smart tokens are secure computers in their own right. Most have the ability to securely store and process information such as biometric templates and cryptographic keys. Implementing a two-factor policy can increase security under the right conditions.

Identity Protection

This function group deals with the protection of an individual's identity and the integrity of the scheme, as encryption of biometric data at the earliest possible time during both enrollment and verification is critical. Furthermore, all participants in the transaction—the client, the authentication server, and the requesting application—should digitally sign their transmissions to ensure component integrity and nonrepudiation. Man-in-the-middle attacks should be guarded against using challenge–response protocols and digital signatures.

It is important that biometric templates and images be protected in any ID management system. Organizations wishing to deploy a solution must ensure that data is encrypted securely. Consider the damage that would be done to any organization should the biometrics of its enrollment database be published on the Internet. There are ways to guard against this happening. Best-in-class deployments use a tamper-resistant *hardware security module (HSM)* to perform all cryptography. Keys are generated on board the HSM itself to ensure effective key generation and management protocols. These keys should never be exported in the clear outside the HSM. The U.S. federal government has published the Federal Information Processing Standard (FIPS) 140 regarding the HSM. Organizations should look for systems that use FIPS 140-accredited HSMs to obtain a level of assurance regarding the security of stored biometrics. Be aware, however, that using an HSM does not necessarily make a system secure. It is how the ID management system is designed and put together that ensures the security of the system.

Privacy is obviously a concern of every individual, and organizations must act appropriately to protect personal privacy if they expect to obtain acceptance from employees, partners, and customers. Measures that should be implemented include giving participants a written guarantee that the organization will not share or divulge their biometric, will implement best-in-class security and management procedures for the system, and will immediately remove any personal information upon the request of the owner.

A secure and complete audit trail must also be maintained for every operation the system performs. No matter how trivial, operations should be logged in a tamper-proof manner for later interrogation, if required. This secure audit trail guards against system administrators being able to tamper with or amend records. For example, the use of secure FIPS-certified HSMs to generate symmetric *message authentication codes (MACs)* on each transaction is desirable, as the keys required for MAC generation are available only within the HSM.

Identity Issuance

Security can be increased further through the combination of multiple authentication factors in a single authentication transaction. This group of functions deals with issuing a credential and possibly biometric identifiers on tokens, such as smart cards.

A solution should support the issuance of various token types. Older tokens, such as magnetic stripe cards, are being phased out by organizations in favor of smart cards (both contact and contactless varieties). A token issuance function should support open standards, such as *PKCS#11*, for management of the token information. The system should also support a range of smart cards to ensure that an organization is not limited to a single card supplier. The issuance of the token should be recorded in the database and a secure audit entry generated. When the smart card contains cryptographic keys, the generation of these keys should be done in a secure manner, such as on the smart card itself or in a secure HSM. Tokens are very often lost, broken, or stolen. When this happens, the ID management solution must be able to securely reissue the token and deliver it to the person requiring it. Audit trail and reporting functions must be available to allow an organization to track token reissuance statistics.

There is also a growing interest in authenticating an individual to a biometric stored on the smart card, as opposed to the central server. This model works in certain situations, such as when it is not feasible to connect a biometric reader to a central server. However, the performance and accuracy of authentication on a smart card is not equivalent to that which can be achieved on a server. Neither is the flexibility of a server-centric, policy-based approach available. Furthermore, the match-at-server model supports easy upgrades to authentication algorithms or the introduction of new modes as required. Management of card hot lists also becomes an issue in offline environments. The authentication terminal/reader is generally not powerful enough to maintain a complete list of all blocked or hot-listed cards. As stated earlier, smart cards have a finite life span; if they are broken, lost, or stolen, they need to be reissued to individuals immediately. Unlike traditional PIN-protected smart cards, in which a new PIN is simply generated, the issuance of biometrically protected smart cards is more complex because of their physiology. To reissue a PIN-protected card does not require any knowledge of the individual's biometric characteristics, nor does it require the user to provide any personal information to the issuance system. The situation changes with biometric cards. The issuance of the card involves the secure storage of an individual's personal biometric data on the card. This requires the proper capturing of this data (enrollment) and the subsequent secure storage (management).

Identity Life Cycle Management

A biometric is unique to an individual, but that does not mean it is static. As people grow older, a biometric element—such as an individual's fingerprints—changes. For example, a person accumulates damage to fingers, such as cuts and scratches, and the brittleness of their skin increases.

Some algorithms support the automatic aging of biometric templates. For example, through a process known as progressive enrollment, each verification operation slightly modifies the enrollment template to allow for aging or cracked/damaged fingers. BIM engines must support this function in situations where the organization has chosen an algorithm that does include progressive enrollment. In situations where progressive enrollment is not supported, an organization must be able to determine which enrollees are in danger of not being accepted by the system (generating a false reject) because the enrollment record is out-of-date compared to the current state of the individual's biometric. To do this, a biometric ID management system must support the generation of reports based on time of enrollment, time of last verification, verification frequency, etc. Armed with this information, the organization can ensure that steps are taken to contact the individual and revalidate the enrollment prior to the expiration of the effective period for the first enrollment expiring.

System Management

As organizations deploy a single ID management solution in an attempt to move away from retrofitted solutions, it is important that they look to deploy a solution that integrates with existing system management technologies. This leads to a reduced cost of administration and allows the effectiveness of the system to be constantly managed within the existing critical systems infrastructure. Organizations must be able to report against the activity of their ID management systems to ensure effectiveness of the system, analyze activity, and spot trends, as well as provide data for conflict resolution, employee disputes, etc. Organizations deploying a large-scale ID management system must ensure that the data in the audit trail is complete and that a method exists to query this information in an efficient manner. For example, audit trail logs should be stored securely and should be accessible through a SQL interface to a relational database. In a large-scale system, searching audit logs in text files is not sustainable.

ID management is now entering mainstream utility within organizations. Biometric algorithms have reached the levels of accuracy required, and the cost-per-user has declined significantly in recent years, yielding a form of authentication that is most cost-effective and secure. This is clear from the explicit requirements in recent legislation requiring the deployment of biometric technologies as well as recent biometric standards adopted by the International Civil Aviation Organization (ICAO). Organizations looking to deploy a BIM solution need to look at the bigger, long-term picture and deploy a centralized, policy management solution that enables multifactor, multimodal, and flexible authentication and authorization policies, while maintaining individual privacy. [4]

Governance

Boards and executive management have long known the need for enterprise and corporate governance. However, most are beginning to realize that there is a need to extend governance to information technology as well, and to provide the leadership, organizational structures, and processes that ensure that the enterprise's IT sustains and extends the enterprise's strategies and objectives.

ISACA recognized this shift in emphasis in 1998, and formed the *IT Governance Institute (ITGI)* to focus on original research, publications, resources, and symposia on IT governance and related topics. In addition to the work carried out by the ITGI, ISACA addresses the topic through a regular column in and occasional dedicated issues of the *Information Systems Control Journal*, conference sessions and tracks, and education courses.

Security Guidelines

Designing and deploying security architectures that incorporate biometrics is not an easy task. Questions must be answered, such as:

- Which is the most suitable biometric technology to protect the specific application?
- What performance and security levels are chosen?
- What policies and procedures will protect the system?
- Will the biometric templates be stored in a central database or in smart cards?
- How will the components of the system communicate with each other, and what technologies can protect the communication channels?

The IS auditors who are responsible for ensuring the existence of adequate security controls during all phases of the systems development life cycle, or for its security evaluation, can deploy general audit techniques. Security-related standards provide excellent guidelines for organizing and ensuring system security, and include:

- ANSI X9.84—Biometric Information Management and Security
- Best Practices in Testing and Reporting Performance of Biometric Devices
- Control Objectives for Information and Related Technology (COBIT)
- ISO 17799—Code of Practice for Information Security Management
- ISO/IEC 27001—Information Security Management Systems
- ISO/IEC 15408—Evaluation Criteria for IT Security

These standards are good starting points for secure systems that incorporate biometrics, with respect to the protection of personal biometric data. However, at some point, IS/IT auditors will be forced to conduct a risk analysis to ensure that security controls implemented by the security administrator are adequate. To complete this task successfully, the auditors will need access to a database of the common risks of biometrics, followed by the corresponding controls. These risks and countermeasures include:

- Enrollment, administration, and system use risks
- Impostor trial and error
- Noise and power loss risks
- Power and timing analysis risks
- Residual characteristic risk
- Similar template and characteristics risk
- Spoofing and mimicry attacks
- Fake template risks on the server side
- Communication links risks
- Cross-system risk
- Component alteration risks

Enrollment, Administration, and System Use Risks

Poor enrollment, system administration, and system use procedures increase the system's risk factor. During the enrollment phase, raw biometric data and biometric templates can be compromised, and databases can be altered or filled with imprecise data. Poor system administration procedures might lead to altered system configuration files with increased FAR, making false acceptance easier and security weaker. Similarly, users might exceed authority, threatening the system. IS auditors should ensure the establishment of detailed procedures and controls as extensions of the system's security policy, forcing, for example, segregation of duties, job rotation procedures, logging facilities, alteration, or anomaly detection mechanisms.

Impostor Trial and Error

This type of attack is based on trial and error. The impostor is continuously attempting to enter the system by sending incrementally increased matching data to the matching function until a successful score is accomplished. This method is most effective in systems that implement identification rather than verification, since the biometric measurement is compared to a great number of templates, making the system weaker (as the number of users increases), due to the increased probability of the existence of similar templates or characteristics among the population. Biometrics, however, is more resistant to this attack than are traditional systems, since the impostor has to insert the data to the system. IS auditors should ensure that traditional controls are in place, such as the automatic locking of the user's account after a specified number of attempts.

Noise and Power Loss Risks

Off-limit power fluctuation or flooding of a biometric sensor with noise data—for example, flashing light on an optical sensor, changing the temperature or humidity of the fingerprint sensor, spraying materials on the surface of a sensor, or vibrating the sensor outside its limits—might cause biometric devices to fail. IS auditors should ensure security policies that incorporate security controls will make the system environment as controlled as possible. These controls depend on the nature of the application.

Power and Timing Analysis Risks

Capturing the power consumption of a chip can reveal the software code running on the chip, and even the actual command. Simple power analysis and differential power analysis techniques are deployed for such purposes and are capable of breaking cryptographic algorithms, such as DES, by using statistical software. The same strategy can be followed for breaking the matching mechanism of the biometric system. The secret key or biometric template will appear as the peaks of a diagram projecting the result of applying the appropriate software to the power consumption measurement. Timing attacks are similar and measure the processing time instead of the power consumption.

IS auditors should ensure that all necessary controls are in place. These include the use of microcontrollers with lower power consumption and noise generators for power blurring. Regarding timing attacks, the algorithm and program code have to be designed as time-neutral. These technological countermeasures must be included in the biometric system.

Residual Characteristic Risk

The residual biometric characteristic of a previous user on the sensor may be sufficient to allow access to an impostor. The attack can be realized on a fingerprint sensor with a residual fingerprint from the previous measurement by pressing a thin plastic bag of warm water on the sensor; by breathing on the sensor; or by using dust with graphite, attaching a tape to the dust and pressing the sensor. Even when a specific rule in the login algorithm is in place for declining exactly the same measurement, repositioning the tape to provide a slightly different input would deceive the system. IS auditors should conduct a technology assessment. Some fingerprint sensors require the user to sweep the finger on the sensor, thus avoiding this risk. In general, deploying interactive authentication is an adequate control for this type of risk.

Similar Template and Characteristics Risk

A fraudulent user who has a similar template or characteristic to a legitimate user might deceive the system, especially in identification applications where there is a one-to-many template comparison. IS auditors should ensure the maturity of the encoding algorithm, which should deploy functions that produce unique outputs from different inputs, as well as the FAR of the biometric device. Auditors should examine the results of independent testing based on biometric performance evaluation standards or rely on the evaluation assurance level (EAL) of a certified product by the common criteria. For security applications, the biometric system should be calibrated to produce a FAR less than or equal to 0.001 percent.

Spoofing and Mimicry Attacks

Poor biometric implementations are vulnerable to spoofing and mimicry attacks. An artificial finger made of commercially available silicon or gelatin, can deceive a fingerprint biometric sensor. Pictures and speech synthesis tools can deceive face and voice recognition systems. Auditors must ensure that vitality detection features, such as the relative dielectric constant, conductivity, heartbeat, temperature, blood pressure, detection of vitality under the epidermis, or spontaneous dilation and constriction of the pupil, are integrated in the biometric device. If these features are not present, compensating controls must be applied, such as the deployment of multimodal biometrics (e.g., combination of face and lips movement recognition) or the implementation of interactive techniques (e.g., the request for the user to say a specific phrase).

Fake Template Risks on the Server Side

Server-based architectures, where the biometric templates are stored centrally, foster the threat that an impostor can insert a template in the system under someone else's name. Distributed architectures (template storage in a smart card) are preferred. In that case, the template is stored in a tamper-resistant memory module that is write-once and erased or destroyed if altered. When this scenario is not an option, strong security controls must protect the server, including encryption of the templates, system and network security controls, and a strong security policy followed by detailed procedures based on the standards mentioned previously.

Communication Links Risks

Data could be captured from the communication channel, between the sensor and the feature extractor, the feature extractor and the matching algorithm, and the matching algorithm and the application to be replayed at another time to gain access. This is also called electronic impersonation. During system development, the IS auditor should request that the parts of the system be integrated into a hardware security module, decreasing this type of risk. An example is the biometric smart card that has an embedded fingerprint sensor and matching mechanism. Similar security levels are addressed in integrated terminal devices, such as personal digital assistants (PDAs) or mobile phones. If this is not an option, challenge and response can address this risk. An additional control is the introduction of a rule to discard a signal when it is identical to the stored template or to the last measurement that was conducted.

Cross-System Risk

Using a template in two or more applications with different security levels tends to equalize security levels by decreasing the higher security level to the lower one. Depending on the criticality of application, IS auditors should verify the deployment of custom encoding algorithms to ensure the creation of custom templates per user for each application. Another option is to combine existing biometric-encoding algorithms with one-way hash functions to ensure that templates produced for a specific user in the specific system are unique. This feature also provides for revocation to the system in case an impostor compromises a template.

Component Alteration Risks

An attack can be realized with a Trojan horse on the feature extractor, the matching algorithm, or the decision algorithm of the system, acting as a manipulator of each component's output. IS auditors should define security controls, such as write-once memory units that host the feature extraction program and the matching algorithm, and integrate systems to a hardware security module. Additional controls include developing a strong security policy that controls the operation of the system to protect it from exposure to manipulating attempts. [5]

Electronic Access Control

Electronic access control uses computers to solve the limitations of mechanical locks and keys. A wide range of credentials can be used to replace mechanical keys. The electronic access control system grants access based on the credential presented. When access is granted, the door is unlocked for a predetermined time and the transaction is recorded. When access is refused, the door remains locked and the attempted access is recorded. The system will also monitor the door and alarm if the door is forced open or held open too long after being unlocked.

Access Control System Operation

When a credential is presented to a reader, the reader sends the credential information, usually a number, to a control panel, which contains a highly reliable processor. The control panel compares the credential's number to an access control list, grants or denies the presented request, and sends a transaction log to a database. When access is denied based on the access control list, the door remains locked. If there is a match between the credential and the access control list, the control panel operates a relay that in turn, unlocks the door. The control panel also ignores a door-open signal to prevent an alarm. Often the reader provides feedback, such as a flashing red LED for an access denied and a flashing green LED for an access granted.

The above description is a single factor transaction. Credentials can be passed around, thus subverting the access control list. For example, Alice has access rights to the server room but Bob does not. Alice either gives Bob her credential or Bob takes it—he now has access to the server room. To prevent this, two-factor authentication can be used. In a two-factor transaction, the presented credential and a second factor are needed for access to be granted. The second factor can be a PIN, a second credential, operator intervention, or a biometric input. As previously described, the factors are characterized as something possessed (a credential), something known (a PIN), or something physiological (a biometric input).

Access Control System Components

An access control point can be a door, turnstile, parking gate, elevator, or other physical barrier where granting access can be electrically controlled. Typically, the access point is a door. An electronic access control door can contain several elements. At its most basic, there is an electric lock. The lock is unlocked by an operator with a switch. To automate this, operator intervention is replaced by a reader. The reader could be a keypad where a code is entered, it could be a card reader, or it could be a biometric reader. Readers do not usually make an access decision but send a card number to an access control panel that verifies the number against an access list. To monitor the door position, a magnetic door switch is used. In concept, the door switch is not unlike those on refrigerators or car doors. Generally, only entry is controlled and exit is uncontrolled. In cases where exit is also controlled, a second reader is used on the opposite side of the door. In cases where exit is not controlled, free exit, a device called a request-to-exit (REX), is used. Request-to-exit devices can be a pushbutton or a motion detector. When the button is pushed or the motion detector detects motion at the door, the door alarm is temporarily ignored while the door is opened. Exiting a door without having to electrically unlock the door is called mechanical free egress. This is an important safety feature. In cases where the lock must be electrically unlocked on exit, the request-to-exit device also unlocks the door.

Five Steps to Improve Cyber Security

Both the National Cyber Security Alliance (NCSA) and Capitol One Small Business recommend the following five steps to improve the organization's cyber security.

1. Conduct a Risk Assessment

To protect customer information, small business owners need to conduct an initial risk assessment of their online and operating systems. This includes determining if any sensitive information critical to the bottom line, such as the customer database, is attached to the Internet.

Another area of importance is physical security. Small desktop servers and other portable devices should be locked in a closet or office. It is too easy for someone to walk in and take them.

There are several components of a comprehensive risk assessment. Most importantly, small business owners should install updated antivirus programs, antispyware programs, and firewall software on all computers. Make sure to keep these programs, along with the operating system and applications, up-to-date with the most current patches. In addition, ensure that all employees use effective, complex passwords. Passwords should be changed every 60 to 90 days.

2. Educate Employees

It is essential that managers and employees have a basic understanding of cyber security, including company-specific procedures and overall best practices.

Employees should be taught to avoid risks. This includes basic Cyber Security 101, which involves opening attachments from an unknown source. These issues and concepts must be refreshed more than once a year. Threats evolve and managers need to keep employees alert to those threats.

Small business owners need to integrate a cyber security rollout plan within the yearly business plan. This plan should also include steps for measuring success.

3. Back Up Critical Information

Make regular, weekly, backup copies of all important data and information. Creating backups on a regular basis ensures that critical data is not lost in the event of a cyber attack or natural disaster.

Store all backup copies offsite, such as on an external hard drive, and use encryption to protect any sensitive information about the company and customers from thieves and hackers. Encryption programs encode data, making it unreadable until the user enters a password or encryption key to unlock it.

It is easier now than ever to keep data backups offsite. Managers will need that data in the event of a fire, flood, or other disaster. Just having data onsite is not helpful.

4. Create a Contingency Plan

Small business owners should have a contingency plan in place in case the business suffers a cyber security attack. The contingency plan should include steps on how to continue business operations at an alternate location when necessary. This plan should be tested annually.

Managers should understand how to preserve evidence of the attack, who to contact—the police, the bank, customers, and attorneys. Think things out ahead of time because employees and users will not be effective in the heat of the moment.

5. Sign a Security Agreement

Have all employees sign a security agreement to demonstrate that they are taking cyber security seriously and are active participants in helping to maintain a secure online environment. This agreement should also require employees to report any suspicious online activity or known Internet crime to the proper authorities.

If fraud or criminal intent is suspected, it should be reported to the local law enforcement agencies, the local FBI, Secret Service, FTC, or state attorney general's offices. Moreover, some states require business owners to notify their customers if hackers or thieves could have had access to unencrypted personal information. One way to prevent Internet crime is by erasing and destroying all data on a hard drive before recycling or throwing away a computer.

Maintaining a culture of security is important. Employees need to understand that they need to protect their PC data, company information, and laptops with encryption. Having and enforcing an agreement like this shows how seriously management takes cyber security. [6]

Privacy Concerns

By properly addressing privacy issues, a successful deployment of a system is possible. Just the mere suggestion of scanning a person's iris to gain entrance to a building can send many privacy-conscious individuals running scared. The use of biometrics—that is, simply the automatic recognition of a person using distinguishing traits—raises privacy concerns galore, yet, it can be argued that biometric technologies are not that different from individual identifiers, such as Social Security numbers. In many cases, biometrics can enhance the protection of one's identity. [7]

Considering the controversy, before entertaining thoughts of employing biometric tools, the DOD had to conduct research. The DOD is required by many laws, regulations, and policies to provide privacy protections. The DOD Biometrics Management Office fully realizes that its biometric activities must comply with the law.

Understanding the Issues

To help it understand the privacy issues involved, in 1999 the military asked RAND Corporation to complete a study on the concerns raised by the DOD's use of biometrics. The RAND study, titled *Army Biometric Applications: Identifying and Addressing Sociocultural Concerns*, is available to the public. The study examines concerns such as an individual's ability to control personal information, freedom from contact with other people or monitoring agents, and an individual's objection to the use of biometrics based on sincerely held religious beliefs. Delving into these individual concerns more thoroughly helps shed some light on the confusion that has often plagued biometrics.

The most significant informational privacy concerns relate to the threat of function creep and the tracking capabilities of biometrics. Function creep is when the original purpose for which information is obtained is broadened. This can occur with or without the knowledge or agreement of the person providing the data.

Depending on whom it affects and how it affects them, function creep may be seen as desirable or undesirable. For instance, using a Social Security number to search for a parent receiving a federal entitlement who is delinquent with child support payments may be seen as highly desirable. On the other hand, having a person's digitized state Department of Motor Vehicles (DMV) photograph sold to a commercial firm to create a national photo ID database might be seen as unacceptable.

Tracking refers to the ability to monitor an individual's actions or to search databases that contain information about these actions. The use of databases containing detailed personal information, in both the public and private sectors, has raised concerns about individuals' abilities to maintain their anonymity. Some people fear a "Big Brother" government that is able to track every individual. If an individual must use a standard biometric for multiple governmental, business, and leisure transactions of everyday life, it becomes possible for records of these transactions to be linked. This in turn could allow an entity, such as the government, to compile a comprehensive profile of the individual's actions.

The possibility of clandestine capture of biometric data increases "Big Brother" concerns. Facial recognition systems can track individuals secretly without the individual's knowledge or permission. Moreover, the information from tracking, combined with other personal data acquired through biometrics or other means, can be used to provide even more insight into an individual's private life.

Misuse of personal information, including the stealing of identities, has become more of a threat as information technology, including electronic commerce, has become ubiquitous. Where biometrics is authenticated remotely, for instance, by transmission of data from a sensor to a centralized data repository, there is a chance that a hacker can steal, copy, or reverse-engineer the biometric. This misappropriation could also come about through insider misuse. Without good safeguards, the fear is that files could be misappropriated and transactions could be performed using other people's identities.

Physical Privacy

The use of biometrics may raise physical privacy concerns. These concerns are threefold: the stigma associated with some biometrics, such as fingerprints; the possibility of actual harm that might be caused to participants by the technology itself; and the concern that the devices used to obtain or "read" the biometric may be unhygienic.

Some individuals—and segments of society—associate fingerprinting with law enforcement, acts of criminal behavior, and oppressive government. However, these concerns seem to be held by a minority and can generally be overcome through education about the protections in place for using and safeguarding data. Moreover, all military members and federal government employees are required to provide copies of their fingerprints.

Concerns about actual harm that could be caused by biometrics are primarily perceptual. There are no documented scientific or legal cases of biometrics causing injury. Others may be concerned that a dismembered limb could be stolen and used to fool a system. At its Biometrics Fusion Center, the DOD performs tests and evaluates biometric technologies to ensure that performance standards are met. There are objections to biometrics based on concerns about the cleanliness of sensors; however, users have no more contact with the sensor than with doorknobs or ATM keypads.

Religious Objections

Some segments of American society have religious objections to the use of biometrics. Among these objections, individuals oppose being compelled to participate in a government-mandated biometric application.

The New York Department of Social Services and the Connecticut Department of Social Services (DSS) have encountered legal challenges based on religious concerns from entitlement program recipients who refused to provide a biometric identifier. In New York, the state court ruled that mandating a biometric for the receipt of benefits did not violate the recipient's religious beliefs. In Connecticut, a decision by the DSS commissioner to grant a recipient exemption from the digital imaging requirement was made before the state court could render a decision on the legal challenge.

Privacy Enhancement

As a new technology, biometrics raises privacy concerns. However, these concerns are not uniquely associated with biometrics. In fact, they are very similar to the concerns expressed over the use of many identifiers, and are related to the fact that we live in an age where information has value as a commodity.

Also, when evaluating biometrics, the privacy-enhancing aspects of this technology should not be overlooked. Used in many ways, biometrics provides greater security because the identifier is much harder to steal or counterfeit compared to passwords or other identifiers. Using a biometric is also more convenient than having to remember a password.

Another privacy-enhancing aspect is that a biometric identifier discloses little personal information. Unlike information on a driver's license, a biometric template contains no information about an individual's name, address, SSN, race, height, weight, or medical condition. To the extent biometrics protects ID; it acts as privacy's friend.

To ensure that privacy concerns are properly addressed in its work, the DOD Biometrics Management Office has established a biometric policy working group consisting of members from various defense components. A legal working group has also been created and consists of attorneys from DOD and the armed services. Both groups were initiated to explore legal and privacy issues associated with the technology.

The DOD also has procedures in place to deal with service members' religious concerns. A senior coordinating group oversees the work of the management office, and can also weigh in on any privacy matters. Moreover, DOD policy concerning biometrics must go through an extensive coordination process where many different organizations get an opportunity to voice their opinions. In this way, DOD is building a program that protects the privacy and related concerns of all users.

Federal Information Security Management Act

The Federal Information Security Management Act (FISMA) implementation project is concerned with protecting the nation's critical information infrastructure. [8] The objective is to promote the development of key security standards and guidelines to support the implementation of and compliance with the FISMA, including:

- Standards for categorizing information and information systems by mission impact
- Standards for minimum security requirements for information and information systems
- Guidance for selecting appropriate security controls for information systems
- Guidance for assessing security controls in information systems and determining security control effectiveness
- Guidance for certifying and accrediting information systems

This activity should lead to:

- The implementation of cost-effective, risk-based information security programs
- The establishment of a level of security due diligence for federal agencies and contractors supporting the federal government
- More consistent and cost-effective application of security controls across the federal information technology infrastructure
- More consistent, comparable, and repeatable security control assessments
- A better understanding of enterprise-wide mission risks resulting from the operation of information systems
- More complete, reliable, and trustworthy information for authorizing officials—facilitating more informed security accreditation decisions
- More secure information systems within the federal government, including the critical infrastructure of the United States

Committee on National Security Systems

Under Executive Order (E.O.) 13231 of October 16, 2001, Critical Infrastructure Protection in the Information Age, the president redesignated the National Security Telecommunications and Information Systems Security Committee (NSTISSC) as the **Committee on National Security Systems** (CNSS). The CNSS provides a forum for the discussion of policy issues, sets national policy, and promulgates direction, operational procedures, and guidance for the security of national security systems. National security systems are information systems operated by the U.S. government, its contractors, or agents that contain classified information.

The CNSS policy states that all automated information systems which are accessed by more than one user shall provide automated controlled access protection for all classified and sensitive unclassified information.

Summary

- A corporate security policy is the gateway to a company's intellectual property. In today's world of information technology, the main threat to information security within a company is its employees. The overall program policy is usually a compilation of issue-specific policies. These policies are used to address specific parts of the network such as the Internet and e-mail, antivirus programs, data backup and recovery, and software security. Network security policy development involves identifying assets, risk assessment, management of access to information, data backup and restoration, incident handling, and maintenance.
- ID management is a broad administrative area that deals with identifying individuals in a system, such as a country, network, or enterprise, and controlling their access to

resources within that system by associating user rights and restrictions with the established identity.

- A biometric policy is a set of rules that indicate the applicability of a biometric reference template for use by some community or class of applications that have common security requirements. Biometric policy establishes the foundation, mandate, and authority for any organization and is an essential component to deploying biometrics effectively. Policy content will include privacy, planning guidance, acquisition, enterprise organization capabilities, and outreach and community education.
- A biometric information security management system includes the biometric policy, security practices, operational security procedures, organizational structure, assigned responsibilities, and resources needed to protect biometric assets. A Biometric Practice Statement describes the security practices that an organization follows during the biometric reference template life cycle, including the business, legal, regulatory, and technical considerations.
- Using a biometric identifier to authenticate employees, partners, or customers/consumers allows an organization to have the highest level of assurance in the other participant's authenticity. By properly addressing privacy issues, a successful deployment of a system is possible.

Key Terms

Biometric information security management system (ISMS) Includes the biometric policy, security practices, operational security procedures, organizational structure, assigned responsibilities, and resources needed to protect biometric assets.

Biometric policy (BP) A set of rules that indicate the applicability of a biometric reference template for use by some community or class of applications that have common security requirements.

Biometric practice statement (BPS) Describes the security practices that an organization follows during the biometric reference template life cycle, including the business, legal, regulatory, and technical considerations.

Committee on National Security Systems (CNSS) This organization provides a forum for the discussion of policy issues, sets national policy, and promulgates direction, operational procedures, and guidance for the security of national security systems.

Electronic access control Uses computers to solve the limitations of mechanical locks and keys.

Federal Information Security Management Act (FISMA) This act requires each federal agency to develop, document, and implement an agency-wide program to provide information security for the information and information systems that support the operations and assets of the agency, including those provided or managed by another agency, contractor, or other source.

Guideline Typically, a collection of system-specific or procedural-specific "suggestions" for best practice.

Identity (ID) management A broad administrative area that deals with identifying individuals in a system, such as a country, network, or enterprise.

ISO 19092 Provides a technology-specific extension to the ISO/IEC 17799/27001 Code of practice for information security management.

Policy Typically, a document that outlines specific requirements or rules that must be met.

Standard Typically, a collection of system-specific or procedural-specific requirements that must be met by everyone.

Review Questions

1. A ________________ is typically a document that outlines specific requirements or rules that must be met.
2. A ________________ is typically a collection of system-specific or procedural-specific requirements that must be met by everyone.
3. A ________________ is typically a collection of system-specific or procedural-specific "suggestions" for best practice. These are not requirements that must be met, but are strongly recommended.
4. A ________________ is the gateway to a company's intellectual property.
5. Policy establishes the ________________, ________________, and ________________ for any organization and is an essential component to deploying biometrics effectively.
6. ________________ is a broad administrative area that deals with identifying individuals in a system, such as a country, network, or enterprise, and controlling their access to resources within that system by associating user rights and restrictions with the established ID.
7. ________________ will directly correlate to the combination of effective technology and effective policy. It will deliver capabilities in order to contribute to the enhancement of the biometric community.
8. A ________________ includes the biometric policy, security practices, operational security procedures, organizational structure, assigned responsibilities, and resources needed to protect biometric assets.
9. A new biometric security standard, ________________, provides a technology-specific extension to the ISO/IEC 17799/27001 Code of practice for information security management.
10. A ________________ is a set of rules that indicate the applicability of a biometric reference template for use by some community or class of applications that have common security requirements.
11. A ________________ describes the security practices that an organization follows during the biometric reference template life cycle, including the business, legal, regulatory, and technical considerations.

12. Management of biometric information, including its registration, storage, protection and verification, is known as ________________.
13. ________________ consists of a group of functions that facilitate the inclusion of users and their biometric data into the system.
14. The three main functions within the ID capture functional group are ________________, ________________, and ________________.
15. The ________________ is responsible for focusing on original research, publications, resources and symposia on IT governance and related topics.
16. The ________________ standard concerns the Evaluation Criteria for IT Security.
17. For security applications, the biometric system should be calibrated to produce a FAR less than or equal to ________________ percent.
18. ________________ uses computers to solve the limitations of mechanical locks and keys. This control system grants access based on the credential presented.
19. The five steps recommended by the NCSA for improving an organization's cyber security are ________________, ________________, ________________, ________________, and ________________.
20. ________________ implementation project is concerned with protecting the nation's critical information infrastructure.

Discussion Exercises

1. Provide an overview of the X.509 ITU-T standard.
2. Describe the responsibilities of organizational management as it relates to system design and implementation.
3. Describe the policies that impact the use of biometrics in an organization.
4. Provide an overview of the responsibilities of the Biometrics Task Force.
5. Identify the ethical concerns that have been expressed by user and user organizations.
6. Provide an overview of the ISO 19092 biometric security standard.
7. Describe the differences between ISO 19092 and ISO/IEC 27001.
8. Provide an overview of biometric ID management.
9. Provide an overview of the PKCS# 11 standard.
10. Review the Information Systems Control Journal and prepare an overview of some security article.

Hands-On Projects

7-1: Biometric Standards and Procedures Awareness

Time required—30 minutes

Objective—Identify standards and procedures that have been developed for the biometric environment.

Description—Use various Web tools, Web sites, and other tools to identify biometric standards and procedures that are in development or production for both commercial and government organizations.

1. Develop a compendium of standards that relate to biometrics policy and program management.
2. Identify privacy organizations that have expressed concerns with the use of biometrics. Describe the positions taken by these organizations.
3. Provide an overview of the Federal Information Security Management Act. What is the impact of this standard on biometric usage, systems, and implementations?

7-2: Identify Policy and Program Management Requirements

Time required—45 minutes

Objective—Identify policy and program management requirements for biometrics solutions that can provide some level of access control and security for OnLine Access assets and resources.

Description—The owners and managers of OnLine Access have identified a multimodal biometric access control solution. They must now look at the policies and procedures necessary to implement a biometric solution. The daily operational requirements must be identified before undertaking this biometric project.

1. Research the topics of policy and program management for access control and security.
2. Identify other organizations that have implemented a biometric solution.
3. Produce a report describing issues that would impact a successful implementation of a biometric solution.
4. Identify the project management steps involved in biometric solution implementations.
5. Identify help desk and technical support requirements and issues.

chapter 8

Security and Access Technologies

After completing this chapter, you should be able to do the following:

- Understand the basics of computer and network security issues
- Look at access control issues and security implications for resources and assets
- Identify hardware and software components used in the biometric environment for access control
- Become familiar with the biometric terms and definitions associated with biometric security and access capabilities
- Identify biometric access and security technologies and issues
- See how biometrics can be utilized in providing access control

Selecting and implementing a biometric solution requires an understanding of the current legacy operation and the impact of the new security and access control solution. The selection of biometric hardware and software components can have a significant impact on the successful operation of the solution. User acceptability and the issue of "user-friendliness" are significant considerations that must be addressed. Users may find ways to circumvent the new biometric solution if it is difficult to use.

Introduction

This chapter primarily discusses the various techniques associated with computer and network access issues that impact security. Organizational assets must be protected from unauthorized access; therefore, this discussion includes the issues of identification, authentication, and authorization processes. Sections of this chapter are devoted to the implementation of digital signatures and certificates, as well as to public and private keys and their relationship to asymmetrical and symmetrical security systems. The chapter also covers trust and trust relationships, which are closely aligned with authentication and authorization.

Biometric access control systems consist of a reader or scanning device, software that converts the gathered information into digital form, and a database that stores the information for comparison with previous records. These readers, or scanning devices, can scan for a fingerprint, hand geometry, a signature, the iris/retina, facial recognition, a voice print, and vascular pattern.

This technology can be used for a number of applications, including time and attendance reporting, building access control, verification of signatures, point-of-sale identity verification, process control security, and cellular phone security. Information presented in this chapter looks at the hardware, software, and network components utilized in providing security and access control.

This chapter also presents an overview of the e-commerce, computer networking environment, and the various hardware and software components and electronic devices that might be employed as security devices. It provides examples of hardware and software systems that might be considered in a search for a biometric security system, and compares features and functions available on security surveillance and management systems for access control. Finally, the chapter presents options for hardware and software system solutions that address network and computer security issues.

Access into computer and network facilities and buildings has become an issue with most organizations, especially those with large asset bases. It is essential that only authorized personnel be allowed (unattended) into a facility. Not only is ingress an issue, egress can be just as important, as hardware, software, documentation, and other items can "walk" if not protected. The chapter concludes with a discussion of access control for personnel, contractors, visitors, and others.

Authentication and Access Control Mechanisms

The various methods, procedures, and vehicles that address access control issues include biometrics, passwords, firewalls, filters, tokens, and access control lists. In networked and distributed computer systems, the communications traffic between and among clients and servers is a point of attack for intruders. Mechanisms and services, which are part of communications security, can counteract vulnerabilities introduced by insecure communications. Access control mechanisms fall into the following categories:

- Access control by authentication servers
- Access control lists (ACL)
- Intrusion detection
- Physical access control
- Policy filters
- Traffic filters

Authentication

Authentication is a security measure designed to establish the validity of a transmission, message, or originator. It can also include a means of verifying an individual's authorization to receive specific categories of information. In private and public computer networks, including the Internet, authentication is commonly done through the use of logon passwords. Knowledge of the password process is assumed. There are numerous terms associated with authentication. These include:

- Two-factor authentication
- Three-factor authentication
- Access authentication
- Access codes
- Carrier identification codes
- Carrier access codes
- Password authentication
- Network authentication
- Remote authentication
- User authentication
- User ID

Authentication mechanisms include the following:

- Biometric techniques
- Certificates
- Challenge–response handshakes
- Kerberos authentication
- One-time passwords
- Passwords and personal identification numbers (PINs)
- Remote Authentication Dial In User Service (RADIUS)
- Security tokens

Hardware and software products and services provide security of an organization's assets and resources. These products would be implemented in physical settings, computer installations, and networking systems. Discussions are directed toward various software, hardware, and networking security technologies and configurations.

Software

Software includes the programs or other "instructions" a computer needs to perform specific tasks. Examples of software include word processors, e-mail clients, Web browsers, video games, spreadsheets, accounting tools, and operating systems. A basic background of the software environment is important to understand the issues associated with biometric solutions. An *application* software system is usually implemented to satisfy a set of requirements. In this context, an application incorporates a biometric system to satisfy a subset of requirements related to the verification or identification of an end user's identity so the end user's identifier can be used to facilitate their interaction with the system. An application profile includes conforming subsets or combinations of base standards used to provide specific functions. Application profiles identify the use of particular options available in base standards, and provide a basis for the interchange of data between applications and interoperability of systems.

An *application program interface (API)* is a set of services or instructions used to standardize an application. An API is computer code used by an application developer. Any biometric system that is compatible with the API can be added or interchanged by the application developer. APIs are often described by the degree to which they are high level or low level. High level means that the interface is close to the application; low level means that the interface is close to the device. An *algorithm* is a sequence of instructions that tell a biometric system how to solve a particular problem. An algorithm will have a finite number of steps and is typically used by the biometric engine to compute whether a biometric sample and template is a match. An example of a biometric software solution is FaceIt®.

FaceIt®

FaceIt® is facial recognition, software engine that allows computers to rapidly and accurately detect and recognize faces. FaceIt® enables the broadest range of applications in the marketplace; from ID Solutions, to information security, to banking, and e-commerce. The capabilities of FaceIt® include: [1]

- Face detection—Detects single or multiple faces, even in complex scenes
- Face recognition—Functions in either of the following: Verification (one-to-one matching) or Identification (one-to-many matching)
- Image quality—Evaluates quality of image for face recognition and, if needed, prompts for improved image
- Segmentation—Crops faces from background
- Face print—Generates digital code or internal template, unique to an individual
- Tracking—Follows faces over time
- Compression—Compresses facial images down to 84 bytes in size

Hardware

This section contains information about both physical security hardware devices and computer hardware components and devices. Computer hardware is generally used to support the software that resides on the system hard drive storage, whereas physical hardware is used to support physical access control into some facility or work area.

Computer Hardware Devices

Computer **hardware** is the physical part of a computer, including the digital circuitry, as distinguished from the computer software that runs within the hardware. The hardware of a computer is infrequently changed, in comparison with software and data, which are "soft" in the sense that they can be readily created, modified, or erased from the computer. A **microcontroller** is a highly integrated computer chip that contains all of the components that comprise a controller. Typically this includes a CPU, RAM, some form of ROM, I/O ports, and timers. Unlike a general-purpose computer used in information technology, a microcontroller is designed to operate in a restricted environment.

A **head-end system** is the physical access control server, software, and database(s) used in a physical access control system.

A *smart card* is a device that includes an embedded integrated circuit that can be either a secure microcontroller or equivalent intelligence with internal memory or a memory chip alone. The card connects to a reader with direct physical contact or with a remote contactless radio frequency interface. With an embedded microcontroller, a smart card has the unique ability to store large amounts of data, carry out its own on-card functions, such as encryption and mutual authentication, and interact intelligently with a smart card reader. Smart card technology conforms to international standards (ISO/IEC 7816 and ISO/IEC 14443) and is available in a variety of form factors, including plastic cards, subscriber identification modules used in GSM mobile phones, and USB-based tokens. There are a variety of cards. These include:

- Contact smart card
- Memory card
- Contactless smart card
- Hybrid card
- Microprocessor card

A **contact smart card** is a smart card that connects to the reading device through direct physical contact between the smart card chip and the smart card reader. A *memory card* is typically a smart card or any pocket-sized card with an embedded integrated circuit or circuits containing nonvolatile memory storage components and perhaps some specific security logic. A **contactless smart card** is a smart card that communicates with a reader through a radio frequency interface. A **hybrid card** is a smart card that contains two smart card chips (both contact and contactless chips) that are not interconnected. Lastly, a *microprocessor card* is typically a smart card or any pocket-sized card with an embedded integrated circuit or circuits containing memory and microprocessor components. Additionally, an integrated circuit card typically refers to a plastic (or other material) card containing an integrated circuit which is compatible to ISO/IEC 7816.

The physical device that contains the smart card chip is called a **form factor**. Smart chip-based devices can come in a variety of form factors, including plastic cards, key fobs, wristbands, wristwatches, PDAs, and mobile phones.

Physical Security Devices

Efforts to secure a computer, a database, and network resources will be in vain if the physical plant which houses them is not protected. This section provides an overview of alternatives to counter such situations, as well as an in-depth analysis of building, campus, and facility security. Every organization should assume that it will be the target of some kind, and as many barriers as feasible should be erected to counter these threats. Although sometimes expensive and troublesome, it is fairly easy to erect barriers against physical access. Equipment rooms and network demarcation closets can be kept locked. A **demarcation** is the physical point where a regulated network service is provided to the user. Access can be restricted to only those with a specific need to be there.

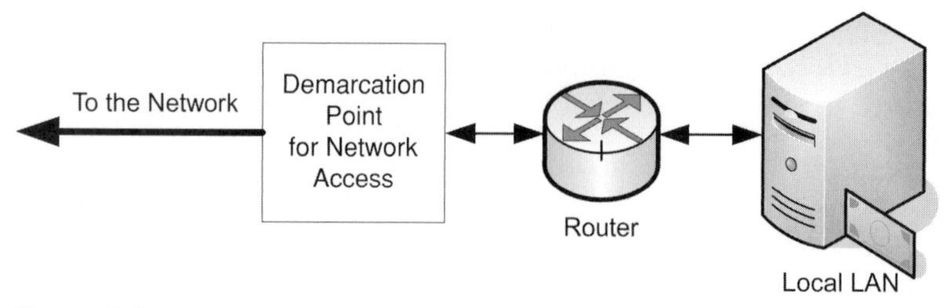

Figure 8-1 Demarcation Point

Course Technology/Cengage Learning

As concentration points for large numbers of circuits, equipment rooms and telecommunications closets are particularly vulnerable. They should be kept locked, and the keys given only to authorized personnel. Circuit records and wiring schematics and other communication master records should not be kept in these rooms. As cabling and wiring comes closer to a workstation or PC, it becomes increasingly easy to determine which one to tap. It is essential that these connections—which attach to sensitive computer and network devices—not show conspicuous tags. Most organizations are vulnerable to malicious activities in these locations, which mean access to these areas must be monitored and restricted. Physical access to a network link would allow someone to tap into that link, jam it, or inject network traffic into it. In the network world, it is important to avoid the "single point of failure." A saboteur could disrupt an entire organization's network by destroying or incapacitating its network demarcation point. Figure 8-1 illustrates a common demarcation point for network access.

In addition to accessing networks through computers and terminals, intruders can physically break into network cabling systems. Fiber-optic cable is considerably more difficult to tap than copper wire because it does not radiate signals that can be collected through surreptitious means. An intruder might resort to tapping copper connectors in a wiring closet or at cable outlets. Physical security measures at plants and offices are the primary means to prevent such break-ins.

Physical access to a building or data center can be controlled via locks and access control mechanisms. Enterprise physical security can be based on security guards, closed circuit television, and card-key entry systems. Data centers and limited-access buildings may have a double-door system, which requires that anyone entering the facility pass through two sets of security checkpoint doors, one door at a time. With these security measures in place, organizations can feel confident that their assets are protected and high user productivity is maintained.

A **physical access control system** (PACS) is a system made up of hardware and software components that controls access to physical facilities such as buildings, rooms, airports, and warehouses.

Traditional access management is meant to allow authorized users, while blocking unauthorized or malicious ones. Hardware authentication does the same thing for devices. An authorized device, like an authorized user:

- Is trusted
- Is confirmed to be virus and malware free

- Is patched with up-to-date software
- Will not bring anything harmful in
- Will not take anything unauthorized out

Identity-Enabled Devices

The heart of identity-enabled device technology is an embedded chip called the *trusted platform module (TPM)* that is preinstalled with the usual accoutrements for authentication, such as passwords, encryption keys, and digital certificates. The organization behind this hardware-authentication initiative is the **Trusted Computing Group** (TCG), a vendor consortium founded in 2003 to promote the use of vendor-neutral specifications.

The key to the TPM is its internal firmware, which can't be easily accessed or manipulated, and doesn't need to be programmed again after installation. If a device is lost or stolen, the certificate or other authentication credentials can be revoked, like any other authentication credential, and it won't connect to the network.

A number of hardware components are required when installing a physical access control solution, including a control panel and components installed at the entrance location, such as a door reader and a Faraday cage.

A **control panel** is the access control system component that connects to all door readers, door locks, and the access control server. The control panel validates the reader and accepts data. Depending on the overall system design, the control panel may then send the data to the access control server or may have enough local intelligence to determine the user's rights and make the final access authorization.

The **door reader** is the device on each door that communicates with a card or credential and sends data from the card to the controller for decision on access rights. A **door strike** is the electronic lock on each door that is connected to the controller. An **excite field** is the RF field or electromagnetic field constantly transmitted by a contactless door reader. When a contactless card is within the range of the excite field, the internal antenna on the card converts the field energy into electricity that powers the chip. The chip then uses the antenna to transmit data to the reader.

A **Faraday cage** is an enclosure formed by conducting material, or by a mesh of such material, that blocks out external static electrical fields. Any electric field will cause the charges to rearrange so as to completely cancel the field's (RF signal) effects in the cage's interior. A Faraday device is usually used to protect sensitive electronic evidence for later prosecution or legal action.

According to the TCG, every enterprise device shipped from the top 20 vendors would have a TPM. This covers an estimated 20 million devices shipped by vendors that include Lenovo Group Ltd., Hewlett-Packard, Dell Inc., Gateway Inc., Fujitsu, Toshiba, Acer Inc., and Panasonic Corp. The TCG also cites usage examples from companies in a number of industries, including pharmaceuticals, food and car rental companies, plus government institutions, such as the National Security Agency (NSA). Its success to date has been attributed in part to the cross-platform nature of the TCG's initiative. It's seen as an alternative to Microsoft's Network Access Protection (NAP) and Cisco Systems Inc.'s Network Admission Control (NAC), which are strongly tied to its networking hardware.

Any organization considering implementation of identity-enabled devices needs to thoroughly study its network architecture—as it would for any new deployment—to determine if the TPM is compatible with its existing infrastructure. Implementing trusted computing can only be done in phases as TPM-enabled hardware is installed.

Identity-enabled technology is still developing and growing; however, it is already a part of most new hardware. The question is whether it will become a widely adopted part of enterprise authentication systems, or go largely ignored. [2]

Security Elements and Components

The field of identity-enabled network devices goes under different names. It's been called trusted computing, endpoint security, or **network access control**. But network access control is actually a little different, as it is a process, or a series of technologies that may include software, hardware, or servers that monitor devices accessing a network, or a combination of all of the above. Network access control systems verify that devices connecting to the network are authorized. Identity-enabled network devices are locked down to a greater degree, as they are authenticated and secure endpoints in themselves. Several questions that should be answered include:

- Do they deliver on the promise of increasing security and locking down a network any tighter than other network access control systems?
- Are they realistic?
- Who are the players?
- How do they plan to roll out the technology to the technical masses?

For identity-enabled devices to work, they must be cross-platform. It must be possible to move them around the network, just like any other piece of hardware. Obviously, the chips embedded in the hardware must be standardized.

A virtual private network (VPN) or wireless client with a TPM can use its self-contained digital certificates to authenticate. In fact, TPMs can be combined with other authentication methods, like smart cards, one-time passwords tokens, and biometrics for a multilayered approach to securing network access. *Note: These multibiometric systems were explored in Chapter 6.*

Computer and Network Resource Access Control

A basic requirement of a security system is to identify the user who is trying to gain access to the network. The owner of the network must determine whether this user is a friend or foe. It is essential that networks, computers, and information be protected from unauthorized disclosure. This concept is called *confidentiality*. In a computer system or network there must be a method of identifying anyone who is allowed into the environment. This takes the form of identification, authentication, and authorization.

Identification

This section examines techniques for identifying those active entities that are responsible for initiating specific actions on a computer system or network. This class of techniques is called identification. **Identification** consists of procedures and mechanisms which allow an entity external to a system resource to notify that system of its identity. Identification techniques occur when it is necessary to associate an action with some user or entity that causes each action. Computer systems and network devices can determine who invoked an operation by examining the reported identity of the entity that initiated the session in which the operation was invoked. This identity is usually established via a login sequence.

A system administrator may have assigned this login. The login is the simplest type of mechanism that will comprise identification. Once users are logged in, they are allowed to access various resources based on the rights and privileges assigned to their user accounts or the objects they possess. Identification is usually combined with another procedure, called authentication, which allows a system to determine if the identification sequence was correct.

Authentication

Authentication is the process whereby a user proves he is who he claims to be. Additionally, to authenticate means to establish that a transmission attempt is authorized and valid. There are several techniques for authenticating a user. The first is to enable network logon, which ensures that only valid users are allowed to access the network. The next step is to provide user authentication on servers. Authentication at the server level operates independently of the network logon. Authentication is also required for message exchange to verify that a particular message has not been fabricated or altered in transit.

A user name/password is the most common form of identification for network security. This authentication method can be used as either a manual or dynamic entry.

A manual user name/password allows users to choose their own passwords. The password remains the same until the user changes it. When using the dynamic user name/password, the password continually changes. A device such as a security card can be used to implement the dynamic technique.

Passwords

The **password** is a form of secret authentication data used to control access to a resource such as a location, database, or file storage. The password is kept secret from those not allowed access, and those wishing to gain access are tested on whether or not they know the password and are granted or denied access accordingly. A password is typically referred to as "something you know" for single-factor authentication. It is usually the first line of defense in network and computer security. A successful password system depends on the passwords being kept secret. Security is often compromised because users tend to violate security procedures by posting the password on the computer monitor or committing other violations of security policies.

Passwords are the most widely used network security feature and almost all organizations employ some password scheme. Office applications and databases are routinely secured with login passwords. They provide a minimum level of security and may be used to restrict user access to specific network facilities during designated shifts or working hours.

An area of vulnerability is the database that contains the password file. This database must be protected from those that would attempt to illegally gain access. Many systems are susceptible to unanticipated break-ins. Once the attacker has gained access, it is possible to obtain a collection of passwords in order to use different accounts for different logon sessions, which could decrease the chances of detection. It is also possible that a password file could become readable, which could compromise all accounts.

There are several actions that would be prudent when managing passwords:

- When an employee leaves the organization, this password must be disabled.
- Login attempts must be limited to three before access is denied.
- An audit trail needs to be created that tracks logins and denials.
- An alert needs to be activated when password violations occur.
- The last login date stamp can be displayed when a user logs in. This will flag the user that someone else has been trying the user's password.
- Multiple users are not allowed to know the password for a single user ID.

Some users with accounts on multiple computers in other protection domains might use the same password. Thus, if duplicate passwords are used and are compromised, multiple computers may be affected. Numerous organizations employ single sign-on passwords for all system access.

The use of a password is commonly referred to as first-function authentication. A second-function authentication is based on something the user possesses. This could include a disk, token, or unique card. A more personal form of authentication could include a thumbprint, retinal scan, or a voiceprint. A second-function authentication might be used to control the access to individual offices within a floor, whereas a password might allow access to common areas on a building.

An alternative password system uses a *one-time password* that is used once and then discarded. Each time the user authenticates to a system, a different password is used, after which that password is no longer valid. The password is computed either by software on the logon computer or by tokens in the user's possession that are coordinated through a trusted system.

Authorization

Authorization refers to securing the network by specifying which area of the network—whether it is an application, a device, or a system—a user is allowed to access. The authorization level varies from user to user. Access is based on the need to know; for example, marketing personnel have access to sales systems and programmers have access to design systems. This also means that access to system resources and services available to users or processes, previously authenticated, are selectively granted. It is possible to explicitly grant or deny authorizations. By explicitly denying access, only access to confidential information is restricted. This compares with explicitly granting access, which does not allow access to any information without specifically granting an access. Authorization can be defined as system-wide or limited to specific data elements, and can be based on the type of access, which includes read, write, delete, or execute. There are different authorization processes for routers, gateways, servers, and switches. Figure 8-2 shows the authentication and authorization flow of a login session.

Routers, gateways, and switches require authorization so unauthorized personnel do not change configurations. Access is limited to a server because directory integrity must be preserved. Often a user is allowed a maximum of three attempts to enter the correct authorization before the access attempt is terminated.

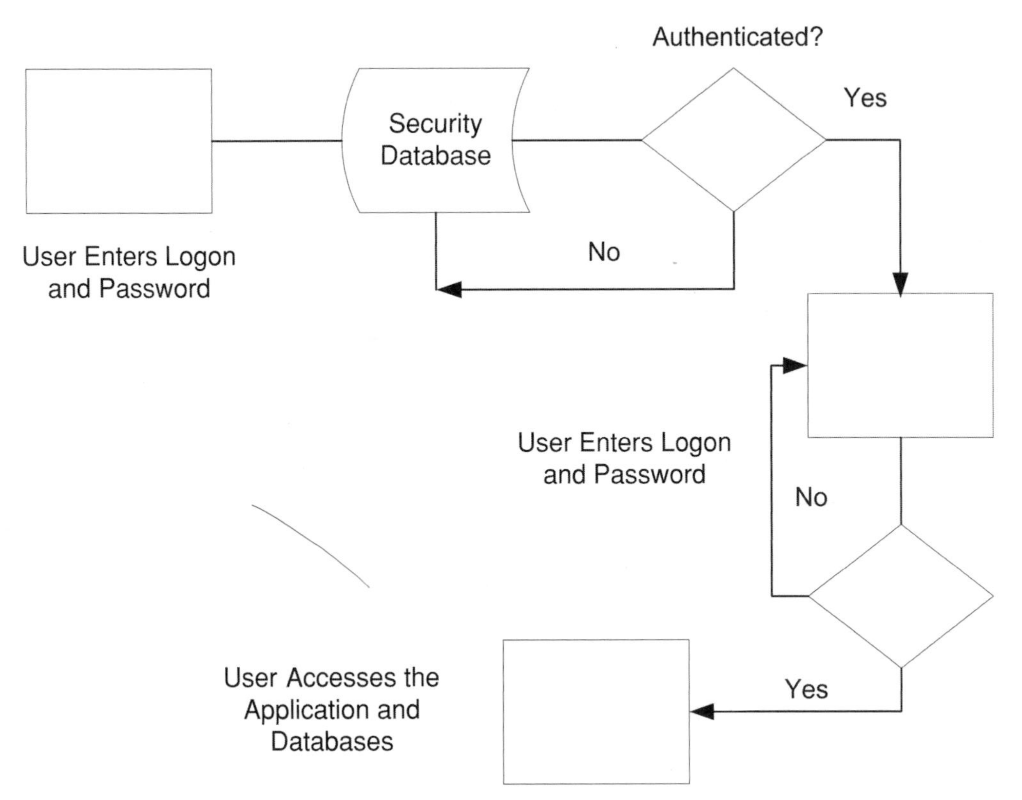

Figure 8-2 Authentication and Authorization Flow of a Login Session

Course Technology/Cengage Learning

A *message authentication code (MAC)* is a short piece of information used to support authentication of a message. A MAC algorithm accepts as input a secret key and an arbitrary-length message to be authenticated, and outputs a MAC (also known as a tag or checksum). The MAC value protects both a message's integrity as well as its authenticity, by allowing verifiers, who also possess the secret key, to detect any changes to the message content. MACs are computed and verified with the same key, unlike digital signatures.

A concept closely related to authentication and authorization is trust. *Trust* is the composite of security, availability, and performance. It includes the ability to execute processes with integrity, keep confidential information private, and perform the required functions without interference or interruption. It follows that a *trusted system* is one in which a computer, network, or software can be verified to implement a given security policy. A framework for an organizational secure systems policy that addresses resource authorization is called a **trust model.**

A **trust relationship** is the link between two entities on a network (e.g., two servers) that enables a user with an account in one domain to have access to resources in another domain. Trust is essential among organizations that engage in business transactions. Internet users are often required to trust someone they have never met. Likewise, commercial transactions between diverse computer systems often contain sensitive and proprietary information. Trust must be established in both cases before secure transactions can occur.

Trust relationships are established between systems so those systems can exchange information without the need for an administrator to actively monitor and authorize these exchanges.

Security and system administrators establish relationships for trust in a cooperative way. Access control techniques that support trust systems will be presented in subsequent sections. *Note: Technical solutions that can provide support to an organization's trust systems were presented in Chapter 6.*

The **chain of trust** is an attribute of a secure ID system that encompasses all of the system's components and processes and assures that the system as a whole is worthy of trust. A chain of trust should guarantee the authenticity of the people, issuing organizations, devices, equipment, networks, and other components of a secure ID system. The chain of trust must also ensure that information within the system is verified, authenticated, protected, and used appropriately.

Privileges and Roles

In the computer environment, a *privilege* is defined as a collection of related computer system operations that can be performed by users of that system. These include the access rights to a directory, file, or program, which typically includes read, write, delete, create, and execute. This is closely related to the role or job-related function that the user is trying to accomplish. These ***roles* are defined as a collection of related privileges. Least privilege is a principle that may be used to guide the design and administration of a computer system or enterprise network. This suggests that users only be granted the privilege for some activity if there is a justifiable need for its associated authorizations.**

There is a relationship between roles and privileges. A role might contain only one privilege or many. This means a role might be defined to include only one privilege for one activity or it might be defined to include many different privileges to perform a number of system operations. Privileges can also exist in multiple roles. This implies that a number of privileges can be shared by multiple different roles. Some privileges are more powerful than others, due in part to the level of access required to perform the role. "Normal" users who are assigned non-administrative roles typically include less powerful privileges. The security administrator must decide the appropriate privilege for each role.

A **role** consists of the actions and activities assigned to or required or expected of a person or group. **Role-based access control (RBAC)** provides access to resources based on a user's assigned role. Access permissions, which determine accessible resources and the privileges in the context of that resource, are administratively associated with roles, and users are administratively assigned appropriate roles. Roles can be granted new permissions as new resources are incorporated, permissions can be revoked from roles as needed, and role assignments for users can be modified or removed as needed. Since users are not assigned permissions directly, but only acquire them through a role, or roles, management of individual user rights becomes a matter of simply assigning the appropriate roles to the user, which simplifies common operations such as adding a user, or changing a user's department.

Device authentication doesn't replace user authentication, it augments it. Hence the idea behind using identity-enabled network devices is to authenticate a device, rather than a user, before it can access a network. This emerging authentication paradigm is intended to add an extra layer of security to any kind of network device, be it a workstation, desktop or laptop, PDA, cell phone, or even a wireless access point. A user would still have to use a user ID and password, or some other authentication mechanism, to logon to a network, but the device itself would also have to authenticate.

Anonymous Access

Many computer systems and network servers provide *anonymous access* or guest access. An account is established with the name "anonymous" or guest," which do not require a password. Multiple users can login to the account simultaneously just by entering a user name. This account usually has very restricted access to the system and is allowed access only to public files. These accounts can be used for File Transfer Protocol (FTP) and Web servers on the Internet.

Digital Certificates and Signatures

A **certificate** is a unique collection of information, transformed into an unforgeable form and used for authentication of users. It is a digital representation of information that at least:

- Identifies the certification authority issuing it
- Names or identifies its subscriber
- Contains the subscriber's public key
- Identifies its operational period
- Is digitally signed by the certification authority issuing it

It follows that a Web server can have a certificate to prove its authenticity to users who access it. Users on these servers must be assured that malicious individuals are not collecting personal information or distributing infected documents. A **certificate authority** (CA) is a trusted third party who is responsible for issuing and revoking digital certificates within the public key infrastructure. A **certificate revocation list** (CRL) is a list of certificates that have been revoked before their expiration by a certificate authority.

Both certificates and signatures are mechanisms that provide an identity of an individual or user. A *digital certificate* is a password-protected file that contains identification information about its holder. This file includes a public key and a unique private key. The electronic exchange of keys and certificates allow two parties to verify their identities before communicating. The *Online Certificate Status Protocol (OCSP)* is an online protocol used to determine the status of a public key certificate.

A **digital signature** is an authentication mechanism that enables the creator of a message to attach a code that acts as a signature. The signature guarantees the source and integrity of the transmitted message. A message can be digitally signed by a system by including a header, a body, and a signature as part of the message. These components are described as follows.

- The header describes the identity of the sender.
- The body contains the message to be sent.
- The signature contains a computed checksum of the message contents, encrypted with the secret key of the sender.

The receiver can decrypt the checksum using the sender's public key and ensure the checksum matches a computed checksum of the transmitted message. The *checksum* process provides a sum for a group of items in the message, which is used for error checking.

The risk of message forgery is an important reason to use a digital signature. Forged messages can be used to provide false information about a person or some event. Message forgery can also occur on the Internet if someone wants to smear another person or an organization. Another issue related to this subject is called message *repudiation* or denial that a message was transmitted.

Nonrepudiation is the ability to ensure and show evidence that a specific action occurred in an electronic transaction. The message originator cannot deny sending a message or the party in a transaction cannot deny the authenticity of a signature or approval. This feature provides assurance that the sender is provided with proof of delivery and that the recipient is provided with proof of the sender's identity so that neither can later deny having processed the data.

Certificates and Certification Systems

Certification is the process of verifying a statement or claim and then issuing a certificate as to its correctness. Remember that a digital certificate is a personal digital identifier or digital signature that can be used for authentication, to ensure that the sender is validated, and to ensure the message has not been altered in transit. This authenticity verification extends to the server, which can prove its authenticity to users who access it. Server users must be assured that malicious people, who may be trying to collect credit card and personal information or distribute bogus and virus-infected copies of software, are not spoofing (impersonating) a site. Certificates can be used in place of credit card numbers for online buying transactions. The secure electronic transaction (SET) scheme, developed by major credit card companies, is designed to hide credit card numbers from merchants by substituting the card number with a digital certificate. *Note: SET will be discussed in Chapter 12.*

A basic certificate contains a public key, an expiration date, serial number, certifying authority, and a name. More importantly, it contains the digital signature of the certificate issuer. The International Telecommunication Union (ITU) X.509 certificate contains the following information:

- User name
- User organization
- Certificate start date
- Certificate end date
- User public key parameter
- Certificate authority name
- Certificate authority signature on certificate

Corporations, government agencies, and universities issue their own certificates, which are signed and issued to clients by a trusted certificate server. A certificate granted to an individual is a signed recognition on the individual's identification and authenticity. Checking the existing certificate against the public key of the certificate server and the public key of the individual can validate it.

Access Control

Security cards keep the network secure by preventing network access to users who don't possess one. This card requires a permanent user name and a temporary password. This password changes after a specific amount of time, often 60 seconds. Passwords belonging to a specific

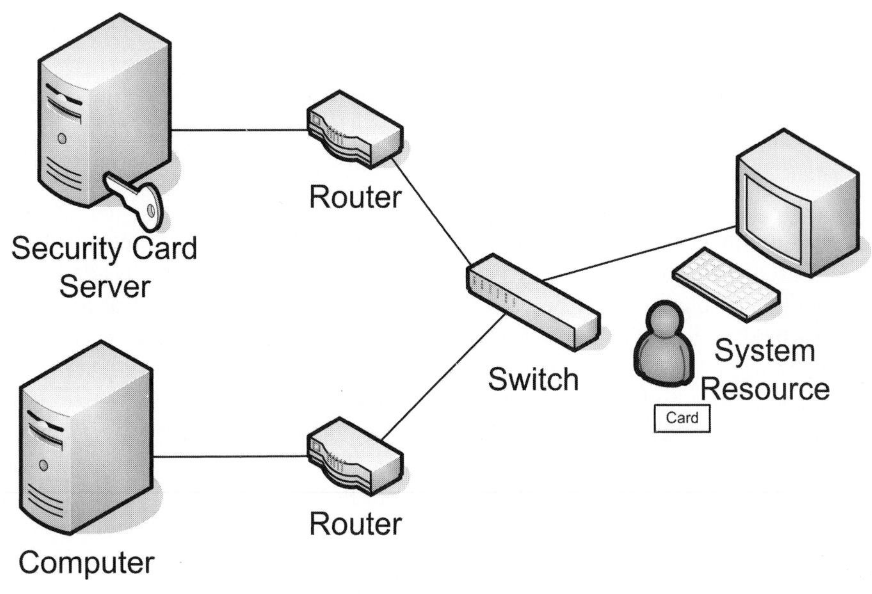

Figure 8-3 Security Card Environment

Course Technology/Cengage Learning

security card are synchronized with a security card server. Figure 8-3 depicts the security card environment. The steps involved with using this dynamic user name/password pairing is as follows:

- The user turns on the card and types in the user name.
- The card displays a password for the user to enter for network access.
- The user types in this dynamic password.
- This password is validated through a central security card server.
- If the password is correct, access is allowed to the resource; if not, the user must restart the process.

Smart card systems allow users to access data by inserting a card into a card reader. The card system provides access for authorized users to any reader-equipped workstation or computer. Automatic teller machines (ATMs) use this type of security combined with a *PIN*. Smart cards containing microprocessor chips can perform such tasks as online encryption and user-activity logging, as well as perform user identification. Some smart card setups require two individuals to perform an entry procedure before admittance is granted. Smart cards are more difficult to administer than passwords and more inconvenient for end users.

An *open ID* is a decentralized digital identity system, in which any user's online identity is given by, for example, a URL (such as for a blog or a home page) and can be verified by any server running the protocol. Users are able to clearly control what pieces of information can be shared, such as their name, address, or phone number. With open ID, users create a single user name and password, along with some personal data. Users can then log into an increasing number of sites without registering each time. Open ID also provides users with a place to store a digital identity where a user can easily be found on the Web.

Data Security and Encryption

An important issue that must be addressed after a user has successfully gained access to the network and the server is *data security*. As more and more organizations rely on a network to transport sensitive information, data security is rapidly becoming an area of significant concern. Networks that require a high level of security can use a method called encryption.

Encryption is the conversion of plaintext or data into unintelligible form by means of a reversible translation, based on some algorithm or translation table. It is sometimes called *enciphering*. This process of scrambling cannot be read by anyone except the intended receiver. This means the data must be decoded or decrypted back to its original form before the receiver can understand it. This encryption process maintains data integrity on the network and assures the confidentiality of the sender and the receiver from other users on the network. Data security and encryption information presented show how the various encryption/decryption methods are implemented in the enterprise network environment. This section compares hash functions to asymmetric and symmetric methods. Another section discusses key management for encryption systems.

Hash Functions

A **hash function** is a reproducible method of turning some kind of data into a relatively small number that may serve as a digital "fingerprint" of the data. The hash algorithm substitutes or transposes the data to create such fingerprints. A cryptographic hash function is a transformation that takes an input and returns a fixed-size string, which is called the hash value. Hash functions with this property are used for a variety of computational purposes, including cryptography. The hash value is a concise representation of the longer message or document from which it was computed. The main role of a cryptographic hash function is in the provision of digital signatures. Since hash functions are generally faster than digital signature algorithms, it is typical to compute the digital signature to some document by computing the signature on the document's hash value, which is small compared to the document itself. Additionally, a digest (digital fingerprint) can be made public without revealing the contents of the document from which it is derived. A *digest algorithm* is a signature algorithm that provides cryptographic protection against inversion. The *hash algorithm* is a software algorithm that computes a value (hash) from a particular data unit in a manner that enables detection of intentional/unauthorized or unintentional/accidental data modification by the recipient of the data.

Asymmetric and Symmetric Encryption

Encryption is the encoding of data in order to hide its content from everyone except its intended viewer. Encryption converts readable data, called plaintext, into a seemingly random sequence of characters, called ciphertext. The level of protection provided by encryption is determined by an encryption algorithm. *Decryption* is the conversion of an encrypted message to a form that is readable.

Symmetric encryption is a fast-performing algorithm typically used for bulk (large volume of data) encryption. The same key is applied to the encryption and decryption processes. As the name suggests, the key is kept secret between sender and receiver. Widely used examples are DES, Triple DES, and RC4 standards. This technique is also called private-key encryption. *Asymmetric encryption* and decryption include the use of two different keys, one of which is referred to as the public key and the other is referred to as the private key. This technique is also known as public-key encryption.

Properly applied, encryption can provide a secure communication channel even when the underlying system and network infrastructure are not secure. This is particularly important when data passes through shared systems or network segments where multiple users may have access to the information. In these situations, sensitive data, especially passwords, should be encrypted in order to protect it from unintended disclosure or modification.

Pattern Recognition

Pattern recognition aims to classify data (patterns) based on either *a priori* knowledge or on statistical information extracted from the patterns. The patterns to be classified are usually groups of measurements or observations, defining points in an appropriate multidimensional space. This is in contrast to *pattern matching*, where the pattern is rigidly specified as in exact password matching.

A complete pattern recognition system consists of a sensor that gathers the observations to be classified or described; a feature extraction mechanism that computes numeric or symbolic information from the observations; and a classification or description scheme that does the actual job of classifying or describing observations, relying on the extracted features.

The classification or description scheme is usually based on the availability of a set of patterns that have already been classified or described. This set of patterns is known as the training set, and the resulting learning strategy is characterized as supervised learning. Learning can also be unsupervised, in the sense that the system is not given an *a priori* labeling of patterns; instead, it establishes the classes itself based on the statistical regularities of the patterns.

The classification or description scheme usually uses one of the following approaches:

- Statistical (or decision theoretic)
- Syntactic (or structural)

Statistical pattern recognition is based on statistical characterizations of patterns, assuming the patterns are generated by a probabilistic system. Syntactical pattern recognition is based on the structural interrelationships of features. A wide range of algorithms can be applied for pattern recognition, from very simple Bayesian classifiers to much more powerful neural networks.

Typical applications are automatic speech recognition, classification of text into several categories (e.g., spam/nonspam e-mail messages), the automatic recognition of handwritten postal codes on postal envelopes, or the automatic recognition of images of human faces. The last two examples form the subtopic image analysis of pattern recognition that deals with digital images as input to pattern recognition systems.

Pattern recognition is more complex when templates are used to generate variants.

Access Control of Digital Assets and Resources

As was stated in Chapter 1, a priority issue for many organizations has been access control into computer and network facilities and buildings. This is particularly true if the digital assets and physical resources are expensive and replacing or recreating them would be difficult and cost-prohibitive. This section looks at the issues relating to egress and ingress to buildings and facilities that house computer and network resources.

Mandatory access control (MAC) is an access control technique that assigns a security level to all resources (e.g., information, parts of a building), assigns a clearance level to all potential users requiring access, and ensures that only users with the appropriate clearance level can access a requested resource.

It is essential that only authorized personnel be allowed, unattended, into a facility. Not only is ingress an issue, egress can be just as important, as hardware, software, and documentation tend to disappear if management controls are not present. Unauthorized people can cause several problems, including theft of equipment or data, destruction of equipment or data, and viewing of sensitive data and information. There are a number of devices and systems that can provide access control and management. As previously stated, a program must be in place that addresses access control for both external individuals and internal personnel who are not authorized access. Smart cards with PINs can be utilized in the organization's access control program.

A number of support organizations, vendors, and suppliers might need access to the physical facility, including:

- Maintenance personnel, who will probably require frequent access to the computer and network components
- Software engineers and other specialists, who may also require access during all hours of the day
- Communication suppliers and vendors, who often have a requirement to access the demarcation closet

The facility that houses the digital assets and resources in any organization is a critical location and access must be adequately controlled. Different groups may require a different level of access control. Very few of these people should be allowed unescorted access into sensitive areas. Someone must be responsible for providing authorization for these personnel to access computer equipment and networking facilities.

Visitors into a facility can fall into a number of different categories, including:

- Relatives of employees
- Tour groups
- Competitors
- Sales representatives
- Trainers
- Others

None of these visitors should be allowed free access throughout an organization. Trainers and education providers fit in a different category than the others, as they may require access to both computer and networking components as part of a training program.

Techniques for Theft Prevention

The most successful method of preventing theft of computer and networking resources is to control access. Access control devices can prevent access by unauthorized individuals and record access by those authorized. Mainframes, client/servers, and other large devices are difficult to get past the front door; however, there are a number of devices and software that are easy to conceal. Several approaches can be taken to reduce the incidence of theft of these resources, including preventing access, restricting portability, and detecting exit. These approaches and solutions include:

- Guard
- Lock
- Electronic

Guard Solution

A common approach is to employ a human security guard; however, there are situations where guard dogs are deployed. Guards are traditional, well understood, and adequate in many situations. There are a number of security services that offer guard solutions. To be effective, human guards must be on duty continuously, which implies seven days a week, 24 hours a day. A guard must personally recognize a person or some identification, such as a badge. People often lose or forget badges, and badges can be forged. Employees and contractors who no longer have authorized access may still have their badges. The guard must make a record of everyone entering and leaving a facility, if any tracking is required. This guard service can be expensive. It is also essential to establish the credibility of the guards, who can be a source of theft.

Lock Solution

The simplest technique to prevent theft is to lock the room or closet that contains the equipment or resource. The lock is easier, cheaper, and simpler to manage than employing a guard. However, a lock does not provide a record of entries and exits, and there are difficulties attributed to lost and duplicated keys. An unauthorized person may walk through a door which has been unlocked by another. There is also the inconvenience of having to fumble for a key while carrying an armload of something. A lock and key solution requires someone be the "keeper of the keys." More often than not, this person is unavailable when someone needs access to a locked room.

Electronic Solution

More sophisticated access control devices employ magnetic strip cards, cards with radio transmitters, and cards with electronic circuitry that makes them difficult to duplicate. Because each of these cards interfaces with a computer, lists can be produced of those entering and exiting a secured location, along with date and time stamps. Some devices operate by proximity, so that a person can carry it in a pocket or on a neck rope. It is also easy to update the access list of authorized personnel when someone is added or deleted from the list. New employees, dismissed employees, and lost or stolen cards can be readily noted on the access list.

An electronic solution—a label or a tag—can also alert when a protected resource is removed from a secure location. This label is small and unobtrusive, and is used by a number of libraries. It looks like a normal pressure-sensitive label; however, sensors at the entrance and exit can detect the security tag. These tags are available for vehicles, machinery, equipment, electronics, software, and other documents. The detector sounds an alarm, and someone must approach the person trying to leave with a protected object. This may require the addition of a human guard, which can add to the security cost.

Biometric Access Control

Biometric access control systems consist of a reader or scanning device—software that converts the gathered information into digital form—and a database that stores the information for comparison with previous records. These readers (or scanning devices) can scan for a fingerprint, hand geometry, a signature, an iris/retina, facial recognition, a voice print, or a vascular pattern. This technology can be used for a number of applications, including time and attendance reporting, building access control, verification of signatures, point-of-sale identity verification, process-control security, and cellular phone security.

Authentication is the process of establishing confidence in the truth of some claim. The claim could be any declarative statement, for example: "This individual's name is 'Joseph K.'" or "This child is more than 5 feet tall." In biometrics, "authentication" is sometimes used as a generic synonym for verification.

Verification is a simple process for users. A PIN is entered into a keypad, a magnetic stripe/barcode card is swiped, or a proximity card is used to touch the biometric reader. As a result, the reader accesses a template taken of the person's biometric data at the time of enrollment. Depending on the type of reader, the user places a hand on the unit or a finger on the window or provides another type of biometric input. If the resulting template matches the stored template, the person is verified. Although this can be considerably more convenient than current access methods, such as passwords and cards, many think of the technology as confined to heightened security applications. It is true that biometrics is used to check employees coming into almost every airport and to guard almost every nuclear plant. These access systems are also the mode of entry at embassies around the world. However, the majority of implementations are used in common, everyday locales, including hair salons and restaurants.

Embedded Biometric Devices

Biometric sensors are now being embedded into systems such as cell phones or PDAs to customize features or settings and provide increased comfort and convenience. The stand-alone biometric reader, although seldom thought of as an embedded system, has been on the market for some time. A newer alternative features template management performed by a smart card. Breaches of security can be minimized in many ways. One method limits the opportunity for infringement. For instance, if the database of those authorized for access to a facility does not rely on networked hardware systems, the chance of someone infiltrating the system is reduced, as are the possibilities of downtime. Although new methods and attributes, such as vascular recognition and DNA matching, are promising, they are not yet mature enough for mainstream usage. Alternatives that are further along in development already exist, and a variety of common methods are commercially available today. A comparison of biometric technologies is depicted in Table 8-1. [3]

Selection of the appropriate biometric access control technology depends on a number of application-specific factors, including the environment in which the identity verification process is carried out, the user profile, accuracy requirements, and overall system cost.

Success Factors and Design Challenges

The main goal of any access control system is to keep some people out and allow others to get in. Several key factors must be considered early on when designing a biometric application.

Requirement	Face	Fingerprints	Hand Geometry	Iris	Retina	Signature	Voice
Ease of use	Medium	High	High	Medium	Low	High	High
Accuracy	High	High	High	V. High	V. High	High	High
User acceptance	Medium	Medium	Medium	Medium	Medium	High	High
Error factors incidence	Light, age, hair, glasses	Dryness, dirt, age	Hand injury, age	Lighting	Glasses	Changing signatures	Noise, colds
Stability	Medium	High	Medium	High	High	Medium	Medium

Table 8-1 The Effectiveness of Access Control Technologies

These factors include user acceptance, throughput, accuracy, encryption, and identity-theft aversion.

User acceptance of the access control device is one of the most critical factors in the success of a biometric-based implementation. In order to prevent improper use, which can cause access errors, the device should not cause discomfort or concern and must be easy to use. Throughput, which is application-dependent, is the total time required to use the device. The elapsed time from presentation to identity verification is known as verification time. Most readers can verify identity within one second.

However, when considering the use of biometrics for access control, the total time it takes a person to use the reader must be considered. This includes the time it takes to enter the ID number and the time required to get into the right position for scanning. The total time required for each person varies. The use of smart cards can help avoid identity theft. Since the card stores the user's ID and biometric template, there is no database that can be hacked. When the smart card is presented to the smart card reader embedded in a biometric reader, the user's hand is placed on the unit. The hand geometry reader compares length, width, thickness, and surface area measurements against the template stored in the smart card to verify identity. This process takes approximately one second and is virtually foolproof. [4]

Biometric identification devices use individual physical attributes such as fingerprints, palm prints, voice attributes, and retina patterns to identify authorized users (see Figure 8-4). Fingerprint identification is the most popular biometrics technique and retinal scanning is the most accurate. Biometrics is currently used as a network security technique in a limited number of U.S. companies.

Biometric identification has been expensive to implement, but costs are decreasing. There is considerable end-user resistance to fingerprinting, which is associated with criminal activity. While biometrics promises more secure identification of authorized users than that of passwords, it also can be circumvented. Several terms need to be defined concerning biometric identification before proceeding.

A **sensor** includes hardware found on a biometric device that converts biometric input into a digital signal and conveys this information to the processing device. **Sensor aging** is the gradual degradation in performance of a sensor over time. A *fingerprint sensor* is part of a biometric device used to capture a fingerprint image for subsequent processing. *Live capture* is

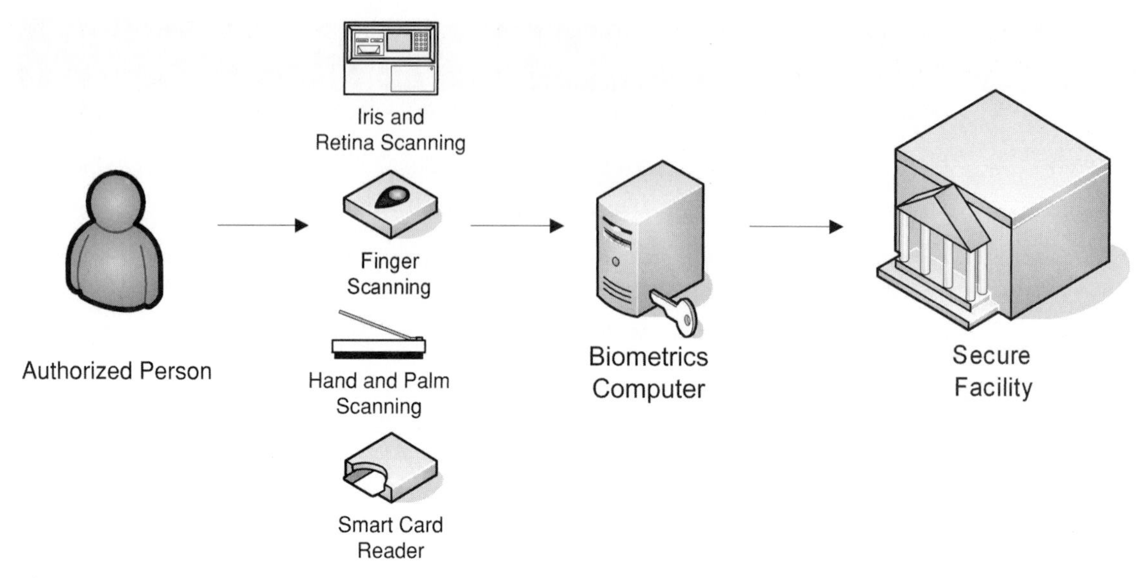

Figure 8-4 Biometric Access Methods

Course Technology/Cengage Learning

the process of capturing a biometric sample by an interaction between an end user and a biometric system and *live scan* occurs when taking a fingerprint or palm print directly from a subject's hand.

- Accessibility means the biometric feature is easily presented to an imaging sensor.
- Acceptability means the biometric feature is perceived as nonintrusive by the user.
- Availability is the amount of time that a computing resource is available for use by a user or system.

Role of Biometrics in Physical Access Control

Biometrics can play a major role in physical access control. It is essential to determine if biometrics can be integrated into access control applications. It is necessary to consider a number of key issues when using a biometric device. Biometrics identifies a person via a unique human characteristic: the size and shape of a hand, a fingerprint, one's face, or several aspects of the eye. If the goal of an access control system is to control where people, not credentials, can and cannot go, then only a biometric device truly provides this capability to the end user. [5]

As a result, biometrics is used at the front doors of thousands of businesses around the world, at the entrances to the tarmacs of major airports, and at the entrances of other facilities where the combination of security and convenience are desired. Many biometric hand readers control client and employee access to special areas of banks and prisons across the world. Biometric applications are useful for both prisoner and visitor tracking. Universities use hand readers for on-campus meal programs and to safeguard access to dormitories and protect computer centers. Hospitals utilize the biometric devices for access control and payroll accuracy. There are biometric systems available today which economically meet the needs of almost any commercial or governmental access control application. And, as costs continue to decline, justifying the use of a biometric is becoming a reality and necessity for more and more organizations.

Today, it is surprising to many that biometric access control is actually found in applications requiring minimal security. For instance, health clubs are major users because biometric readers easily let customers into the club. The customer doesn't need to carry another card, and the club doesn't have to issue and administrate a card system.

Biometric systems have produced millions of verifications at San Francisco International Airport (SFO), with more than 50,000 produced on high-volume days. Hand readers span the entire airport, securing more than 180 doors and verifying the identity of more than 18,000 employees. The use of biometrics at San Francisco is airport-wide and fully integrated into the primary access control system.

The Benefits of Biometrics in Access Control

The goal of any access control system is to let authorized people, not just their credentials, into specific places. Only with the use of a biometric device can this goal be achieved. A card-based access system will control the access of authorized pieces of plastic, but not who is in possession of the card. Systems using PINs require that an individual only know a specific number to gain entry, but who actually entered the code cannot be determined. On the contrary, biometric devices verify who a person is by what he is, whether by hand, eye, fingerprint, or voice characteristic.

The SFO access control regulations expect the airport to validate people gaining access to the airfield, not a piece of plastic or an ID card. Unless a biometric component is being utilized, the only thing being validated is the ID card. It's a system that begins when an employee shows up for the first day of work and creates a template of the hand geometry at the license and permit bureau. When an employee at the airport is terminated, regardless of whether her ID badge is recovered, the template can be instantaneously removed from the system, and she will never be able to get through one of the access control doors. In addition to its proven performance and reliability, one of the most attractive features of this biometric identification is its efficiency. A hand cannot (usually) be lost, forgotten, or stolen like a password or ID card.

Biometrics eliminates the need for cards. While dramatic price reductions have lowered the capital cost of the cards in recent years, the true benefit of eliminating them is realized through reduced administrative efforts. For instance, a lost card must be replaced and reissued by someone. Just as there is a price associated with the time spent to complete this seemingly simple task, when added together, the overall administration of a card system is costly.

Integration Tailored to the Application and Organization Served

Access control requires the ability to identify the person plus unlock a door, grant or deny access based on time restrictions, and monitor door alarms. There are a variety of systems where biometrics accomplishes these tasks. Generally, these include:

- Stand-alone systems
- Networked systems
- Smart card systems
- Third-party system integration

Stand-Alone Systems

Many biometric devices are available in a stand-alone configuration. Such devices are not only a biometric, but also a complete controller for a single door. Users are enrolled at the unit and their biometric template is stored locally for subsequent comparison. The actual comparison is accomplished within the unit and a lock output is energized depending on the outcome.

Networked Systems

Many access control applications have a need to manage more than one door. While multiple stand-alone units could be employed, a network of biometric readers is much more feasible. By networking the systems together and then connecting them to a computer, several advantages are available to users. The most obvious is centralized monitoring of the system. Alarm conditions and activity for all the doors in the system are reported back to the PC. All transactions are stored on the computer's disk drive and can be recalled for a variety of user-customized reports. Figure 8-5 depicts a networked biometric system linked to a remote computer and database system.

Networked systems also provide convenient template management. Although a user enrolls at one location, the template is available at other authorized locations. Deletion of a user or changes in an access profile is simply entered at the PC. Some biometric systems store all information in the PC where template comparisons are also performed. Others distribute template information to the individual readers at each door. Either way, the net effect of template management is the same.

Smart Card Systems

Smart cards raise the bar even higher, providing additional capabilities and flexibilities. As costs begin to come down and usage is more widespread, biometric devices can leverage their secure data storage. For example, a smart card can store both the user's ID number and hand geometry template on the card. Because of this, there is no need to distribute hand templates across a network of hand readers or require the access control system to manage biometric templates. This means integration to any existing access control application is greatly simplified and additional network infrastructure costs are eliminated. Since the template only resides on the card, the solution also eases individual privacy concerns.

Providing the best of smart cards and biometrics, the solution provides dual authentication by requesting both the right card and the right person. A smart card reader is embedded into the

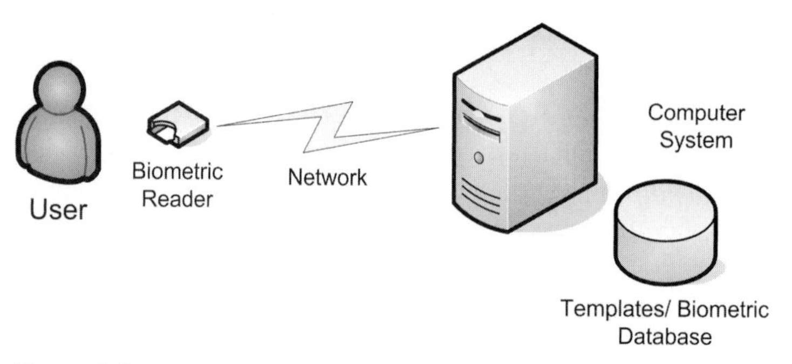

Figure 8-5 Networked Biometric System

Course Technology/Cengage Learning

biometric reader. A plastic cardholder is affixed to the side of the unit. The verification process takes approximately one second and is virtually foolproof.

Third-Party System Integration

Manufacturers offer a variety of different methods to integrate biometrics into conventional access control systems. The most common way is card reader emulation. This method is very effective when integrating into existing card-based systems to bring extra security to the front entrance or server room. The wiring is identical to the card readers wiring.

In this mode, the biometric device essentially works with the access control panel in the same way that a card reader does. The biometric device is connected to the panel's card reader port. The unit outputs the ID number of an individual if, and only if, the user is verified. The format of the output is consistent with the card technology used by the access control panel. Once an ID number reaches the panel, it is handled as if it came from a card reader. The determination of granting access is made by the panel. The access control panel, not the biometric, handles door control and monitoring.

As an alternative to a keypad, some biometric readers also have card reader input capability—the most common being proximity and smart cards, although other technologies are also supported. At the biometric unit, the user swipes the card, which contains an ID number. If verified, that card number is sent to the panel for a decision.

Typical Device/Systems Process Flow

While individual biometric devices and systems have their own operating methodology, there are some generalizations one can make as to what typically happens within a biometric systems implementation.

Figure 8-6 depicts the process pictorially and the accompanying notes provide a more detailed explanation.

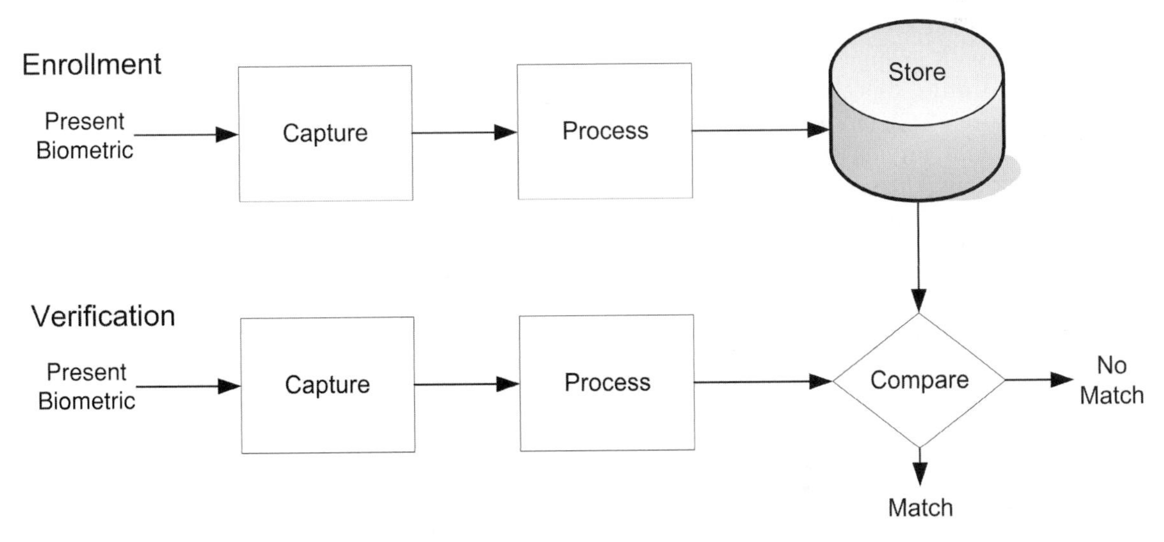

Figure 8-6 Systems Process Flow

Course Technology/Cengage Learning

A. *Enrollment:* The first priority is to capture a sample of the chosen biometric to verify an individual's identity. This sample is referred to as a *biometric template* and is the reference data against which subsequent samples provided at verification time are compared. A number of samples are usually captured during enrollment (typically three) in order to arrive at a truly representative template via an averaging process. A biometric sample includes information or computer data obtained from a biometric sensor device. Examples are images of a face or fingerprint. The template is then referenced against an identifier (typically a PIN or card number if used in conjunction with existing access control tokens) to recall and make it ready for comparison with a live sample at the transaction point. The enrollment procedure and quality of the resulting template are critical factors in the overall success of a biometric application. A poor-quality template will often cause considerable problems for the user, often resulting in a reenrollment.

B. *Template storage:* This is an area of interest, particularly with large-scale applications which may accommodate many thousands of individuals. The possible options are:

1. Store the template within the biometric reader device.
2. Store the template remotely in a central repository.
3. Store the template on a portable token such as a chip card.

Option 1, storing the template within the biometric device, has both advantages and disadvantages, depending on how it is implemented. The advantage is a potentially fast operation as a relatively small number of templates may be stored and manipulated efficiently within the device. In addition, users are not relying on an external process or data link in order to access the template. In some cases, where devices may be networked together directly, it is possible to share templates across the network. The potential disadvantage is that the templates are somewhat vulnerable and dependent upon the device being both present and functioning correctly. If something happens to the device, users may need to reinstall the template database or possibly reenroll the user base.

Option 2, storing the templates in a central repository, is the option which will occur to many IT systems engineers. This may work well in a secure networked environment where there is sufficient operational speed for template retrieval to be invisible to the user. However, with a large number of readers working simultaneously, there could be significant data traffic, especially if users are impatient and submit multiple verification attempts. The size of the biometric template itself will have some impact on this, with popular methodologies varying between 9 bytes and 1.5k. Another aspect to consider is that if the network fails, the system effectively stops unless there is some sort of additional local storage. This may be possible to implement with some devices, using the internal storage for recent users and instructing the system to search the central repository if the template cannot be found locally.

Option 3, storing the template on a token, is an attractive option for two reasons: first, it requires no local or central storage of templates; and second, the user carries his template with him and can use it at any authorized reader position. However, there are still considerations. If the user is attracted to the scheme because he believes that there is effective personal control and ownership of the template, then also storing the template elsewhere in the system would not be acceptable. If the user subsequently loses or damages the token, then there will be a need to reenroll. Another consideration may be unit cost and system complexity if users need to combine chip card readers and biometric readers at each enrollment and verification position.

If the user base has no objection, administrators may wish to consider both on-token and central storage of templates (options 2 and 3). This could provide fast local operation with a fallback position if the chip card reading process fails for any reason, or if a genuine user loses his token and can provide suitable identity information. The choice of template storage may be dictated to some extent by the choice of biometric devices, as some devices offer greater flexibility than others.

C. *The network:* There are numerous possible options with regard to networks. Some devices have integral networking functionality, often via RS485 or RS422 with a proprietary protocol. This may enable administrators to network a number of devices together with no additional equipment involved, or maybe with a monitoring PC connected at one end of the network. In this case, reliance will be on the vendor's systems design and message functionality, together with the user's own software.

Alternatively administrators may design the networking, message passing and monitoring system, taking advantage of the recent generic biometric APIs and accessing the reader functions directly. This will provide absolute flexibility and control over systems design, if the chosen device supports it.

Another option may be to use the vendor's network for message passing and primary interconnection, coupled to custom software at the monitoring point, which may, in turn, interface with other systems being controlled. In some cases, the vendor may possess an existing network and control interface into which the biometric devices may be integrated via a common security standard. In this case, they will appear as just another device, although template storage and access must be separately considered.

D. *Verification:* This process requires the user to claim an identity by either entering a PIN or presenting a token, and then to verify this claim by providing a live biometric to be compared against the claimed reference template. There will be a resulting match or no match, accordingly. A record of this transaction will then be generated and stored, either locally within the device or remotely via a network and host (or both).

With certain devices, administrators may give the user a number of attempts at verification before finally rejecting them if the templates do not match. Setting this parameter requires some thought. Administrators may want to provide every opportunity for a valid user—who may be having difficulty using the system—to be recognized. An *attempt* is the submission of a single set of biometric samples to a biometric system for identification or verification. Some biometric systems permit more than one attempt to identify or verify an individual. However, administrators do not want impostors to have too much opportunity to experiment. With some systems, the reference template is automatically updated upon each valid transaction. This allows the system to accommodate minor changes to the user's live sample as a result of aging, local abrasions, etc., and may be a useful feature when dealing with large user databases.

E. *Transaction storage:* This is an important area as there must be a secure audit trail with respect to the use of the system. Some devices will store a limited number of transactions internally, scrolling over as new transactions are received. This is acceptable as long as all such transactions can be retrieved before the buffer fills up and transactions are lost. In practice, this is unlikely to be a problem unless there are severe network errors. In some cases, each biometric device is connected directly to a local PC, which may in turn be polled periodically (overnight, for example) to download transactions to a central point. In either case, adoption of a local procedure to deal with error and exceptional conditions could be required, which will require some sort of local messaging. This may be as simple as a relay closure in the event of a failed transaction activating an enunciator of some description. What happens with this transaction data is another matter. It can be analyzed via an existing reporting tool, or perhaps technicians can write a custom application to show transactions in real time as well as write them to a central database.

F. *The network:* How the network handles transactions may be of critical importance in some applications. For example, multiple terminals may be distributed within a large facility, each of which requires a real-time display of information. This will require fast and reliable message transmission. Each terminal user may wish to "hold" a displayed transaction until a response has been initiated. This will require a separate local message buffer and possibly a message prioritization methodology to ensure that critical messages are dealt with promptly.

Terminal/host software may be required according to the user and core function. These details will need to be accommodated within the overall network in a secure and efficient manner. There are many potential issues to consider in this respect and the overall system design should reflect this.

Issues to Consider

The most critical factor in the success of a biometric system is user acceptance of the device. There are several factors that impact acceptance. These include:

- Cause no discomfort to the user
- Easy to enroll
- Easy to use
- Work correctly

Biometric devices must cause no discomfort or concern for the user. This may be a subjective issue, but it is important to fully explain any concerns that users may have. If people are afraid to use a device, they most likely will not use it properly, which may result in not being granted access. The system must enroll people easily. Many people will become frustrated if they have to go through the process over and over again. Potential users are predisposed to reject the system. The biometric device must be easy to use. People like things that are simple and intuitive. How many times are users frustrated with a card reader that gives no indication of which way to swipe the card? The biometric device must work correctly, even though no device or system is perfect. In the biometric world, the two errors a unit can make are letting a bad guy in and keeping a good guy out. The probability of one of these errors happening is characterized by the false acceptance and false rejection error rates, explained in previous chapters.

Throughput is a logistical issue that should be considered when using a biometric: this is the total time it takes for a person to use the device. It is difficult for manufacturers to specify a throughput, since it is application-dependent. Most manufacturers specify the verification time for the reader, but that is only part of the equation.

When a person uses a biometric reader, she typically enters an ID number on a keypad. The reader prompts her to position the hand, finger, or eye where the device can scan physical details. The elapsed time from presentation to identity verification is the verification time. Most biometric readers verify ID in less than two seconds.

However, when considering the use of biometrics for access control, one must look beyond the verification time and consider the total time it takes for a person to use the reader. This includes the time it takes to enter the ID number, if required, and the time necessary to be in position to be scanned. If ID numbers must be entered, they should be kept as short as possible. If a long ID number must be used, some devices can obtain the number by reading a card that contains the ID number in the card code.

The total time required for a person to use the reader will vary between biometric devices, depending on their ease of use and verification time. A card-based access system may seem faster. However, as one hand geometry user points out, the speed difference between a card and the hand reader is about two seconds, but users make up for the difference since the hand is right in front, verses fumbling around looking for a card.

Proven Biometric Technologies

There are a wide variety of human characteristics used by biometric devices to confirm a person's identity. The biometric industry is constantly finding new attributes and ways to measure their uniqueness. Biometric devices that utilize one or more of these characteristics are commercially available today. A proper assessment is built not only on a general understanding of biometrics, but also on an understanding of specific technologies. An understanding of both biometrics in general and specific biometric technologies is a necessary condition for a solid understanding of the larger social implications of biometrics. A recap of these biometric modalities, described in Chapter 2, is repeated here to refresh the reader's awareness.

Hand Geometry

A hand reader uses the size and shape of the hand and fingers to verify a person's identity. Hand geometry evaluates a three-dimensional image of the four fingers and part of the hand. This was the technology used in the very first commercially available biometric device, which came to market in 1976. It continues to be the most widely used biometric device for access control applications. Frost & Sullivan's *World Biometric Report* [6] determined that hand geometry readers continue to be the dominant biometric technology for access control and time and attendance applications. In fact, more people order hand readers for these applications than all fingerprint and face systems combined.

Fingerprint

Law enforcement agencies have used fingerprints for decades to identify individuals, and businesses continue to do so today when undergoing background checks. However, relatively inexpensive fingerprint access control readers are different than these devices. The FBI fingerprint ID system takes images of all ten fingers, while an access control fingerprint product may only capture one or two fingers for verification. The system then creates a template in a process similar to hand geometry readers for local comparison. Due to throughput concerns, fingerprint access control may be best applied in smaller user populations. Because of cost and size, fingerprint readers are a perfect choice for single-person verification applications, such as in logical access control, where they are used to log onto PCs or computer networks. This is certainly a fast-growing application for this technology.

Facial Systems

The shape of the face, determined by distances between the eyes, ears, and nose, and other facial characteristics, is put into a template. When viewed via a CCTV camera, the image is matched against the template to verify a person. This is a technology many are counting on in the fight against terrorism, as the system could scan large crowds or many people waiting in line. The system can select individuals who should be further scrutinized. In many instances, facial systems are combined with another technology.

For instance, IR Recognition Systems and several partners have developed a dual biometric technology, which combines hand geometry and facial recognition technologies. The system works well in conjunction with travel documents and provides unparalleled convenience for three-factor authentication of face, card, and hand.

After extensive testing, the Israeli government determined the combination of the hand reader along with face technology best fit the security, convenience, and efficiency requirements of a new high-volume, border-crossing control system application.

Eye

The iris of each eye of each person is absolutely unique. In the entire human population, no two irises are alike in their mathematical detail. This even applies to identical twins.

This iris of each eye is protected from the external environment. It is clearly visible from a distance, making it ideal for a biometric solution. Image acquisition for enrollment and recognition is easily accomplished and most importantly is nonintrusive. No other biometric technology can rival the combined attributes of mathematical certainty, speed, and noninvasive operation offered by iris recognition.

The IrisCode™ creation process starts with video-based image acquisition. This is a purely passive process achieved using CCD video cameras. This image is then processed and encoded into an IrisCode™ record, which is stored in an IrisCode™ database. This stored record is then used for identification in any live transaction when an iris is presented for comparison.

Considerations for Access Control Systems

If the goal of an access control system is to control where people, not credentials, can and cannot go, only a biometric device truly provides this capability to the end user. There are biometric systems available today that economically meet the needs of almost any commercial access control application. Justifying the use of a biometric is a reality for more and more organizations. Several questions should be answered before allocating resources to access control. These include: [7]

- How is biometrics integrated into access control applications?
- What are the key issues to be considered when using a biometric device?

Summary

- There are numerous legacy methods, procedures, and vehicles that address the various access control issues, including passwords, firewalls, filters, tokens, and ACLs. Hardware and software products and services provide access security for the organization's assets and resources. These products would be implemented in physical settings, computer installations, and networking systems. As concentration points for large numbers of circuits, equipment rooms, demarcation points, and telecommunications closets are particularly vulnerable and must be kept locked and access controlled.
- In a computer system or network there must be a method of identifying anyone who is allowed into the environment. This takes the form of identification, authentication, and authorization. The password is a form of secret authentication data that is used to control access to a resource, such as a location, database, or file storage. Authorization refers to securing the network by specifying which area of the network, whether it is an application, a device, or a system, a user is allowed to access.
- It is essential that only authorized personnel be allowed—unattended—into a facility. Not only is ingress an issue, egress can be just as important, as hardware, software, and documentation tend to disappear if management controls are not present. The facility that houses the digital assets and resources in any enterprise is a critical location and access must be adequately controlled. Different groups may require a different level of access control.
- Biometric access control systems consist of a reader or scanning device, software that converts the gathered information into digital form, and a database that stores the information for comparison with previous records. These readers—or scanning devices—can scan for a fingerprint, hand geometry, a signature, an iris/retina, facial recognition, a voice print, and a vascular pattern. This technology is particularly effective in facility, computer, and network access control, ID verification, and authorization.
- Biometrics can play a major role in physical access control. However, it is essential to determine if it can be integrated into access control applications. It is also necessary to consider a number of key issues when using a biometric device. If the goal of an access control system is to control where people, not credentials, can and cannot go, then only a biometric device truly provides this capability to the end user.

Key Terms

Certificate authority (CA) A trusted third party that is responsible for issuing and revoking digital certificates within the public key infrastructure.

Certificate revocation list (CRL) A list of certificates that have been revoked before their expiration by a certificate authority.

Certification The process of verifying the correctness of a statement or claim and issuing a certificate as to its correctness.

Chain of trust An attribute of a secure ID system that encompasses all of the system's components and processes and assures that the system as a whole is trustworthy.

Contact smart card A smart card that connects to the reading device through direct physical contact between the smart card chip and the smart card reader.

Contactless smart card A smart card that communicates with a reader through a radio frequency interface.

Control panel The access control system component that connects to all door readers, door locks, and the access control server.

Demarcation The physical point where a regulated network service is provided to the user.

Digital signature The number derived by performing cryptographic operations on the text to be signed.

Door reader The device on each door that communicates with a card or credential and sends data from the card to the controller for decision on access rights.

Door strike The electronic lock on each door that is connected to the controller.

Excite field The RF field or electromagnetic field constantly transmitted by a contactless door reader.

FaceIt® A facial recognition software engine that allows computers to rapidly and accurately detect and recognize faces.

Faraday cage An enclosure formed by conducting material, or by a mesh of such material, that blocks out external static electrical fields.

Form factor The physical device that contains the smart card chip.

Hardware The physical part of a computer, including the digital circuitry, as distinguished from the computer software that executes within the hardware.

Hash function A reproducible method of turning some kind of data into a relatively small number that may serve as a digital "fingerprint" of the data.

Head-end system The physical access control server, software, and database(s) used in a physical access control system.

Hybrid card A smart card that contains two smart card chips—both contact and contactless chips—that are not interconnected.

Identification (2) Procedures and mechanisms which allow an entity external to a system resource to notify that system of its identity.

Least privilege A principle that may be used to guide the design and administration of a computer system or enterprise network.

Microcontroller A highly integrated computer chip that contains all of the components comprising a controller.

Network access control A series of technologies that may include software, hardware, or servers that monitor devices accessing a network.

Nonrepudiation The ability to ensure and have evidence that a specific action occurred in an electronic transaction.

Password A security measure used to restrict access to computer systems and sensitive files. A password is a unique string of characters that a user types in as an identification code.

Physical access control system (PACS) A system composed of hardware and software components that controls access to physical facilities such as buildings, rooms, airports, and warehouses.

Role The actions and activities assigned to or required or expected of a person or group.

Role-based access control (RBAC) Provides access to resources based on a user's assigned role.

Security card A card that requires a permanent user name and a temporary password. This password changes automatically after a specific amount of time, often 60 seconds.

Sensor Hardware found on a biometric device that converts biometric input into a digital signal and conveys this information to the processing device.

Sensor aging The gradual degradation in performance of a sensor over time.

Software Includes the programs or other "instructions" that a computer needs to perform specific tasks.

Trusted Computing Group (TCG) A vendor consortium founded in 2003 to promote the use of vendor-neutral specifications.

Trust model A framework for an organizational secure systems policy that addresses resource authorization.

Trust relationship The link between two entities on a network (e.g., two servers) that enables a user with an account in one domain to have access to resources in another domain.

Review Questions

1. Identify access control mechanisms.
2. Identify the various types of authentication.
3. Describe the smart card and identify the different types of smart cards.
4. What is the relationship between a privilege and a role?
5. What does a digital certificate accomplish?
6. What is the purpose of encryption and enciphering?
7. Why are egress and ingress an access control issue?
8. List the categories of visitors that might require access control techniques.
9. Provide a list of the various techniques for theft prevention.
10. Identify those types of systems that provide biometric controls.
11. An API is a set of services or instructions used to standardize an application. (True or False?)
12. A ________________ is the physical point where a regulated network service is provided to a user or organization.
13. A ________________ is a device on an entrance that communicates with a card and sends data to a controller for an access decision.

14. A ______________ consists of the actions and activities assigned to or required or expected of a person or group.
15. A digital signature is a password-protected file that contains identification about its holder. (True or False?)
16. Repudiation is the ability to ensure and have evidence a specific action occurred in an electronic transaction. (True or False?)
17. Pattern matching is another name for pattern recognition. (True or False?)
18. Three possible solutions for theft prevention include: ______________, ______________, and ______________.
19. A scanner includes hardware found on a biometric device that converts biometric input into a digital signal and transmits it. (True or False?)
20. The sample of a chosen biometric to verify an individual's identity is referred to as a/an ______________.

Discussion Exercises

1. Provide a list of authentication categories and describe each.
2. Describe the various access control mechanisms.
3. Describe the purpose and use of an API.
4. Describe the purpose and use of an algorithm.
5. Identify the hardware and software components of some access control system.
6. Provide an overview of the various categories of smart cards.
7. Provide an overview of the TCG and the TPM.
8. Provide an overview of the use of Faraday technologies.
9. Describe the actions that are required when managing passwords.
10. Identify the currently available certificate authorities.

Hands-On Projects

8-1: Security and Access Technologies Awareness

Time required—30 minutes

Objective—Identify biometric security and access technologies that have been developed and implemented in some organization.

Description—Use various Web tools, Web sites, and other tools to identify biometric security and access technologies that are in development or production for both commercial and government organizations.

1. Use a real-world example to describe an authorization and authentication flow.
2. Develop a graphic that depicts a security card environment for some organization or application.
3. Develop a graphic that depicts a biometric system process and flow.
4. Provide an overview of the FaceIt® facial recognition product. Identify the capabilities and features of this tool.
5. Provide an overview of the responsibilities of the Trusted Computing Group and describe the trusted platform module. Identify the components of this type of implementation.
6. Develop a scenario where a hash function or cryptography techniques can be used as a digital fingerprint.

8-2: Identify Security and Access Control Technologies

Time required—45 minutes

Objective—Identify security and access control technologies that are available for providing some level of access control and security for OnLine Access assets and resources.

Description—The owners and managers of OnLine Access have identified the necessary policy and management requirements for implementing a biometric access control solution. They must now look at the security and access control technologies available for implementing a biometric solution.

1. Research the security and access control technologies that are available for access control and security solutions.
2. Produce a cost–benefit analysis of the various products.
3. Identify security and access control technologies that have been implemented in other biometric implementations.
4. Produce a report describing how these technologies have impacted the implementation of a biometric solution.
5. Identify help desk and technical support provided by the product providers.

chapter 9

System Integrity and Accessibility

After completing this chapter, you should be able to do the following:

- Understand the issues relating to computer and network system integrity and confidentiality
- Look at the requirements of system availability and accessibility
- See how IS/IT systems are protected from various attacks and compromises
- Look at the performance measures used in biometric systems
- Become familiar with the issue of biometric methods' accuracy
- Identify areas of concern that apply to the protection of biometric systems and database assets
- Understand smart card technologies within the access-control environment

Confidentiality, integrity, and accessibility are especially relevant when implementing and operating a networked computer system. An organization's credibility is on the line when the integrity of confidential data and information is compromised. An article in the press or an exposé on the evening news concerning a virus attack on a bank can cause a loss of numerous customer accounts. Incorrect account balances could indicate lax control, fraudulent accounting practices, dishonest employees, or external thieves. The e-commerce network depends upon the ability of buyers and sellers to access products and services promptly, while being assured that the transactions are confidential and accurate.

Introduction

This chapter addresses the issues of confidentiality, integrity, availability, and accessibility. Maintaining IS/IT system integrity is an ongoing effort involving data and system baselines, backup, policies, procedures, and guidelines.

Threats to system resources and assets can occur from both internal and external sources. Both authorized and unauthorized personnel and individuals can cause havoc in corporate systems. Numerous attacks can occur that compromise system resources, particularly those vulnerabilities that exist on most computer and network systems.

Administrators and security managers have many hardware and software tools to effectively counter system threats and attacks. Threats can occur in all parts of the network, including server systems and mainframe computer configurations.

Local area networks, involving intranets and extranets, must be protected from adverse situations. Databases are common sources of compromise that must be protected. Threat sources can include viruses, device failures, natural disasters, and espionage.

Biometric verification and identification techniques and systems are being developed and implemented to counter security issues. Numerous hardware and hardware solutions are used to make these biometric techniques successful. The major issues with these biometric techniques include performance and error rates.

Considerable efforts are being devoted to the use of smart cards for biometric access control. Implementation time frames and cost are major issues with this initiative.

Confidentiality and Integrity

Confidentiality has been defined by the *International Organization for Standardization (ISO)* as ensuring that information is accessible only to those authorized to have access and is one of the cornerstones of information security. Confidentiality is one of the design goals for many cryptosystems, made possible in practice by the techniques of modern cryptography. Confidentiality of information, enforced in an adaptation of the military's classic need-to-know principle, forms the cornerstone of information security in today's corporations.

Confidentiality also refers to an ethical principle associated with several professions, such as medicine, law, and professional psychology. In ethics, some types of communication between a person and one of these professionals are "privileged" and may not be discussed or divulged to third parties. In those jurisdictions in which the law makes provision for such confidentiality, there are usually penalties for its violation.

Integrity implies protection against unauthorized modification or destruction of information. This defines a state in which information has remained unaltered from the point it was produced by a source, during transmission, storage, and eventual receipt by the destination. In computer and network technologies, the term system integrity has the following meanings:

- That condition of a system wherein its mandated operational and technical parameters are within the prescribed limits
- The quality of a system when it performs its intended function in an unimpaired manner, free from deliberate or inadvertent unauthorized manipulation of the system
- The state that exists, under all conditions, when there is complete assurance that a system is based on the logical correctness and reliability of the operating system, the logical completeness of the hardware and software that implement the protection mechanisms, and data integrity

Accessibility and Availability

Accessibility is a general term used to describe the degree to which a device, service, or environment is accessible by as many users as possible. It is not to be confused with usability, which is used to describe the extent to which a device, service, or environment can be used by specified users to achieve specified goals with effectiveness, efficiency, and satisfaction in a specified context of use. Accessibility can also be viewed as the ability to access the functionality—and possible benefit—of some system or entity. Accessibility is an aggregate measure of how reachable locations are from a given location. Common measures of accessibility are distance and cost. There are also new laws that define accessibility for handicapped individuals.

Availability means ensuring that authorized users—people or computer systems—are able to access information without interference or obstruction, and to receive it in the required format. High availability means there are implementations of products and services for ensuring high availability of systems running mission-critical applications. These include instruments for monitoring and reaction, alert management, and consistency checking. Fault-tolerant messaging and effective state and session management also ensure reliable operation.

Maintaining System Integrity

Protecting network assets and operations is a continuing task with unpredictable results. When intelligently applied, protective efforts can reduce, but never completely eliminate, the chances of losses due to security breaches. Although most network security violations take place within corporate networks and are initiated by authorized and internal users, most funding for security programs are allocated to measures that guarantee that only authorized users are allowed to access the network. These funds are also designated for the prevention and detection of external invasions. Management must develop and implement security policies for computer and network assets and resources.

Security Policies

All security policies set forth in a contingency plan should be detailed and followed closely. The security policies will depend on the computer and network infrastructure size, the organization's security standards, and the value and sensitivity of the data. The security plan must include physical security, network, database, and computer security. The five most significant issues for the security of a computer network that should be addressed include:

- Identification/authentication: Users are accurately identified
- Access control/authorization: Only legitimate users can access a resource
- Privacy: Eavesdropping is not an issue and transmissions are private
- Data integrity: Activities on a database are controlled and protected
- Nonrepudiation: Users cannot deny any legitimate transactions

Network security and access control can be enhanced by a number of user name and password requirements and resource access requirements. Standards for user name and passwords include the following suggestions:

- Establish minimum and maximum password lengths for user accounts.
- Provide the users with the details for character restrictions.
- Determine the frequency for changing passwords.
- Decide if and when passwords can be reused.
- Decide if there will be exceptions to the policy.

Resource access is generally granted only to those who specifically require it. It is always easier to grant new access to users than to take it away. For dial-in users, special security arrangements will probably be necessary. Many organizations require a security card, which provides a code that must be entered for dial-up access. It is essential that the number of users who perform network administration tasks be limited to the absolute minimum. The more users with access to administrative functions, the more likely security problems will occur.

Another user issue that can cause a security breach is leaving the work area without logging out of the system. This not only allows anyone to utilize that person's access rights, but also could put that employee's job in jeopardy. The system can assist operators by logging off any user who has not entered a transaction within a certain time period, such as five minutes. Users who leave their workstations for more than five minutes would have to go through the logon process again. This can become a performance issue if transaction-processing time is sporadic and the workstation is frequently logged off. A transaction log can be utilized to identify the optimum automatic logoff value.

Security Threats

Security is concerned with protecting data and data systems and includes security techniques for both physical facilities and software. It is also concerned with ensuring the integrity of the *network operating system (NOS)*. This system contains the software that runs on a server and controls access to files and other resources to and from multiple users. Physical security is concerned primarily with providing secure access, while software security includes authentication, authorization, access controls, and user logon. This section addresses security controls for operating and networking systems.

Vulnerabilities and threats are the reason it is necessary to be concerned about security. These include the following destructive activities:

- Internal users may try to access unauthorized data systems.
- Internet users may try to attack systems available to the network.

- Unauthorized users may try to access personal user accounts.
- Attackers may modify data values or destroy data.
- Proprietary information may be copied for resale to competitors.
- Attackers may gain access to an account and lockout or restrict the legitimate users.

A number of these security attacks can be attributed to hackers. Historically, a **hacker** was defined as a person who "hacks" away at a computer until access is gained. Students initially performed these activities, primarily for fun and sport. Today, *hacker* has a more positive meaning and has been replaced with the term "*cracker*," which is reserved for the individual who willfully breaks into computer systems with the purpose of wreaking havoc and destruction.

The most common security breach is access to unauthorized user accounts. This activity occurs when the attacker impersonates or masquerades as the legitimate user by obtaining a user name and password. These can often be obtained from posted notes, "shoulder surfing," or from a network monitor. Because some users may use some personal object or a relative's name as a password, the possibility exists for a *password cracker* to guess user names and passwords. Another successful technique is the *dictionary attack*, where the cracker has access to a very large dictionary of words commonly used for passwords. This threat can be countered by locking a user account after a number of failed logon attempts, usually three.

The most serious attack, however, is when someone gains access to an administrator or superuser account. A *superuser* is a UNIX system administrator who has high-level access privileges. With this access, attackers may lock out the legitimate account owner and perform destructive activities on the system or databases.

Another technique that can have significant repercussions is *eavesdropping*. Network traffic can be monitored with sniffers or wiretaps. A **sniffer** is a LAN protocol analyzer that supports a variety of hardware to include Ethernet and Token Ring LAN topologies. The attacker may monitor the network for long periods of time and record valuable or sensitive information that can be used for future attacks. Information captured can be re-sent in a *replay attack*, which results when a service already authorized and completed is forged by another duplicate request in an attempt to repeat authorized commands. It is possible for an attacker to replay an authentication routine to gain illegal access to a system or network. An eavesdropper might capture packets, modify them, and reinsert them into the data stream, which can be unknown to the legitimate sender and receiver. Sequencing and time-stamping packets can avoid these situations.

The last threat discussed in this section concerns the *denial of service* attack. This involves an attack that attempts to deny corporate computing resources to legitimate users. The attacker can cause a server or processor to slow or stop operations using techniques that overwhelm it with some useless task. The common denial of service attack is implemented by flooding ports on the server with a massive number of transmissions. It is also possible to corrupt the operating system of the processor. This type of attack is becoming commonplace across the Internet.

Securing Data Systems

Network and computer security features such as access controls can be used to protect resources from unauthorized access. Many directories and network resources, such as routers, contain *access control lists (ACL)* that contain entries, which specifically identify access parameters. These lists identify which users and groups have object access and the respective permissions for that access. An object is an entity or component, identifiable by the user; that may be distinguished by its properties, operations, or relationships. These objects can have specific access levels, which include read, write, delete, and execute functions.

Most network operating systems have an auditing system designed to record network activities. These systems can be available for real-time or almost real-time network traffic monitoring. The auditing records can be reviewed on a regular basis to determine if the system is being compromised. This auditing system can record activities such as logons, file access, network traffic, and account access. It is essential that auditing records be protected to prohibit an attacker from modifying them, which could effectively cover up an illegal activity.

Enterprise and organization system access for many individuals still occurs via modem dial-up. Because this technique is available to everyone in the world using a dial-up modem, special attention must be devoted to limiting the access to legitimate users. A security policy that is both usable and enforceable is a requirement in all enterprise computer-networking systems. This policy should include some, if not all, of the following elements:

- Provide protection for analog telephone numbers. Don't advertise them to the world. Use numbers that are different from regular enterprise telephone numbers.
- Maintain an inventory of all dial-up telephone lines. If the line is not being used, make sure that it is not connected to a modem.
- Don't have modems scattered around the premises. Home all modems into a central rack where they can be controlled. Use a modem pool to consolidate them into a controlled environment.
- Put dial-up lines and modems in a secure closet and tag them with circuit IDs.
- Monitor all dial-up activity. Look for failed login attempts. Caller ID may be useful for identifying calling telephone numbers.
- Don't display any banner information that is presented after a connection. This is a place where a warning message can be displayed to callers.
- Require a dial-back authentication procedure, if feasible. This may be impractical with roving users.
- Provide help desk and PBX employees with specific instructions about social engineering tactics. Remote access credentials cannot be reset without the approval of a supervisor.
- Connectivity for all analog facility activity, including fax and voicemail, must be controlled by a central authority. This includes the standard telephone (POTS) line.
- Provide for an audit activity that monitors all of the above elements.

Dial-Back Systems

An increasingly mobile workforce often requires remote access to the computer resource. Remote network access can help employees gain access to databases, facilitate a telecommuting system, or help traveling executives maintain contact with corporate headquarters. This type of dial-up access should be granted cautiously and judiciously, and only to those who require it, because network resources are easily compromised by this access method. This type of doorway into the network invites other nonemployees to try the locks on the door. There are many technical solutions that can be implemented to provide the required level of

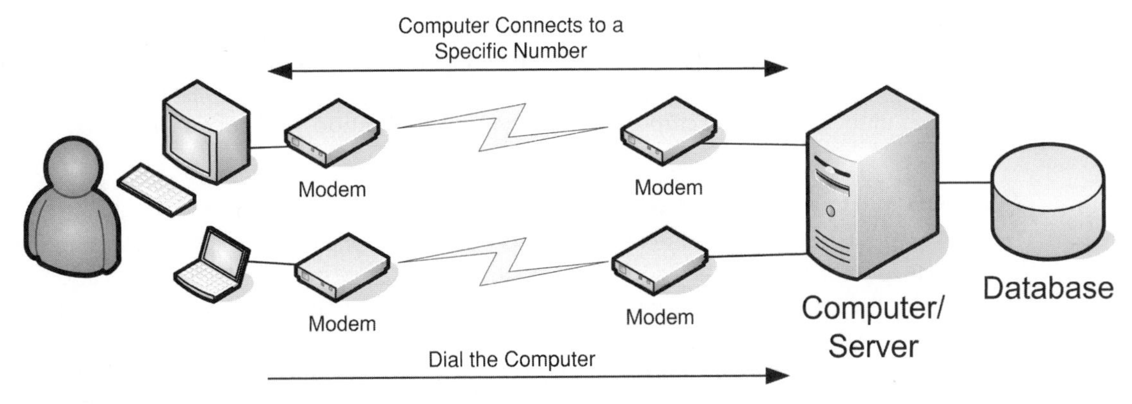

Figure 9-1 Dial-Back Process

Course Technology/Cengage Learning

security and ensure that only authorized individuals gain access to the corporate network using such a door. Figure 9-1 describes the dial-back process.

A *dial-back* system is a solution that can provide some level of protection for these remote dial-in users. Using a dial-back modem, the remote user calls the computer resource and the dial-back modem calls the user back at a predefined telephone number to make the connection. The normal identification and authentication process occurs over this connection. The downside to this technique is that the dial-back number is fixed to a specific location.

Virtual Private Network

A **virtual private network (VPN)** is a private data network that makes use of the public telecommunications infrastructure, which maintains privacy through the use of tunneling and security procedures. The VPN is so named because an individual user shares communications channels with other users. Switches are placed on these channels to give an end user access to multiple end sites. A VPN can be contrasted with a system of owned or leased lines, which can only be used by one company. The connection between sender and receiver acts as if it were completely private, even though it uses a link across a public network to carry information. Figure 9-2 depicts a VPN using a tunnel over the public Internet infrastructure.

Ensuring security is a major issue when implementing virtual networks, and should not be treated lightly. Security problems can be solved with various techniques, such as tunneling, when implementing a VPN. Securing tunnels for private communications between corporate

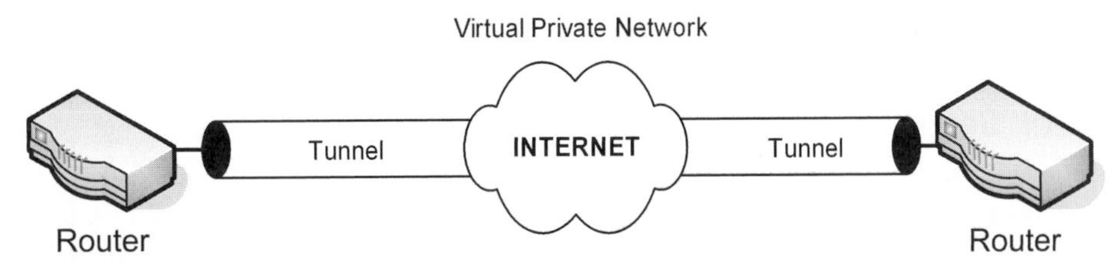

Figure 9-2 VPN Tunnel Across the Internet

Course Technology/Cengage Learning

sites will do little if employee passwords are openly available or if other holes in the security of the network exist. VPN-related security management, of keys and user rights, must be integrated into the rest of the organization's security policies. Security is implemented in VPNs with authentication, encryption, integrity, nonrepudiation, and content filtering.

Authentication Techniques

As presented previously, authentication methods identify users. Once users are identified and authenticated, they can access resources based upon an authorization. An authentication system may be the most important part of an NOS. It requires that a user supply some identification, such as a user name and password. Simple password systems may be sufficient for small systems, but if a higher level is required, then more advanced security may be necessary. These systems may provide dial-back capabilities, or use biometric or token authentication devices.

Token-based authentication is a security technique that authenticates users who are attempting to login to a secure computer or network resource. This method helps eliminate insecure logons—logons that send a user's passwords across the network where it can be observed in the clear. Someone capturing the password could use it to repeatedly masquerade as the legitimate user and access a secure system resource. This situation can be countered by not sending passwords in any form across an insecure channel. An alternative process is to use a *token*, which is a physical object that indicates the identity of its owner and resembles a credit card.

These devices are microprocessor-controlled *smart cards*, which are used to implement *two-factor authentication*. The user supplies the logon password and the one-time value generated by the smart card. The smart card generates one-time passwords that are good for only one logon attempt. Two-factor authentication identifies a user and then authenticates the user. Organizations often assign tokens to remote and mobile users who need to access internal network resources from outside locations.

Security Dynamics offers a token called *SecurID* that uses a time-based technique for displaying a number that changes every minute. This card (token) is synchronized with a security server at the corporate location. When users login, they are prompted for a value from the SecurID card. Because the value changes every minute, someone who manages to capture it online cannot reuse it.

Software-based token devices that run on portable computers are also available that provide similar functionality as hardware tokens. It has been suggested that software tokens are less secure than hardware ones, and can be more easily compromised.

The use of an authentication device, which is protected by a password, is superior to a single password since the loss of one of the authentication factors will not allow system access. The likelihood of losing both factors is remote.

Server Security

Maintaining secure servers is an absolute necessity, since they are the gatekeepers to systems and databases. Computer security involves more than setting up a program of firewalls and passwords. It is an ongoing, organization-wide awareness that someone could break into the network and steal information and other resources. Failing to implement protocols and procedures mitigates the effectiveness of the best-designed security program.

Remember that servers consist of a hardware and software component. The hardware selected will determine the level of performance provided, whereas the software can determine the level of security of a site. Two approaches to server security include purchasing secure server software packages or having in-house development using a variety of packages. Whichever solution is selected, certain elements should be included in this security system, including password protection, encryption, firewalls, and SSL technology. All of these elements offer some degree of security, however, when combined can produce a formidable security scheme.

There are a number of ways to secure a server, given the objectives and requirements. The most common security methods are access control, which includes user passwords, card or key systems, biometrics, and data encryption. Complex security mechanisms usually increase the inconvenience level for authorized users, which could result in employee resistance. The objective is to develop a protection strategy that does not cause employee complacency or circumvention.

A wide range of access control mechanisms has evolved over time, and many are related to the organization's internal policies. Many of these techniques use authentication mechanisms to establish the identity of a user. An alternative mechanism employs security labels to provide input to rules-based access control policy decisions. Rules-based policies could rely on other inputs, such as time and date, when making access control decisions. In addition to controlling access to databases through a secure server, administrators must evaluate all shared resources that are accessible through the network so that their respective access restrictions can be implemented.

Password protection, previously discussed, is the most common means of securing server access. Passwords provide a minimal level of security without requiring special devices such as keys or cards. The downside in implementing only password protection is user ambivalence. If the password method is implemented, several precautions should be mandated:

- Passwords should not contain easily obtainable information.
- Passwords should contain lower and upper case, numbers, and special characters.
- They should be dynamic, with mandatory periodic changes.
- A multilevel password protection scheme should be implemented if hierarchical security is an issue.
- A master password file should be maintained and surveyed periodically.

Magnetic card systems allow users to access data from any available workstation with an attached card reader. The card key system allows selective access for each user as opposed to each workstation. Smart cards offer a wide range of security considerations, including eliminating manual passwords. They are viable for both local and remote access control. Biometric card readers are an important addition to access control and security and discussed in several chapters.

Data encryption is a mathematical method of scrambling data to disguise its original content. This is the most popular security mechanism for protecting sensitive information during transmission between computers. Data confidentiality is achieved by encrypting the message at the sender and decrypting at the receiver. Since attackers, including those using network monitors (sniffers), cannot read encrypted data, it has a high probability of remaining secure.

An additional security technique called inactivity timeout can be utilized for servers that perform remote access. Servers can be configured to automatically terminate calls when there has been no activity for a specified time interval. All of these different approaches in combination provide a high level of security at the server level.

Intrusion Detection

Intrusion detection techniques can be used to monitor network traffic, looking for patterns that suggest known types of computer attacks. An alternative approach, designed to discover previously unknown types of attacks, requires monitoring network statistics for unusual changes that might indicate attack activity. Intrusion detection monitoring may be accomplished using a network host computer, or detection facilities may be distributed at various points throughout the network. These systems often employ both pattern and statistical methods, and use both host-based and distributed monitoring.

The most effective intrusion detection monitoring occurs dynamically as information packets travel through the network. If this activity is in-band, network traffic could experience slow throughput. To meet intrusion detection needs, vendors now offer security chips capable of examining millions of packets per second.

Attackers often turn to scanning systems to gather information on the target network. Scanning tools have the ability to automatically check a target network for possible computer and network vulnerabilities. These tools employ a database of known configuration errors, system bugs, and other problems that can be used to infiltrate a system. Defending against vulnerability scanners require the administrator apply system patches on a regular basis and periodically conduct internal vulnerability scans.

Testing to assess the network intrusion vulnerability may help administrators discover security weaknesses. Such testing can be performed in-house or through a security consultant. When holes in network security are discovered through vulnerability testing, proactive action can be taken to close them. A security scanner consists of software that will audit remotely a given network and determine whether bad guys may break into it, or misuse it in some way. *Nessus* is an example of a vulnerability scanner that might be used to improve the security of the enterprise.

Intrusion Prevention

Intrusion prevention system (IPS) is a preemptive approach to network security used to identify potential threats and respond to them swiftly. Like an IDS, an IPS monitors network traffic. However, because an exploit may be carried out very quickly after the attacker gains access, intrusion prevention systems also have the ability to take immediate action, based on a set of rules established by the network administrator. For example, an IPS might drop a packet that it determines to be malicious and block all further traffic from that IP address or port. Legitimate traffic, meanwhile, should be forwarded to the recipient with no apparent disruption or delay of service.

An intrusion prevention system provides solutions that monitors network and/or system activities for malicious or unwanted behavior and can react, in real time, to block or prevent those activities. Network-based IPS, for example, will operate in-line to monitor all network traffic for malicious code or attacks. When an attack is detected, it can drop the offending packets while still allowing all other traffic to pass. Intrusion prevention technology is considered by some to be an extension of IDS technology.

The role of an IPS in a network is often confused with access control and application-layer firewalls. There are some notable differences in these technologies. While they are similar, how they approach network or system security is fundamentally different.

An IPS is typically designed to operate completely invisibly on a network. IPS products do not have IP addresses for their monitoring segments and do not respond directly to any traffic. Rather, they merely silently monitor traffic as it passes. While some IPS products have the ability to implement firewall rules, this is often a mere convenience and not a core function of the product. Moreover, IPS technology offers deeper insight into network operations, providing information on overly active hosts, bad logons, inappropriate content, and many other network and application layer functions.

Securing Mechanized Transactions

A number of technology companies have developed protocols for transacting business in a secure environment. This is in addition to securing the operating system (OS) and preventing unauthorized access to the server. Some types of communications require protection between the Web browser and the secure server. The most common mechanisms, which will be described in Chapter 12, include the following:

- *Secure Socket Layer (SSL)* provides for server authentication, data encryption, and data integrity
- *Secure HyperText Transfer Protocol (S-HTTP)* allows servers to encrypt responses to browsers, digitally sign replies to browsers, and authenticate the identity of browsers
- *Private Communication Technology (PCT)* provides privacy between a client and server, and to authenticate the client and server
- *Generic Security Service Application Program Interface (GSSAPI)* provides for mutual authentication and data encryption capabilities in Web browsers and servers

There is an option that allows for the combining of HTTP and S-HTTP, which offers the freedom, flexibility, and efficiency of HTTP, while using S-HTTP to protect sensitive parts of a transaction. This is possible because S-HTTP and HTTP are different protocols that use different ports. This allows merchants to offer catalog information to anyone, while providing protection to the server and client during order entry.

Another issue concerns the use of a firewall component between the Internet and the server. Firewalls and routers are used for creating an internal site or private network that is isolated from the general population of Internet users. Using a server for commercial purposes exposes it to the entire untrusted Internet outside the firewall's perimeter of protection. This means that placing a commercial server within the confines of a protected network is self-defeating, unless potential clients are granted accessibility on an individual basis.

Protecting the Data and Database Asset

The issue is not "if" the database will be compromised, but "when." Several questions must be answered:

- Can the organization survive if the database asset is destroyed?
- How much would it cost to recreate the asset?
- What are the repercussions if it cannot be recreated?

These and other questions are pertinent if the organization utilizes a database to conduct its day-to-day activities. The organization must take proactive steps to ensure that a complete recovery can be made in a reasonable time period. It is essential for the administrator to identify the possible causes of a database disaster and the steps that will be required to recover it completely. Enterprise network security is not the same for each file, data item, or database. This means that potential incidents are identified, the ramifications explored, and corrective measures identified to be taken for each occurrence.

Data and Database Integrity and Security

Two issues that must be addressed include data integrity and data security. *Note: Database, repository, and other information relating to data were presented in Chapter 4.* The possibility of errors in data transmission exists in all public and private telecommunication networks. Correcting and preventing these errors is called maintaining data integrity. Even though accurate transmission of data is essential, data security may be even more important. Data security can be assured by implementing the five basic goals of the National Institute of Standards and Technology (NIST). These goals state that secure data messages should be:

- Sealed
- Sequenced
- Secret
- Signed
- Stamped

Data integrity is addressed by utilizing a number of error-control techniques. These include:

- Parity checking
- Cyclical parity checking
- Hamming codes
- Checksums
- Cyclical redundancy checks (CRC)

Enterprise Threats and Vulnerabilities

There are many opportunities for threats to occur in an organization's computer and networking environment. This is particularly true in the Internetworking environment, where attackers, crackers, and hackers abound. A threat from these individuals can have a potential adverse effect on the organization's assets and resources. Threats can be listed in generic terms; however, they usually involve fraud, theft of data, destruction of data, blockage of access, and so on.

As these interlopers continue to create more ingenious methods for penetrating the telecommunication's network, administrators must take a comprehensive approach to security. The use of antivirus programs, firewalls, and the triple-A techniques is a good start, but much more is required to ensure network security. Of major importance are continuous vigilance with a security package of intrusion detection techniques and updating software with the latest patches. When the network is hacked, the response should be quick, thorough, decisive, and effective.

It is essential to identify the various threats and rank-order them as to their importance and impact on the organization. These assignments can be made on the basis of dollar loss, embarrassment created, monetary liability, or probability of occurrence. The most common threats to an organization include the following:

- Virus/worm/Trojan horse
- Device failure
- Internal hacker
- Equipment theft
- External hacker
- Natural disaster
- Industrial espionage

Intranet Security

An **intranet** is an organization's internal computer network. A firewall is implemented in an intranet security system, which is designed to limit Internet access to a company's intranet. The **firewall** permits the flow of e-mail traffic, product and service information, job postings, and so forth, while restricting unauthorized access to private information residing on the intranet. The firewall can exist as either a hardware or software component. Usually implemented with this firewall is a hardware device, called a router. Figure 9-3 depicts an intranet configuration providing connectivity between LAN A and LAN B.

9

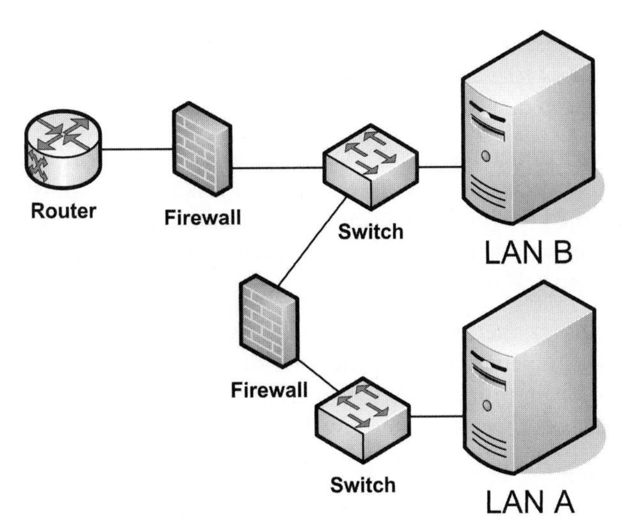

Figure 9-3 Intranet Configuration

Course Technology/Cengage Learning

Firewalls are implemented with routers and proxy servers. A *proxy server* is a firewall component that controls how internal users access the outside world and how Internet users access the internal network. In some cases, the proxy blocks all outside connections and only allows internal users to access the Internet, which means that the only packets allowed back through the proxy are those that return responses to requests from inside the firewall. Each organization must decide the level of Internet access that is acceptable, which can ensure that their IT resources remain fully protected.

Extranet Security

An **extranet** is a private, Internet-like network that a company operates to conduct business with its employees, customers, vendors, and suppliers. They include Web sites that provide corporate information to internal employees and also to external partners, large customers, and particular suppliers through a secure access arrangement. It is called an extranet because it uses the technology of the public Internet (TCP/IP) and customers and suppliers often access it through the Internet via an ISP.

An extranet uses the Internet protocols and the public telecommunications system to securely share part of an organization's information or operations with the aforementioned entities. An extranet has become the means for companies to engage in business-to-business (B2B) e-commerce on a global basis. It is a secure version of the Internet with virtual private network (VPN) features.

An extranet can also be viewed as part of a company's intranet that is extended to users outside the company. It has also been described as a "state of mind," in which the Internet is perceived as a way to do business with other companies as well as to sell products to customers. The same benefits that HTML, HTTP, and SMTP provides for the Internet users can be applied to the extranet users.

An extranet can be extended to the general public for accessing certain types of information. Organizations build an extranet to improve communication among key constituents, facilitate information distribution, and broaden access to each other's resources. It also provides a browser front-end, such as a portal, to various corporate databases to expedite inventory tracking, invoicing, and supply-side management.

One of the issues of collaborative extranet systems is they must access systems and databases behind the firewalls of multiple organizations, and all elements in this network must interoperate. Interoperability is facilitated by a number of standard technologies. The aim of these standards is to enable a comprehensive, interoperable infrastructure that permits secure transmission of information across a network of diverse users.

Sensitive data can be kept private via the use of a firewall. Private information and other resources can be kept off-limits by implementing a number of strategies, including packet filtering and intrusion detection. When implemented properly, extranets provide access to appropriate resources, while securing others on a selective basis.

The real security challenge involving intranets relates to partners whose business relationships are dynamically changing and complex. Today's partners may become tomorrow's competitors and could also be a current competitor. An extranet must permit dynamic changes in access control to guard against the loss of sensitive and private information. Two key security requirements for extranet access include user authentication and authorization. Figure 9-4 depicts an extranet configuration.

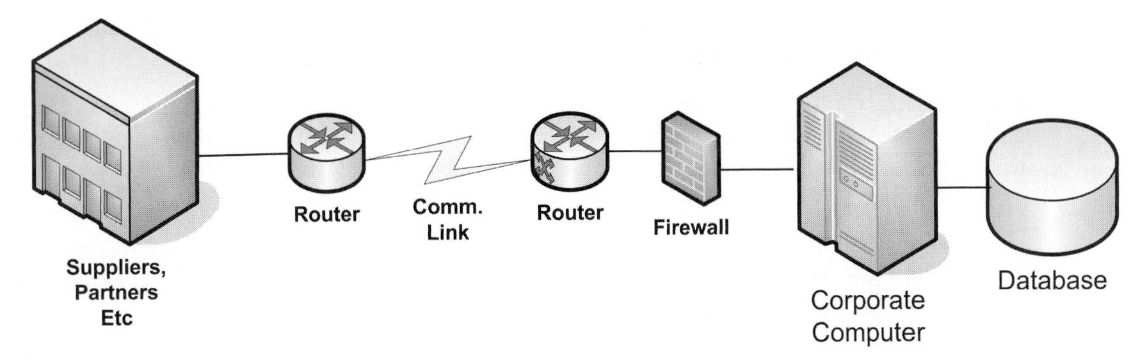

Figure 9-4 Extranet Configuration

Course Technology/Cengage Learning

The identity of a user wishing to access the extranet must be authenticated. This process is complicated when employees or business partners access information from multiple computers and from remote locations over the Internet. This means that remote sites require additional security protocols in order to limit server access. This gives rise to a number of issues, which include the following:

- There are often many Web servers in large organizations and users need access privileges for each server accessed.
- Users must remember passwords for many servers.
- Administrators must manage the access controls for each individual server.
- Many separate entries must be added or removed when a user's access privileges change or when there is user turnover.

A security process that lets the organization manage access controls for all centrally located server elements and present users with a single sign-on to the Web space would simplify security management.

After the user has been identified, access privileges must be determined. Security policies must explicitly grant access rights to Web resources. This access control can be complicated if it is different for each Web server accessed. A centralized authorization framework would simply this server administration.

A security management system must be easy to implement and administer, and must integrate easily into the organization's infrastructure. Complexities in security management increase the possibility of human errors, make the extranet difficult to navigate, and expose it to attack or misuse.

Biometrics Verification and Identification

The terms "verification" and "identification" can be confusing to individuals who are discussing biometrics. The majority of available devices operate in verification mode. This means that an identity is claimed by calling a particular template from storage by the input of a PIN or presentation of a token and then presenting a live sample for comparison, resulting in a match or non match according to predefined parameters. Thus a simple one-to-one match that may be performed quickly can be used to generate a binary yes/no result.

A few devices claim to offer biometric identification, whereby the user submits a live sample and the system attempts identification within a database of templates. A more complex one-to-many match could generate a multiple result according to the number and similarity of stored templates and consume a considerable amount of time.

Imagine a scenario in which many templates are stored in a database. The user presents a live sample and the database engine starts searching. Depending on how tightly defined the likeness-threshold parameter is, the search may result in thousands of possible identities for the user. Additional filters based upon sex, ethnic origin, age, etc. can be applied to reduce this list to a manageable size, if this information can be captured from the user. The end result can still result in a sizeable list of potential identities. Of course, in a smaller database this becomes less of a problem, but it is precisely with large databases this functionality is typically sought.

This situation assumes that the system can function as claimed in identification mode. Certain devices have been demonstrated to work well in this manner with small databases of tens of users, but the situation becomes very complicated with databases of even a few hundred. The mathematical probability of finding an exact match within such a database is extremely slim. A large database, such as might be the case with travelers across borders, for example, would be almost impossible to manage in this manner with current technology. The time taken to search such a database and the impact of multiple concurrent users haven't even been considered. For these and other reasons, one should exercise extreme caution when considering biometric identification systems. While one can readily understand the attraction of this mode of operation, it has often been less than successful in practice, except in small-scale, carefully controlled situations.

Verification systems, alternatively, are straightforward in operation and may easily be deployed within a broad cross section of applications (as has been the case). Errors can occur during the verification or identification process. Two critical measurement factors indicate the level of accuracy, or reliability, of any given biometric: the false reject rate and false accept rate.

False Rejection Rate (FRR)

When a biometric measurement from a live subject is compared to that subject's enrolled template and the system fails to match the two, a "false reject" event occurs. The theoretical probability of this occurring or the actual frequency of it occurring, if there is sufficient historical data available, is known as the *false rejection rate (FRR)*. Although FRRs will vary widely among different biometric systems and technologies, with any single given biometric system, the FRR will be approximately the same whether the process is used for verification or for identification. This is because there is only one authentic template on file against which the live subject can make a match. The FRR will vary widely depending on the situation under which the biometric is used. Factors such as user cooperation and operating conditions can affect the FRR.

False Acceptance Rate (FAR)

There is always a possibility that the measurement from a live subject will be sufficiently similar to a template from another (different) person that a match will be erroneously declared.

This second type of error is called a "false accept" event and the associated probability is called the *false acceptance rate* (FAR). The FAR achieved by a particular biometric directly reflects the fundamental power and specificity of the technology. To achieve a low FAR, the biological entity measured must be absolutely unique to the individual, and the algorithm used to measure the entity must capture this uniqueness very effectively.

Performance Measures

False accepts, false rejects, equal error rates, enrollment, and verification times are the typical performance measures quoted by device vendors. Several questions are relevant:

- What do these measures really mean?
- Are these performance statistics actually realized in real systems implementations?
- Can they be accepted with any degree of confidence?

False accept rates indicate the likelihood that an impostor may be falsely accepted by the system. False reject rates indicate the likelihood that the genuine user may be rejected by the system.

This measure of template matching can often be manipulated by the setting of a threshold which will bias the device toward one situation or the other. Hence, one may bias the device toward a larger number of false accepts, but a smaller number of false rejects (user friendly) or a larger number of false rejects, but a smaller number of false accepts (user unfriendly)—the two parameters being mutually exclusive.

Somewhere between the extremes is the equal error point (*equal error rate, or ERR*) where the two curves cross and which may represent a more realistic measure of performance than either FAR or FRR quoted in isolation. These measures are expressed in percentage of error transactions, with an equal error rate of somewhere near 0.1 percent being a typical figure. Figure 9-5 provides a graphical representation of this relationship.

The quoted figures for a given device may not be realized in practice for a number of reasons. These will include:

- User discipline
- Familiarity with the device
- User stress
- Individual device condition
- The user interface
- Speed of response
- Other variables

Remember that vendor-quoted statistics may be based upon limited tests under controlled laboratory conditions, and supplemented by mathematical theory. They should only be viewed as a rough guide and not relied on for actual system performance expectations.

This situation is not because vendors are trying to be misleading, but because it is almost impossible to give an accurate indication of how a device will perform in a limitless variety of

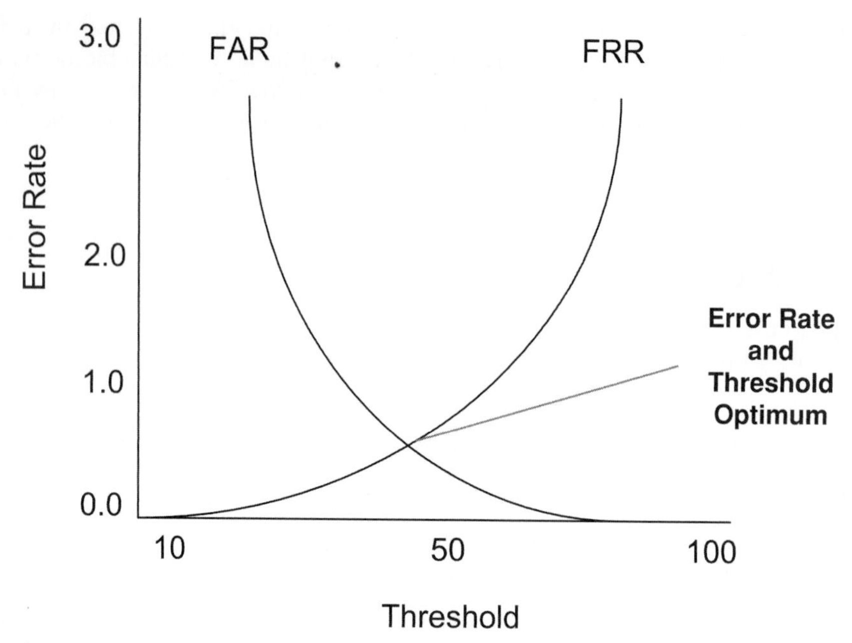

Figure 9-5 FAR and FRR Relationship

Course Technology/Cengage Learning

real-world conditions. Similarly, actual enrollment times will depend upon a number of variables inherent in the enrollment procedure, including the following:

- Are the users preeducated?
- Have they used the device before?
- What information is being gathered?
- Is custom software being used?
- How well trained is the enrolling administrator?
- How many enrollment points will be operating?
- What other processes are involved?

The vendors cannot possibly understand these variables for every system, and their quoted figures will again be based on their own in-house experiences under controlled conditions.

Verification time is often misunderstood, as vendors typically describe the average time taken for the actual verification process, which does not usually include the time taken to present the live sample or undertake other processes, such as the presentation of a token or keying of a PIN. When an average time for user error and system response are also considered, it is apparent that the end-to-end verification transaction time will be greater than the quoted figure.

Thus, it should not be surprising that biometric device performance measures have sometimes become a contentious issue when implementing real systems. To provide an independent view, the National Biometric Test Center has been established in the United States with a similar facility recently announced in Hong Kong. These centers are based at academic institutions and will, over time, provide some interesting views. However, this does not necessarily mean that vendors will rush to conform with regard to their quoted specifications and the method used to derive them. Administrators should continue to view such specifications as a rough guide and rely on their own trials and observations to provide a more meaningful appraisal of overall performance.

There is also a question concerning the uniqueness of biometric parameters, such as fingerprints, irises, voices, hands, faces, and so forth. The degree of individuality or similarity within a user base will naturally affect performance to some degree. It should suffice to say that no one has reliable data for the whole world and cannot therefore say that any biometric is truly unique. However, what can be said is that the probability of finding identical fingerprints, irises, or hands within a typical user base is low enough for the parameter in question to be regarded as a reliable identifier. Beware of claims of absolute uniqueness, because some individuals are similar enough to cause false accepts, even in finely tuned systems.

Biometric methods offer scalability between security and ease of access with a trade-off between the FRR and FAR. What this means is that the more security is desired, the more false rejects are encountered. Alternatively, to reduce false rejects creates a much higher false acceptance rate, leading to unwanted persons being identified erroneously. *Note: A section titled "Biometric Measurements" is included in Chapter 12.*

Smart Card Biometric Access Control

A common authenticator for both physical and logical information technology authentication can provide a single credential that enables rights to corporate resources, timelier user life-cycle management, and improved physical access controls. The benefits are compelling, but implementing a common authenticator requires considerable effort and can take several years. There are various dimensions to the problem, including:

- Smart card personalization and distribution
- Upgrade of physical access control systems
- Emergency access procedures

Smart Card Personalization and Distribution

The process of making the smart card ready for the user is called personalization. The procedure includes printing the user's picture, installing applets and PKI (X.509) certificates, and binding the smart card to both the user and the physical access system.

In most cases, a *smart card management system (CMS)* is a deployment requirement. Important considerations when evaluating a CMS include:

- Platform support: This is necessary for vendor physical access systems, smart card and printer products.
- Remote applet distribution capabilities: Java card applets provide most smart card functionality, and the CMS can deliver new applets after the smart cards have been distributed to the users.
- Key escrow and recovery capabilities: Such features can recreate the user's PKI credentials on a new smart card in the event that the previous card is destroyed or lost.

- Provisioning system integration: Integration provides a single authoritative source of identity information and consistent access rights.
- Administrative delegation and scoping capabilities: These components enable the secure management of smart cards across the organizational hierarchy.

It's easy to see why the distribution of smart cards to an organization's employees is a significant effort. The process can take months or even years, and the many important details require careful planning. The distribution of smart cards to "virtual" employees, those that rarely visit a campus, requires special attention.

Upgrade of Physical Access Control Systems

It's a toss-up as to which activity causes more organizational heartburn: the distribution of smart cards, or the upgrading of the physical access systems across an organization's campuses. Organizations may have a wide spectrum of physical access technologies across their environments, from keys to magnetic stripes to biometric authenticators to contactless tools. As part of the planning process, an organization should take inventory of its facilities' physical access system and determine what upgrades are necessary for the implementation of a common authenticator.

While there is wide variability, the typical components of modern physical access systems include readers, controllers, security servers (hosts), and cards. Even if some of these components are present, the organization may still need to upgrade them. During the gradual migration to a common authenticator, multitechnology readers, or even controllers, may be required to enable the use of various card types. Controllers may also need to be upgraded to function over the Ethernet instead of a serial protocol. Such improvements can better support a more modern architecture and an organization's physical/logical convergence goals.

Emergency Access Procedures

It's a fact of life: users will forget their smart card at home. Without them, they cannot access applications, workstations, buildings, and maybe even the parking lot or bathroom. With proper emergency access measures, such an error should only be a temporary one. The organizational challenge is to implement emergency access procedures that give forgetful, cardless users a timely access to resources. The access processes must also do so in a cost-effective manner. Some tricks of the trade include:

- Self-service kiosks in the building entrance where employees can authenticate and get a temporary smart card
- IT software management tools that temporarily allow the user to authenticate with a password instead of a smart card
- Physical access readers with PIN pads that enable the user to temporarily authenticate with an identification number.

Even in the face of the many details mentioned above, planning for a common authenticator appears more daunting than it really is. If the organization defines achievable milestones and exercises vigilance against the temptation of expanding and redefining the objective of the project, implementation is possible. [1]

Biometric Authentication: What Method Works Best?

There does not appear to be any one method of biometric data gathering and reading that does the "best" job of ensuring secure authentication. Each of the different methods of biometric identification possesses integrity and security features. Some are less invasive, some can be done without the knowledge of the subject, and some are very difficult to fake. Current biometric implementations are using fingerprint and face modalities in multimodal systems.

Fingerprint Identification

Fingerprints remain constant throughout life. In over 140 years of fingerprint comparison worldwide, no two fingerprints have ever been found to be alike, not even those of identical twins. Good fingerprint scanners have been installed in PDAs like the iPaq Pocket PC, so scanner technology is also easy. A scanner might not work in industrial applications since it requires clean hands.

Fingerprint identification involves comparing the pattern of ridges and furrows on the fingertips and the minutiae points (ridge characteristics that occur when a ridge splits into two, or ends) of a specimen print with a database of prints on file.

Face Recognition

Of the various biometric identification methods, face recognition is one of the most flexible, working even when the subject is unaware of being scanned. It also shows promise as a way to search through masses of people who spent only seconds in front of a "scanner," which is an ordinary digital camera.

Face recognition systems work by systematically analyzing specific features that are common to everyone's face: the distance between the eyes, width of the nose, and position of cheekbones, jaw line, chin, and so forth. These numerical quantities are then combined in a single code that uniquely identifies each person.

Hand Geometry Biometrics

Hand geometry readers work in harsh environments, do not require clean conditions, and form a very small dataset. It is not regarded as an intrusive kind of test, and is often the authentication method of choice in industrial environments.

Retinal Scan

There is no known way to replicate a retina. As far as anyone knows, the pattern of the blood vessels at the back of the eye is unique and stays the same for a lifetime. However, it requires about 15 seconds of careful concentration to take a good scan. The retina scan remains a standard in military and government installations.

Iris Scan

Like a retina scan, an iris scan also provides unique biometric data that is very difficult to duplicate and remains the same for a lifetime. The scan is similarly difficult to make and may be difficult for children or the infirm. However, there are ways of encoding the iris scan biometric data in a way that it can be carried around securely in a "barcode" format.

The iris itself is a "good subject" for biometric identification, because it is an internal organ that is well protected, it is mostly flat, and it has a fine texture that is unique even for identical twins. Iris scans can be done regardless of whether the subject is wearing contact lenses or glasses. However, it is necessary for the system to take eye lids and eye lashes into account, as both can obscure the necessary parts of the eye and cause false information to be added into automated systems. Iris scans are extremely accurate. [2]

Signature

A signature is another example of biometric data that is easy to gather and is not physically intrusive. Digitized signatures are sometimes used, but usually have insufficient resolution to ensure authentication.

Voice Analysis

Like face recognition, voice biometrics provides a way to authenticate identity without the subject's knowledge. It is easier to fake by using a tape recording; however, it is not possible to fool an analyst by imitating another person's voice.

Future Developments for Biometrics Data Integrity Assurance

Data encryption is a fundamental tool for the protection of electronically mediated data, yet encryption itself cannot necessarily protect against fraudulent data manipulation when the security of encryption keys cannot be absolutely guaranteed. Efforts are ongoing to integrate biometric security and authentication techniques within generic document models in such a way as to maximize their security potential. Biometric technologies attempt to automate the measurement and comparison of physiological or behavioral characteristics for identifying individuals. Document security might be improved via the generation of encryption keys directly from biometrically based data. This will serve to increase confidence and trust in the integrity of sensitive data both to their producers and users. While encryption technologies can effectively spread security, by themselves they cannot guarantee the security of the identity of the originator of the data.

Biometric technologies allow for security to be introduced at document origination and document retrieval. Such a system will involve the removal of the need to store any form of template for validating the user. The security of the system will be as strong as the biometric and encryption algorithm employed (no back door access). The only ways to gain subsequent access are to provide another sample of the biometric or to break the cipher employed by the encryption technology. The unlikely compromise of a system does not release sensitive biometric template data which would allow unauthorized access to other systems protected by the same biometric or any system protected by any other biometric templates present. A further significant advantage relates to the asymmetric encryption system associated with the proposed technique. Traditional systems require that the private key for decrypting data be stored in some way (memorizing a private key is not feasible). In this system, the key is uniquely associated with the given biometric sample and a further biometric sample will be required to generate the required private key. As there is no physical record of the key, it is not possible to compromise the security of sensitive data via unauthorized access to the key. [3]

Biometrics-Based Information Assurance

Information assurance consists of operations that protect and defend information and information systems by ensuring their availability, integrity, authentication, confidentiality, and nonrepudiation. This includes providing for the restoration of information systems by incorporating protection, detection, and reaction capabilities. Biometric technologies can improve the overall assurance of an information system by incorporating the uniqueness of personal biometric signatures into the security and safety management. The downside of using biometric signature is the problem of the scalability. The problems and issues associated with the widespread use of biometrics are not necessarily technical in nature; the possibility that digital identifiers of a person could be acquired from multiple sites and pooled to build a profile of that person's activity is worrisome. Therefore, it is the responsibility of system designers to assure that the downsides of biometrics technologies cannot be exploited. In the near future, this system-design imperative may become the matter of compliance with the law, if the statutes similar to the Consumer Biometric Privacy Protection Act, introduced into the California Assembly in 1998, become common.

Biometric Security Issues

Various topics relate to system-level design and assessment considerations for applications that use biometrics technology. In an application, a biometric signature is typically used to enhance user authentication capability or to improve the confidence in preserving the integrity of communication. The areas of concern include biometric encryption, secure biometrics-based distributed authentication protocols, and methods for system level availability, reliability, and maintainability assessment of biometric-based systems. Additionally, a number of issues relating to human action and their activities have an impact on the successful implementation of a biometric access control system. These include:

- *Liveness detection* is a technique used to ensure the biometric sample submitted is from an end user. A liveness detection method can help protect the system against some types of spoofing attacks.
- *Spoofing* is the ability to fool a biometric sensor into recognizing an illegitimate user as a legitimate user (verification) or into missing an identification of someone that is in the database.
- *Mimic* is the presentation of a live biometric measure in an attempt to fraudulently impersonate someone other than the submitter.
- *Skimming* is the act of obtaining data from an unknowing end user who is not willingly submitting the sample at that time. An example could be secretly reading data while near a user on a bus.
- *Eavesdropping* is surreptitiously obtaining data from an unknowing end user who is performing a legitimate function. An example involves having a hidden sensor co-located with the legitimate sensor.

Various network cyber attacks can also impact biometric security. **Pharming** is a cyber attack that directs people to a fraudulent website by poisoning the domain name system server. **Phishing** is a cyber attack that directs people to a fraudulent website to collect personal information for identity theft. **Spear phishing** is the latest type of attack to occur and involves a coordinated effort to compromise a single organization or target.

Biometric Encryption Issues

Encryption is the act of transforming data into an unintelligible form so that it cannot be read by unauthorized individuals. A key or a password is used to decrypt (decode) the encrypted data. A **public key** is the public part of an asymmetric key pair that is used to verify signatures created with its corresponding private key. Depending on the algorithm, public keys are also used to encrypt messages, files, or other information that can then be decrypted with the corresponding private key. The user releases this key to the public, who can use it to encrypt messages to be sent to the user and to verify the user's digital signature. This technique is compared with the private-key process. A public key certificate is a digital document that is issued and digitally signed by the private key of a certificate authority (CA) and that binds an attribute of a subject to a public key.

Public (asymmetric) key cryptography is a type of cryptography that uses a pair of mathematically related cryptographic keys. The public key can be made available to anyone and can encrypt information or verify a digital signature. The private key is kept secret by its holder and can decrypt information or generate a digital signature.

Public key infrastructure (PKI) is the architecture, organization, techniques, practices, and procedures that collectively support the implementation and operation of a certificate-based public key cryptographic system. There are four basic components to the PKI:

- The certificate authority (CA) responsible for issuing and verifying digital certificates
- The registration authority (RA), which provides verification to the CA prior to issuance of digital certificates
- One or multiple directories to hold certificates (with public keys)
- A system for managing the certificates

Also included in a PKI are the certificate policies and agreements among parties that document the operating rules, procedural policies, and liabilities of the parties operating within the PKI.

The ability to encrypt a biometric signature is vital for establishing the trust that a computer application will not violate the confidentiality of the users. The encryption is needed to prevent the access of an unauthorized individual to the signatures stored in an appropriate repository, as well as to prevent snooping attacks in cases when signatures are transported over unsecured (public) communication lines. Human biometric signatures are actually digital signals. In most cases they are two-dimensional signals, such as images (face, fingerprint scans, retina scans, etc.) and their compressed representations (eigenfaces, minutia points). Traditional encryption research addresses information assurance concerns in the exchange of textual information, which consists of one-dimensional stream of bits. While these techniques can be used for biometric signals, encryption techniques addressing the specific features of two-dimensional signals in general, and biometric signatures in particular, have not received due research attention.

Biometrics-Based Assurance Protocols

Research efforts in biometric encryption are expected to provide a base knowledge for information assurance of biometric-based systems. However, as with any encryption algorithms, they have to be incorporated into protocols that assure that the context in which they are used does not provide a back door for security breach. As in the case of password-based security, designing biometrics based authentication protocols in a distributed environment requires extra effort. To ensure the integrity and confidentiality of communications in a distributed system, a secure encryption key (session key) must be associated with a biometric signature. One possible scenario assumes that a biometric image and/or the corresponding template is sent (from the client computer) to a secure location (secure server) for template comparison. If the user is verified, then the key is released from the secure location. If the security of the computer performing template comparison cannot be ensured, this solution is not appropriate. Alternatively, the key may be released at the client's side if it is chosen to be a part of the biometric template. This solution calls for the static choice of template features used to derive keys, which is inappropriate in cases when an attack can be launched based on the long-term observation of the key patterns. [4]

Summary

- Accuracy, confidentiality, and integrity are imperative in legacy computer and network systems; however, there is an increased impact with the introduction of biometric techniques and solutions. When system administrators and managers consider these systems, they should address topics such as integrity, accuracy, security, accessibility, availability, performance, and false accepts/rejects.
- Computer and network systems must possess a high degree of accessibility and availability. Systems, devices, data, service, or environments must be accessible by as many users as possible. Authorized users or computing devices must have access to resources and information without interference or obstruction, and receive it in the required format. Products and services must possess high availability for systems running mission critical applications.
- Biometric verification and identification techniques and systems are being developed and implemented to provide for security and access control. Numerous hardware and hardware solutions are being used to make these biometric techniques successful. Considerable efforts are being devoted to the use of smart cards for biometric access control.

Key Terms

Accessibility (2) A general term used to describe the degree to which a device, service, or environment is accessible by as many users as possible.

Availability Ensuring that authorized users—persons or computer systems—are able to access information without interference or obstruction, and to receive it in the required format.

Encryption The scrambling of data so that it becomes difficult to unscramble or decipher.

Extranet A private, Internet-like network that a company operates to conduct business with its employees, customers, vendors, and suppliers.

Firewall Permits the flow of e-mail traffic, product and service information, job postings, and so forth, while restricting unauthorized access to private information residing on the intranet.

Intranet An organization's internal computer network.

Pharming A cyber attack that directs people to a fraudulent Web site by poisoning the domain name system server.

Phishing A cyber attack that directs people to a fraudulent Web site to collect personal information for identity theft.

Public key The public part of an asymmetric key pair that is used to verify signatures created with its corresponding private key.

Public key infrastructure (PKI) The architecture, organization, techniques, practices, and procedures that collectively support the implementation and operation of a certificate-based, public-key cryptographic system.

Sniffer A LAN protocol analyzer that supports a variety of hardware to include Ethernet and Token Ring LAN topologies.

Token-based authentication A security technique that authenticates users who are attempting to login to some secure computer or network resource.

Virtual private network (VPN) A private data network that makes use of the public telecommunications infrastructure, which maintains privacy through the use of tunneling and security procedures.

Review Questions

1. ________________ has been defined as ensuring that information is accessible only to those authorized to have access and is one of the cornerstones of information security.
2. ________________ defines a state in which information has remained unaltered from the point it was produced by a source, during transmission, storage, and eventual receipt by the destination.
3. ________________ is a general term used to describe the degree to which a device, service, or environment is accessible by as many users as possible.
4. ________________ means ensuring authorized users—persons or computer systems—are able to access information without interference or obstruction, and to receive it in the required format.
5. An attack that attempts to deny corporate computing resources to legitimate users is a ________________.
6. These lists, often located in a router, identify which users and groups have object access and the respective permissions for that access. ________________
7. A ________________ is the term reserved for an individual who willfully breaks into computer systems with the purpose of wreaking havoc and destruction.
8. A ________________ system is a solution that can provide some level of protection for remote dial-in computer users.
9. A ________________ is a private data network that makes use of the public telecommunications infrastructure, which maintains privacy through the use of tunneling and security procedures.

10. _______________ techniques can be used to monitor network traffic, looking for patterns that suggest known types of computer attacks.
11. _______________ is a preemptive approach to network security used to identify potential threats and respond to them swiftly.
12. A _______________ permits the flow of e-mail traffic, product and service information, job postings, and so forth, while restricting unauthorized access to private information residing on an intranet.
13. An _______________ is a private, Internet-like network that a company operates to conduct business with its employees, customers, vendors, and suppliers.
14. When a biometric measurement from a live subject is compared to that subject's enrolled template and the system fails to match the two, a _______________ event occurs.
15. The measurement from a live subject that is sufficiently similar to a template from another, different, person, then a match can be erroneously declared. This type of error is called a _______________ event.
16. Somewhere on a performance measurement graph is the point where the two FAR and FRR curves cross. This point is called the _______________.
17. _______________ is a cyber attack that directs people to a fraudulent Web site by poisoning the domain name system server.
18. _______________ is a cyber attack that directs people to a fraudulent Web site to collect personal information for identity theft.
19. _______________ is the latest type of attack that involves a coordinated effort to compromise a single organization or target.
20. _______________ is the architecture, organization, techniques, practices, and procedures that collectively support the implementation and operation of a certificate-based public key cryptographic system.

Discussion Exercises

1. Describe the five most significant issues that should be addressed for the security of a computer network.
2. Provide five suggestions for network user name and password requirements.
3. Identify five destructive activities that network attackers might use to compromise an organization's vulnerabilities.
4. Research the multifactor authentication topic. Provide an example for one-, two-, and three-factor authentication.
5. Provide an overview of the token called SecurID provided by Security Dynamics.

6. Develop a comparison spreadsheet of intrusion detection and intrusion prevention system features.
7. Data integrity is addressed by a number of error-control techniques. Provide an overview of those identified in this chapter.
8. Provide a short description of the most common threats to an organization.
9. Identify terms that apply to biometric measurements and performance. Provide a short description of each and show any relationships with other terms.
10. Create a comparison table that describes the pros and cons of the various biometric modalities used for authentication.

Hands-On Projects

9-1: System Integrity and Accessibility Awareness

Time required—30 minutes

Objective—Research system integrity and accessibility issues that are part of biometric application and system implementations.

Description—Use various Web tools, Web sites, and other tools to identify biometric system integrity and accessibility issues that occur in the development and production in commercial and government organizations.

1. Provide an analysis of modem usage in an organization. Identify policies that can be developed to establish rules for the secure use of modems. Recommend an alternative solution to modem usage.
2. Identify an organization that utilizes a VPN. Describe the applications accessed. Provide an overview of the security aspects and components of this network.
3. Identify vendors and their respective IDP and IPS systems. Prepare a comparison of the various products.
4. Identify organizations that utilize either intranets or extranets, or both. Describe how security is implemented. Produce a logical graphic of these configurations.
5. Provide an overview of the public key infrastructure, its components, and its value. Describe how it would be utilized in a biometric verification and identification system.

9-2: Identify System Integrity and Accessibility Issues

Time required—45 minutes

Objective—Identify system integrity and accessibility issues that are available for providing some level of access control and security for OnLine Access assets and resources.

Description—The owners and managers of OnLine Access have identified the necessary security and access control technology for implementing a biometric access control solution. They must now look at the system integrity and accessibility issues involved when implementing a biometric solution.

1. Research the integrity and accessibility issues that are inherent in access control and security solutions.
2. Identify security and access control system integrity and accessibility issues that have been encountered in other biometric implementations.
3. Produce a report describing how these issues have impacted the implementation of a biometric solution.
4. Describe how these issues have been resolved.

chapter 10

Security and Privacy Issues

After completing this chapter, you should be able to do the following:

- Identify privacy and security issues that can impact an information system
- Look at hardware, network, and software procedures and protocols that can provide security capabilities
- Become familiar with biometric applications and systems that can be used for security and access control
- Understand the issues of privacy, identification, and verification
- Look at the smart card environment as it relates to security and privacy

Management and administrators must address numerous security and privacy issues that arise daily in the computer and network environment. Various privacy rules and regulations are mandated by the government. Security of information systems assets has been required for decades, and continues to be a major issue with data center managers and administrators. The growth of e-commerce systems has raised the bar for security and privacy issues. The opportunities for identity theft and fraud have increased significantly, which means more attention is being given to positively identify valid users of computer and networking systems.

Introduction

Biometrics has found increased adoption over the past few years in a variety of applications, including authentication and identification. However, there are widespread security and privacy concerns about the dangers of using biometric data in ubiquitous and unchecked manners. Security concerns stem from the fact that biometric data cannot be easily revoked or replaced. Once biometric data is compromised, it remains compromised forever. Privacy concerns arise from the fact that biometric data is tightly bound to a person's identity such that it can be used to violate his privacy.

Considerable research has been conducted over the past few decades on developing techniques for capturing and matching biometric data; however, security and privacy issues have received comparably less attention. Unfortunately, traditional cryptographic techniques cannot be easily adapted to protect biometric data. The main difficulty is that biometric samples cannot be reproduced exactly. Efforts are underway to develop simple, practical, and provably effective cryptographic techniques for the security and privacy of biometric data.

Security and privacy issues have been at the forefront of legacy computer and network systems. Of particular importance is the protection of the computer hardware and software systems and the database assets. An overview of this legacy environment is presented at the beginning of this chapter, and includes information about computer security, information, assurance, and privacy.

Information about various societal issues involving legal, privacy, and user psychology is critical to selecting and implementing a successful biometric-based security and identification-management solution. From the Immigration & Naturalization Service (INS) to ATMs, both public and private sectors are making extensive use of biometrics for human recognition. As this technology becomes more economically viable and technically perfected, and thus more commonplace, the field of biometrics will spark legal and policy concerns. Critics inevitably compare biometrics to "Big Brother" and the loss of individual privacy. The probiometric lobby generally stresses the greater security and improved service that the technology provides. This chapter explores the various arguments for and against biometrics and contends that while biometrics may pose legitimate privacy concerns, these issues can be adequately addressed.

Computer Security Basics

To be secure in the information environment means to be free from danger, and not exposed to damage or attack. *Security*, then, provides protection and defense against attack, interference, or espionage and includes procedures to counter these incidents.

Computer security is a branch of technology known as information security as applied to computers. The objective of computer security varies and can include protection of information from theft or corruption, or the preservation of availability, as defined in the security policy.

Computer security imposes requirements on computers that are different from most system requirements because they often take the form of constraints on what computers are not supposed to do. This makes computer security particularly challenging because analysts find it hard enough just to make computer programs do everything they are designed to do correctly. Furthermore, negative requirements are deceptively complicated to satisfy and require exhaustive testing to verify, which is impractical for most computer programs. Computer security provides a technical strategy to convert negative requirements to positive, enforceable rules. For this reason, computer security is often more technical and mathematical than some computer science fields. Typical approaches to computer security, in approximate order of strength, can include the following:

- Physical limits on computer access to only those who will not compromise security
- Hardware mechanisms that impose rules on computer programs, thus avoiding dependence on the computer programs for computer security
- Operating system mechanisms that impose rules on programs to avoid trusting computer programs
- Programming strategies that make computer programs dependable and resist subversion

Information Security Basics

Information security means protecting information and information systems from unauthorized access, use, disclosure, disruption, modification, or destruction. The terms information security, computer security, and information assurance are frequently used interchangeably. These fields are interrelated and share the common goals of protecting the confidentiality, integrity, and availability of information; however, there are some subtle differences between them. These differences lie primarily in the approach to the subject, the methodologies used, and the areas of concentration. Information security is concerned with the confidentiality, integrity, and availability of data regardless of the form the data may take such as electronic, print, or other forms.

Governments, military, financial institutions, hospitals, and private businesses amass a great deal of confidential information about their employees, customers, products, research, and financial status. Most of this information is now collected, processed, and stored on electronic computers and transmitted across networks to other computers. Should confidential information about a business's customers or finances or new product line fall into the hands of a competitor, such a breach of security could lead to lost business, law suits, or even bankruptcy of the business. Protecting confidential information is a business requirement, and in many cases also an ethical and legal requirement. For the individual, information security has a significant effect on privacy, which is viewed very differently in different cultures.

The field of information security has grown and evolved significantly in recent years. As a career choice there are many ways of gaining entry into the field. It offers many areas for specialization including information systems auditing, business continuity planning, and digital forensics science.

When dealing with any of the following areas involving information, there are ramifications with regard to an individual's privacy:

- Personnel records/files
- Credit and financial information
- Education information
- Wire, oral, and electronic communications

- Alcohol and drug abuse diagnosis, treatment, or other information
- Alcohol and drug testing
- Personal history questionnaires
- Surveillance and investigations
- Using an employee's name and/or likeness without her written permission
- Employment application
- Responding to requests for employee information by third parties
- Antitrust law compliance
- Disclosure of union activity
- Off-duty activity or behavior

Security Management

Security management concerns the ability to monitor and control access to network resources, including generating, distributing, and storing encryption keys. Passwords and other authorization or access control information must be maintained and distributed. Security management is involved with the collection, storage, and examination of audit records and security logs.

Security management subsystems work by partitioning network resources into authorized and unauthorized areas. For some users, access to any network resources is inappropriate because such users are usually company outsiders. For internal users, access to information originating from a particular department, such as payroll, is inappropriate.

Security management subsystems perform several functions. They identify sensitive network resources and determine mappings between sensitive network resources and user sets. They also monitor access points to sensitive network resources and log inappropriate access to sensitive network resources.

Information Assurance Basics

The U.S. government's national information assurance glossary defines **information assurance (IA)** as: "Measures that protect and defend information and information systems by ensuring their availability, integrity, authentication, confidentiality, and nonrepudiation." These measures include providing for restoration of information systems by incorporating protection, detection, and reaction capabilities.

IA is closely related to information security and the terms are sometimes used interchangeably. However, IA's broader connotation also includes reliability and emphasizes strategic risk management over tools and tactics. In addition to defending against malicious hackers and malicious code, IA includes other corporate governance issues such as privacy, compliance, audits, business continuity, and disaster recovery. *Note: Malicious code could include viruses, worms, and Trojan horses.* Further, while information security draws primarily from computer science, IA is interdisciplinary and draws from multiple fields, including criminology, fraud examination, forensic science, military science, management science, systems engineering, security engineering, and computer science. Therefore, IA is best thought of as a superset of information security. Information assurance is not just computer security because it includes security issues that do not involve computers.

IA includes the practice of managing information-related risks. More specifically, IA practitioners seek to protect the confidentiality, integrity, and availability of data and their delivery systems. These goals are relevant whether the data is in storage, processing, or transit, and whether it is threatened by accident or malice. IA, therefore, is the process of ensuring that the right people get the right information at the right time. Brief overviews of authentication, confidentiality, and nonrepudiation are provided below.

- Authentication is a security measure designed to establish the validity of a transmission, message, or originator, or a means of verifying an individual's authorization to receive specific categories of information.
- Confidentiality has been defined by the International Organization for Standardization (ISO) as "ensuring that information is accessible only to those authorized to have access" and is one of the cornerstones of information security.
- Nonrepudiation is the ability to ensure and present evidence that a specific action occurred in an electronic transaction. The message originator cannot deny sending a message or that a party in a transaction cannot deny the authenticity of their signature. This process ensures the sender is provided with proof of delivery and the recipient is provided with proof of the sender's identity so that neither can later deny having processed the data.

In regard to digital security, the crypto-logical meaning and application of nonrepudiation shifts to mean:

- A service that provides proof of the integrity and origin of data
- An authentication that with high assurance can be asserted to be genuine

Proof of data integrity is typically the easiest of these requirements to accomplish. A data hash or checksum, such as MD5 or CRC, is usually sufficient to establish that the likelihood of data being undetectably changed is extremely low. Even with this safeguard, it is still possible to tamper with data in transit, either through a man-in-the-middle attack or phishing. Due to this flaw, data integrity is best asserted when the recipient already possesses the necessary verification information.

The most common method of asserting the digital origin of data is through a digital certificate, a form of public key infrastructure to which a digital signature belongs. This can also be used for encryption. It is important to note that the digital origin only means that the certified/signed data can be, with reasonable certainty, trusted to be from somebody who possesses the private key corresponding to the signing certificate. If the key is not properly safeguarded by the original owner, digital forgery can become a major concern.

Computer Privacy Basics

Privacy is the ability of an individual or group to seclude personal information and thereby reveal information selectively. The boundaries and content of what is considered private differs between cultures and individuals, but shares basic common themes. Privacy is sometimes related to anonymity, which is the wish to remain unnoticed or unidentified in the public realm. When a matter is private to a person, it usually means that there is something about it that is considered inherently special or personally sensitive. The degree to which private information is exposed therefore depends on how the public will receive this information, which differs between places and over time. Privacy can be seen as an aspect of security where tradeoffs between the interests of one group and another can become particularly clear.

The right against unsanctioned invasion of privacy by the government, corporations, or individuals is part of many countries' privacy laws. Almost all countries have laws which in some way limit privacy. An example of this is a confidentiality statute concerning taxation, which normally requires the sharing of information about personal income or earnings. In some countries, individual privacy may conflict with freedom-of-speech laws, and some laws may require public disclosure of information which would be considered private in other countries and cultures.

Privacy may be voluntarily sacrificed, normally in exchange for perceived benefits and very often with specific dangers and losses, although this is a very strategic view of human relationships. Academics who are economists, evolutionary theorists, and research psychologists describe revealing privacy as a "voluntary sacrifice" where sweepstakes or competitions are involved. In the business world, a person may give personal details, often for advertising purposes, to enter a content to win a prize. Information, which is voluntarily shared and later stolen or misused, can lead to identity theft.

The term "privacy" means many things in different contexts. Different people, cultures, and nations have a wide variety of expectations about how much privacy a person is entitled to or what constitutes an invasion of privacy. These contexts include the following categories:

- Financial
- Informational
- Internet
- Medical
- Workplace

Financial

Information about a person's financial transactions—including the amount of assets, positions held in stocks or funds, outstanding debts, and purchases—can be sensitive. If criminals gain access to information, such as a person's account or credit card numbers, that person could become the victim of fraud or identity theft. Information about a person's purchases can reveal a great deal about that person's history, such as places he has visited, whom he has had contact with, products he uses, his activities and habits, or medications he has used. In some cases, corporations might wish to use this information to target individuals through marketing that is customized toward those individual's personal preferences, something which that person may or may not approve.

Informational

Data privacy refers to the evolving relationship between technology and the legal right to, or public expectation of privacy in the collection and sharing of personal data. Privacy concerns exist wherever uniquely identifiable data relating to a person or persons is collected and stored, in digital form or otherwise. In some cases, these concerns refer to how data is collected, stored, and associated. In other cases, the issue is who is given access to information. Other issues include whether an individual has any ownership rights to data about them, and/or the right to view, verify, and challenge that information.

Internet

The ability to control personal information revealed over the Internet and who can access that information has become a major issue. These concerns include whether e-mail can be stored or read by third parties without consent, or whether third parties can track the Web sites someone has visited. Another concern is whether Web sites which are visited collect, store, and possibly share personally identifiable information about users. The advent of various search engines and the use of data mining has created a capability for data about individuals to be collected and combined from a wide variety of sources very easily.

Medical

A person may not wish for her medical records to be revealed to others. This may be because she is concerned that it might affect their insurance coverage or employment. Or it may be because she does not wish for others to know about medical or psychological conditions or treatment which could be embarrassing.

Workplace

Employees typically must relinquish some of their privacy while at the workplace, but the amount can be a contentious issue. Employers might choose to monitor employees' activities using surveillance cameras, or may wish to record employees' activities while using company-owned computers or telephones.

A survey of numerous U.S. companies found that over half had disciplined or terminated employees for inappropriate use of the Internet. These incidents included sending an inappropriate e-mail message to a client or supervisor, neglecting work while chatting with friends, or viewing explicit websites during work hours.

Biometric Systems Security and Privacy

Biometric identifiers are becoming widely used by organizations to identify users. Devices with biometric identifiers attempt to automate this process by comparing the information scanned in real time against a template stored digitally in a database. The technology has had several teething problems, but now appears poised to become a common feature in the technological landscape.

The most widely used biometric is the fingerprint identifier. A report by the National Institute of Standards and Technology (NIST) showed that one-fingerprint identification systems had an accuracy rate of 98.6 percent, while the accuracy rate rose to 99.6 percent when two fingerprints were used, and 99.9 percent when 4, 8 and 10 fingerprints were used. The report also showed the accuracy rate for fingerprint identification drops as the age of the person increases.

The United States Visitor and Immigrant Status Indicator Technology (US-VISIT) program has extended its entry/exit biometric capturing system to 50 of the busiest land ports of entry. The system requires two digital index finger scans as well as a digital photograph of the visitor, which are intended to verify identity and are compared to a vast network of government databases.

Several organizations have stated that significant privacy and civil liberty concerns regarding the use of such devices should be addressed before any widespread deployment. Briefly there are six major areas of concern:

- Storage: Is the data stored centrally or is it dispersed? How should scanned data be retained?
- Vulnerability: How vulnerable is the data to theft or abuse?
- Confidence: How much of an error factor is acceptable in the technology's authentication process? What are the implications of false positives and false negatives created by a machine?
- Authenticity: What constitutes authentic information? Can that information be tampered with?
- Linking: Will the data gained from scanning be linked with other information about spending habits, etc.? What limits should be placed on the private use (as contrasted to government use) of such technology?
- Ubiquity: What are the implications of having an electronic trail of every human movement if cameras and other devices become commonplace, and used on every street corner and every means of transportation?

The increasing use of biometric technology raises questions about the technology's impact on privacy in the public sector, in the workplace, and at home.

- Is the use of biometrics compatible with personal and informational privacy?
- What types of protections are necessary to ensure that biometrics are not used in a privacy-invasive fashion?
- Under what circumstances can biometric data be misused?
- In what situations do the potential risks of biometric usage outweigh the benefits?

Addressing these questions is necessary to ensure that biometrics, if and when used, does not undermine or threaten privacy.

Numerous security and privacy issues must be addressed when employing any security or access system or product, including identification and verification issues. Some of the restrictions are government imposed. Several questions are relevant to this discussion:

- How will masses of biometric data be stored?
- How will this information be safeguarded?
- Who will have access to this information?
- Will companies be allowed access to face biometrics, letting them use security cameras to positively identify customers on a routine basis?
- Would customers be concerned about walking into a store, only to be greeted by name by a sales associate who has just read a summary of all recent personal purchases?

A number of terms and definitions relate to individuals who are candidates for a biometric sample. There are also several terms that apply to the security of the biometric process and use. These are categorized as system related and human related.

System Related

- *Eavesdropping* is surreptitiously obtaining data from an unknowing end user who is performing a legitimate function. An example involves having a hidden sensor co-located with the legitimate sensor.
- **Skimming** is the act of obtaining data from an unknowing end user who is not willingly submitting the sample at that time. An example could be secretly reading data while in close proximity to a user on a train.
- An **attempt** is the submission of a single set of biometric sample to a biometric system for identification or verification. Some biometric systems permit more than one attempt to identify or verify an individual.

Human Related

- An **impostor** is a person who submits a biometric sample in either an intentional or inadvertent attempt to claim the identity of another person to a biometric system.
- An **indifferent user** is an individual who knows a biometric sample is being collected and does not attempt to help or hinder the collection of the sample. Example: An individual, aware that a camera is being used for face recognition, looks in the general direction of the sensor, neither avoiding nor directly looking at it.
- A *noncooperative user* is an individual who is not aware that a biometric sample is being collected. Example: A traveler passing through a security line at an airport is unaware that a camera is capturing a face image.
- A **cooperative user** is an individual who willingly provides a biometric to the biometric system for capture. Example: A worker submits a biometric to clock in and out of work.
- An *uncooperative user* is an individual who actively tries to deny the capture of a biometric data. Example: A detainee mutilates a finger upon capture to prevent the recognition of an identity via a fingerprint.

Identification and Verification Privacy

Individuals are currently required to confirm their identity for many diverse purposes, such as verifying eligibility within a health care system, accessing a secure network or facility, or validating their authority to travel. In almost every discussion about implementing personal identification systems to improve identity verification processes, concerns about privacy and the protection of personal information quickly emerge as key issues. Government agencies and private businesses that are implementing ID systems to improve the security of physical or logical access must factor these issues into their system designs. While technologies are available that can provide a higher level of security and privacy than ever before, ID system complexity coupled with increasing public awareness of the risks of privacy intrusion require that organizations focus on privacy and personal information protection throughout the entire ID system design.

Increasing requirements for identity confirmation and for transactions of almost any kind to require personal identification have caused the definition of privacy to change. Modern privacy requires constraints on the collection, use and release of personal information, as well as the imposition of measures to protect such information.

Protecting privacy means protecting individuals' rights to control how personal information is collected and promulgated. Protecting privacy also includes protecting against identity theft; or the use of an individual's personal information for fraudulent purposes. A critical component of protecting privacy is information security—protecting the confidentiality, integrity, and availability of information that identifies or otherwise describes an individual. To be considered privacy-enabled, an identification system must be designed to satisfy these parameters.

Both privacy and security must be considered fundamental design goals for any personal ID system and must be factored into the specification of the ID system's policies, processes, architectures, and technologies. The use of smart cards strengthens the ability of the system to protect individual privacy and secure personal information.

Unlike other identification technologies, smart cards can provide authenticated and authorized information access, implementing a personal firewall for the individual and releasing only the information required when the card is presented.

User Acceptability

User psychology will rarely be discussed in biometric literature, as it would likely open up a sizeable can of worms as far as system performance is concerned. However, it must be considered if a system is to be successfully designed and implemented. If a user is not happy about using a particular biometric device, consistent use is not likely; this will potentially produce a much larger than average error rate. Conversely, if a user is intrigued and enthusiastic about using the device, use is likely to be more consistent and produce relatively low error rates. Additional examples of users who may not be willing to use biometric devices include:

- Those with no particular bias but are nervous or self conscious about using such devices;
- Those who have some physical difficulty in using the device;
- Badly trained users; and
- Those with a poor reference template.

The users' particular temperament, understanding, and current state of mind can have a dramatic impact on actual system performance—much more than the quoted difference between individual devices, for example. Clearly, administrators and managers should aim for well-educated system users who have good-quality reference templates and are happy with the overall system concept and its benefits.

These individuals will have received proper and comprehensive training in the use of the system, been guided carefully and unhurriedly through the enrollment procedure, been invited to ask questions about the system in general, and received some reference documentation with help/support line details included—all within a comfortable, unchallenging environment where individual needs can be addressed. If administrators are not prepared to do this as an absolute minimum, then such a system should not be contemplated.

The larger the potential user base, the more important this becomes as the instances of unusual requirements or misunderstandings increase. In some public applications, such as prison visitor systems or benefit payment systems, there is a tangible and immediate benefit to the user in operating the system correctly, and they are required to do so by the administrating authority. In these situations, where there is a captive audience that needs to be correctly verified, administrators may accept a methodology not normally appropriate.

In a corporate or proprietary system, where the user may have a choice in using the system, this luxury may not exist; designers must then strive to make it an interesting or exciting experience coupled with a clearly defined user benefit. If there is any doubt about this, the system will probably be unsuccessful. The user must, therefore, be the primary focus of the system design—not the technology. The user must be considered at each stage of the process and the technical configuration plan. User requirements and operating experience must also be considered very carefully. Modifications can be made at a later date. [2]

Biometric Security and Business Ethics

A variety of ethical concerns regarding biometric identification methods have been registered by users:

- Some biometric identification methods, like retinal scans, are relatively intrusive.
- The gathering of biometric information (like fingerprints) is associated with criminal behavior in the minds of many people.
- Traditionally, detailed biometric information has been gathered by large institutions, such as the military or police; people may feel a loss of privacy or personal dignity.
- People feel embarrassed when rejected by a public sensor.
- Automated face recognition in public places could be used to track everyone's movements without their knowledge or consent.

Identity Theft and Privacy Issues

Concerns about ID theft through biometrics use have not been resolved. If a person's credit card number is stolen, this situation can cause him great difficulty because information can be used where the security system requires only "single-factor" authentication. Just knowing the credit card number and its expiration date can sometimes be enough to use a stolen credit card successfully. "Two-factor" security solutions require something known plus something possessed; for example, a debit card and a personal identification number (PIN) or a biometric.

Some argue that if a person's biometric data is stolen it might allow someone else to access personal information or financial accounts, in which case the damage could be irreversible. But this argument ignores a key operational factor that is intrinsic to all biometrics-based security solutions. Biometric solutions are based on matching, at the point of transaction, the information obtained by the scan of a "live" biometric sample to a prestored, static "match template" created when the user originally enrolled in the security system. Most of the commercially available biometric systems address the issues of ensuring that the static enrollment sample has not been tampered by using hash codes and encryption. The problem, therefore, is effectively limited to cases where the scanned "live" biometric data is hacked. Even then, most competently designed solutions contain antihacking routines. For example, the scanned "live" image is virtually never the same from scan-to-scan owing to the inherent plasticity of biometrics. Ironically, a "replay" attack using the stored biometric is easily detected because it is too perfect a match.

Major Concerns

Biometric technology is inherently individuating and interfaces easily to database technology, making privacy violations easier and more damaging. If such systems are to be deployed, privacy must be designed into them from the beginning, as it is hard to retrofit complex systems for privacy. Six issues need to be included in systems development and design to ensure privacy concerns are addressed.

- **Biometric systems are useless without a well-considered threat model.**

  Before deploying any such system in a production mode, a realistic threat model is required, specifying the categories of people such systems are supposed to target, and the threat they pose in light of their abilities, resources, motivations, and goals. Any such system will also need to map out clearly in advance how the system is to work, in both in its successes and in its failures.

- **Biometrics is no substitute for quality data about potential risks.**

  It does not matter how accurately a person is identified, identification alone reveals nothing about whether a person is a terrorist, for example. Such information is completely external to any biometric ID system.

- **Biometric identification is only as good as the initial identification.**

  The quality of the initial enrollment or registration is crucial. Biometric systems are only as good as the initial identification, which in any foreseeable system will be based on the exact document-based methods of identification upon which biometrics are supposed to be an improvement. A terrorist with a fake passport would be issued a U.S. visa with the personal biometric attached to the name on the phony passport. Unless the terrorist A) has already entered biometrics data into the database, and B) has garnered enough suspicion at the border to merit a full database search, biometrics will not block entrance at the border.

- **Biometric identification is often overkill for the task at hand.**

  It is not necessary to identify a person—by creating a record of their presence at a certain place and time—if the information sought is the knowledge of whether they are entitled to do something or be somewhere. When in a bar, customers use IDs to prove they are old enough to drink, not to prove who they are, or to create a record of their presence.

- **Some biometric technologies are discriminatory.**

  A nontrivial percentage of the population cannot present suitable features to participate in certain biometric systems. Many people have fingers that simply do not print well. Even if people with "bad prints" represent 1 percent of the population, this would mean massive inconvenience and suspicion for that minority. And scale matters. The INS, for example, handles about 1 billion distinct entries and exits every year. Even a seemingly low error rate of 0.1 percent means 1 million errors, each of which translates to INS resources lost following a false lead.

- **The accuracy of biometric systems is impossible to assess before deployment.**

  Accuracy and error rates published by biometric technology vendors are not trustworthy, as biometric error rates can be manipulated. Biometric systems fail in two ways: false match (incorrectly matching a subject with someone else's reference sample) and false non-match (failing to match a subject with her own reference sample). There is a tradeoff between these two types of error, and biometric systems may be "tuned" to favor one error type over another. When subjected to real-world testing in the proposed operating environment, biometric systems frequently fall short of the performance promised by vendors.

The cost of failure is high. A lost credit card can be canceled and renewed. A lost biometric is lost for life. Any biometric system must be built to the highest levels of data security, including transmission that prevents interception, storage that prevents theft, and system-wide architecture to prevent both intrusion and compromise by corrupt or deceitful agents within the organization.

Despite these concerns, political pressure for increasing the use of biometrics appears to be informed and driven more by marketing from the biometrics industry than by scientists. Considerable federal attention is devoted to deploying biometrics for border security. This is an easy sell, because immigrants and foreigners are, politically speaking, easy targets. But once a system is created, new uses are usually found for it, and those uses will not likely stop at the border.

With biometric ID systems, as with national ID systems, administrators must be wary of getting the worst of two worlds. These could include a system that enables greater social surveillance of the population in general, but does not provide increased protection against terrorists.

Major Privacy Concerns

Why be concerned about biometrics? Organizations such as the Electronic Frontier Foundation (EFF) state that: [3]

- Biometrics isn't dangerous because all the real dangers are associated with the database behind the biometric information, which is little different from problems of *person-identifying information (PII)* databases generally.
- Biometrics actually promotes privacy by enabling more reliable identification and thus frustrating identity fraud.

But biometric systems have many components. Only by analyzing a system as a whole can one understand its costs and benefits. Moreover, analysts must understand the unspoken commitments any such system imposes. EFF concerns are categorized into four major areas:

- Surveillance
- Databases
- Linking
- Tracking

Surveillance

The continuous capture of biometric data is useful for surveillance purposes. Biometric systems entail repeat surveillance, requiring an initial capture and then later captures.

Another major issue relates to how voluntary the capture is. Some biometrics, like faces, voices, and fingerprints, can be easily captured. Other biometrics, at least under present technology, must be consciously provided. It is difficult, for instance, to capture a scan of a person's retina or to gather a hand geometry image without the subject's cooperation. Easily captured biometrics is a problem because people can't control when they're being put into the system or when they're being tracked. But even hard-to-capture biometrics involves a trust issue in the biometric capture device and the overall system architecture.

Databases

To be effective, a biometric system must compare the captured biometric data to that stored in a a biometric database. The static issues concerning databases are mainly about safeguarding large and valuable collections of personally identifying information. If these databases are part of an important security system, then they, and the channels used to share PII, are natural targets for attack, theft, compromise, and malicious or fraudulent use.

The dynamic issues surrounding databases mainly concern the need to maintain reliable, up-to-date information. Databases that seek to maintain accurate residence information must be updated whenever someone moves. Databases that are used to establish eligibility for benefits must be updated so as to exclude persons no longer eligible. The broader the function of the system, the more and broader the updating that is required, increasing the role of general social surveillance in the system.

Logically it may seem that one of the issues that plague token-based ID systems like ID cards, which is the security or integrity of the token itself, does not apply to biometric systems, because "you are your ID." But the question of the reliability of the token is really a question about trust. In an ID card system, the question is whether the system can trust the card. In biometric systems, the question is whether the individual can trust the system. If someone else captures a person's signature, fingerprint, or voice, for instance, nothing prevents it from being used by others. Any use of biometrics with a scanner run by someone else involves trusting someone's claim about what the scanner does and how the captured information will be used.

Vendors and scanner operators may say they protect privacy in some way, perhaps by hashing the biometric data or designing the database to enforce a privacy policy. But the end user typically has no way to verify whether such technical protections are effective or implemented properly. End users should be able to verify any such claims, and to leave the system completely if they are not satisfied. Exiting the system, of course, should at least include the expunging of the end user's biometric data and records.

Linking

A known risk of biometric systems is the use of biometrics as a linking identifier. This risk, of course, depends to some extent on standardization. Consider, for instance, the use of a Social Security number (SSN) as a linker across disparate databases. While the private sector would not have been able to develop anything like the SSN on its own, once the government created this identifier, it became a standard way of identifying individuals. Standardization therefore creates new privacy risks because information gathered for one purpose can be used for completely unrelated, nonconsented purposes.

Currently, automated fingerprint identification systems are heavily used by the government in connection with law enforcement, but there is, at present, little standardization within the AFIS industry. If law enforcement and private industry were to unify their fingerprint databases conforming to one common standard, such as under a national ID system, this would potentially put an individual's entire life history in interoperating databases that are only a fingerprint away.

Tracking

By far the most significant negative aspect of biometric ID systems is the potential to locate and track people physically. While many surveillance systems seek to do this, biometric systems present the greatest danger precisely because they promise extremely high accuracy. Whether a specific biometric system actually poses a risk of such tracking depends on how it is designed.

Why should citizens care about perfect tracking? Some quarters believe that perfect tracking does not fit in a free society. A society in which everyone's actions are tracked is not, in principle, free. It may be a livable society, but it would not be the American society.

Perfect surveillance, even without any deliberate abuse, would have an extraordinary effect on artistic and scientific inventiveness and on political expression. This concern underlies constitutional protection for anonymity, both as an aspect of First Amendment freedoms of speech and association, and as an aspect of Fourth Amendment privacy.

Implemented improperly, biometric systems could increase the visibility of individual behavior, making it easier for measures to be taken against individuals by agents of the government, by corporations, and by peers. It could result in politically damaging and personally embarrassing disclosures, blackmail, and extortion. This would impact democracy, because it would reduce the willingness of competent people to participate in public life. It could also increase the circumstantial evidence available for criminal prosecution. This might dramatically affect the existing balance of plausible-sounding evidence available to prosecutors, and hence increase the incidence of wrongful conviction. Many criminal cases are decided by plea bargaining, a process that is sensitive to the perceived quality of evidence. Even ambiguous or spurious evidence generated by complex technical systems may be difficult for overburdened public defenders to challenge.

Biometric systems could enable the matching of people's behavior against predetermined patterns. This could be used by the government to generate suspicion, or by the private sector to classify individuals into micromarkets, the better to manipulate consumer behavior. It could aid in repressing readily locatable and traceable individuals. While the public's concern is usually focused on the exercise of state power, these technologies may also greatly empower corporations. If proper privacy safeguards are not constructed into such systems, they could become factors in dealing with such opponents as competitors, regulators, union organizers, whistleblowers, and lobbyists, as well as employees, consumer activists, customers, and suppliers.

Security Requirements Assessment

As with any assessment program, a formalized process must be followed to accomplish the stated goals. This usually takes the form of an iterative process where a number of steps are executed many times until the process has been refined. Each of these steps must be clearly defined, including the roles and responsibilities for computer operations processes and network-related activities. This means that process definition and setting of the organization's security goals and standards must precede technology evaluation, selection, and implementation.

There is a simple and straightforward life cycle approach that can be utilized to accomplish this task. The application of this structured process ensures that all potential user group and information combinations have been considered. When successfully completed, this implies that appropriate security processes and technology have been determined that allows legitimate access into any of the organization's computer and network resources. Figure 10-1 depicts a security requirement development life cycle process. The following steps are processed repetitively until an evaluation is successful:

- Identify the organization's security issues.
- Analyze security risks, threats, and vulnerabilities.
- Design the security architecture and the associated processes.
- Audit the impact of the security technology and processes.
- Evaluate the effectiveness of current architectures and policies.

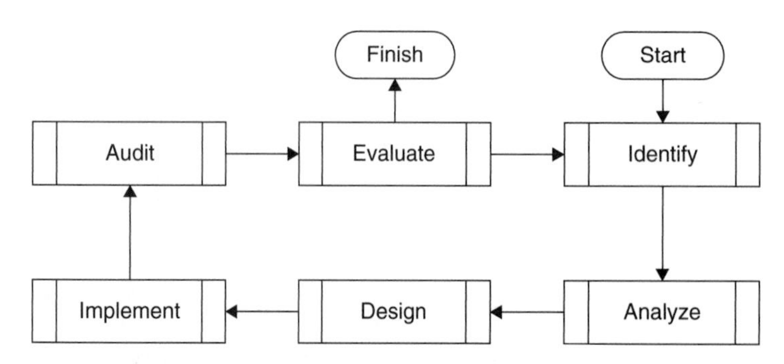

Figure 10-1 Security System Development Life Cycle

Course Technology/Cengage Learning

Note that the evaluation processes validate the effectiveness of the original analysis steps, and feedback from these evaluation steps causes a renewed analysis with possible ripple effects of changes in architecture or implemented technology.

Barriers to Using Biometrics

Fundamental barriers to using biometrics can be categorized into the following system issues:

- Accuracy
- Scale
- Security
- Privacy

System Accuracy

The critical promise of the ideal biometrics is that when a biometric identifier sample is presented to the biometric system, it will offer the correct decision. Unlike password or token-based system, a practical biometric system does not make perfect match decisions and can make two basic types of errors:

- A false match: The biometric system incorrectly declares a successful match between the input pattern and a nonmatching pattern in the database (in the case of identification/screening) or the pattern associated with an incorrectly claimed identity (in the case of verification).
- A false non-match: The biometric system incorrectly declares failure of match between the input pattern and a matching pattern in the database (identification/screening) or the pattern associated with the correctly claimed identity (verification).

The technologies may not be directly comparable in the extent of automation possible or sensing-at-a-distance capability. It is important to realize that, perhaps more than other pattern recognition systems, the false rejection of a user's claim by a biometric system is not a desirable outcome, because the response will then be made to manual identification, which is usually neither effective (verifying enrollment) nor feasible (large-scale identification). Practical biometric systems also have significant failures both in terms of *failure to acquire (FTA)* and *failure to enroll (FTE)*.

FTA is a failure of a biometric system to capture and extract data; the failure to acquire rate is the frequency of an FTA. FTE is any irrecoverable failure in the enrollment process; the failure to enroll rate is the probability that a biometric system will have an FTE.

Performance Requirements It is assumed that a large-scale identification may consist of 1 million identities and that a large-scale screening may involve 500 identities. FTA and FTE are assumed to be zero. These numbers are based on what researchers believe to be an order of magnitude estimate of the performance needed for the viability of a typical application. There are three primary reasons underlying the imperfect accuracy performance of a biometric system:

- Information limitation
- Representation limitation
- Invariance limitation

Information Limitation The invariant and distinctive information content in pattern samples may be inherently limited due to the intrinsic signal capacity, such as the individuality information limitation of the biometric identifier. For instance, the distinctive information in hand geometry is less than that in fingerprints. Consequently, hand geometry measurements can differentiate fewer identities than the fingerprint signal can even under ideal conditions. Information limitation may also be due to a poorly controlled biometric presentation by the user or by an inconsistent signal acquisition. Differently acquired measurements of a biometric identifier limit the invariance across different samples of the pattern. For example, information limitation occurs when there is very little overlap between the enrolled and sample fingerprints, such as with the left and right half of the finger. In such a situation, even a perfect matcher cannot offer a correct matching decision. An extreme example of information limitation occurs when the person does not possess or cannot present the particular biometric needed by the identification system.

Representation Limitation The ideal representation scheme should be designed to retain all invariance and discriminatory information in the sensed measurements. Practical feature extraction systems, typically based on simplistic models of biometric signal, fail to capture the richness of information in a realistic biometric signal, resulting in the inclusion of erroneous features and exclusion of true features. Consequently, a significant fraction of legitimate pattern space cannot be handled by the biometric system, which results in high failure rates. Conventional representations and feature extraction methods limit the effective discrimination among the prints.

Invariance Limitation Finally, given a representation scheme, the design of an ideal matcher should perfectly model the invariance relationship in different patterns from the same class, even when imaged under different presentation conditions. In practice, because there aren't enough training samples, uncontrolled or unexpected variance in the collection conditions, a matcher may not correctly model the invariance relationship, resulting in poor matcher accuracy.

The design challenge is to arrive at a realistic representational/invariance model of the identifier from a few samples acquired under inconsistent conditions, and then, formally estimate the discriminatory information in the signal from the samples. This is especially difficult in a large-scale identification system where the number of classes/identities is large (millions). Administrators would also like to understand how to seamlessly integrate multiple biometric cues to provide effective identification across the entire population. A fingerprint matching algorithm that assumes a rigid transformation between the two fingerprint representations cannot successfully match these two prints. Screening systems are severely information limited. First, the conventional biometric traits available for unobtrusive covert capture from a distance, such as face and gait, offer limited discriminating ability. Second, the lack of user cooperation as well as lack of environmental control typically results in inconsistent presentation. Consequently, a two-pronged approach is necessary to offer an effective identification in screening systems:

- Exploring effective methods of spatiotemporally utilizing weak biometric cues (also called soft biometrics), such as height, gait, hair color, coarse facial features, etc., to reliably identify people against the watch-list
- Engineered approach to signal acquisition: Significant innovation in designing active/purposive vision techniques to obtain higher resolution images

Although higher resolution creates a new set of problems, such as image registration inconsistency, increased processing and storage requirements, there is evidence that it can lead to better discrimination. Both of these approaches have not received much research attention and are fundamental barriers for the success of screening systems.

System Scale

Does the number of identities in the enrolled database affect the speed and accuracy of the system? In the case of verification systems, the size of the database does not really matter since it essentially involves a 1:1 match, comparing one set of submitted samples to one set of enrollment records. In the case of large-scale identification and screening systems containing a total of N identities, sequentially performing N 1:1 matches is not effective. There is a need for efficiently scaling the system to control throughput and false-match error rates with an increase in the size of the database. The fingerprint screening assumes use of two fingers and fingerprint identification performance reflects state-of-the-art AFIS performance based on 10 fingers.

Typical approaches to scaling include using multiple hardware units, coarse pattern classification, and extensive use of "exogenous" data, such as gender, age, and geographical location, which is supplied by human operators. This includes first classifying a fingerprint into major classes (such as arch, tented arch, whorl, left loop, and right loop). Although these approaches perform well in practice, they come at a price. Using hardware that is linearly proportional to the database size is expensive. Coarse pattern classification offers substantial scaling advantage only when multiple measures are available, such as fingerprints from multiple fingers, and can add to the non-match error rates. Use of exogenous information creates a mechanism for intentionally avoiding identification, such as cross-dressing or appearing older.

Ideally, administrators would like to index the patterns in some way similar to that used in conventional databases. However, due to large intraclass variations in biometric data caused by variation in collection conditions and human anatomy/behaviors, it is not obvious how to ensure that the samples from the same pattern fall into the same index bin. There have been very few published studies on reliably indexing biometric patterns. Efficient indexing algorithms would need to be developed for each technology, and it is unlikely that any generic approach would be applicable to all biometric measures.

False match error rates generally increase with the number of required comparisons in a large-scale identification or watch list system. As most comparisons are "false," where a submitted sample is compared to the enrollment pattern of another person, increasing the size of the database increases the number of opportunities for a "false match." Because of the non-independence of sequential comparisons using the same sample data, and the architectural design issues required to sustain the throughput rate while limiting active memory, making multiple passes through the enrollment data, combining parametric and nonparametric measures, the relationship between the number of false matches and database size is a poorly understood issue.

Although the size of the watch list database in a screening system is significantly smaller than that in a large-scale identification, the number of continuous/active comparisons conducted may be huge. Therefore, as in large-scale applications, the throughput and error rate issues are also critical in screening applications. Computationally, scaling of large-scale systems for almost real-time applications involving one million identities or screening the traffic for 500 recognized identities is becoming feasible. However, designing and building a real-time identification system involving 100 million identities is beyond the reach of current understanding.

System Security

The integrity of biometric systems depends on assuring that the input biometric sample was presented by its legitimate owner, and that the system matched the input pattern with genuinely enrolled pattern samples. While there are a number of ways in which a perpetrator may attack a biometric system, there are two very serious criticisms against biometric technology that have not been addressed satisfactorily:

- Biometric identifiers are not secrets
- Biometric patterns are not revocable

The first fact implies that the attacker has a ready knowledge of the information in the legitimate biometric identifier and, therefore, could fraudulently inject it into the biometric system to gain access. The second fact implies that when biometric identifiers have been compromised, the legitimate user has no recourse to revoking the identifiers to switch to another set of uncompromised identifiers. The knowledge of biometric identifier(s) does not necessarily imply the ability of the attacker to inject the identifier measurements into the system. The challenge then is to design a secure biometric system that will accept only the legitimate presentation of the biometric identifiers without being fooled by the doctored or spoofed measurements injected into the system. Development of such a system would obviate the need for revoking the compromised identifiers.

Someone could attempt various strategies to thwart fraudulent insertion of spoofed measurements into the system. For example, someone could use liveness detection to ensure the input measurements are not originating from an inanimate object. The other strategy to consider is multibiometrics which includes data from multiple and independent biometric identifiers being fused. This solution of reinforcing the identity of a subject offers increasingly irrefutable proof that the biometric data is being presented by its legitimate owner and not being fraudulently presented by an impostor. *Note: Multibiometrics was described in Chapter 6.* While these different strategies can be stipulated, it remains a formidable challenge to concretely combine these component blocks to arrive at a foolproof biometric system that does not accept fraudulent data.

System Privacy

A reliable biometric system provides an irrefutable proof of identity of the person. Consequently, users may have multiple concerns:

- Will the undeniable proof of biometrics-based access be used to track the individuals that may infringe upon an individual's right to privacy and anonymity?

- Will the biometric data be abused for an unintended purpose?
- What ensures that fingerprints provided for access control will not be matched against fingerprints in a criminal database?
- Will the biometric data be used to cross-link independent records from the same person, for health insurance and grocery purchases?
- How would administrators ensure and assure the users that the biometric system is being used only for the intended purpose and none other?

The problem of designing information systems whose functionality is verifiable at their deployed instantiation is very difficult. Perhaps someone should devise a system that meticulously records authentication decisions and the people who accessed the logged decisions using a biometric-based access control system. Such a system can automatically generate alarms to the users upon observing a suspicious pattern in the system administrator's access of users' logs. One promising research direction may be biometric cryptosystems; that includes a generation of cryptographic keys based on biometric samples. There are also radical approaches such as total transparency that attempt to solve the privacy issues in a very novel way. While one could stipulate some ingredients of the successful strategy, there are no satisfactory solutions on the horizon for this fundamental privacy problem. Four terms that apply to system privacy are:

- Privacy-protective: A privacy-protective system is one used to protect or limit access to personal information, or which provide a means for an individual to establish a trusted identity.
- Privacy-sympathetic: A privacy-sympathetic system is one that limits access to and usage of personal data, and in which decisions regarding design issues such as storage and transmission of biometric data are informed, if not driven, by privacy concerns.
- Privacy-neutral: A privacy-neutral system is one in which privacy is not an issue, or in which the potential privacy impact is slight. These systems are difficult to misuse from a privacy perspective, but do not have the capability to protect personal privacy.
- Privacy-invasive: A privacy-invasive system facilitates or enables the usage of personal data in a fashion inconsistent with generally accepted privacy principles.

Biometrics and Cryptography

Biometrics and cryptography have long been seen as competing technologies. A new school of thought which sees both technologies as being symbiotic rather than competitive is picking up momentum. An initiative to bridge the divide between areas of biometrics and cryptography is being progressed by several groups. This work is progressing by researching methodologies to fuse biometrics with cryptography while keeping in mind the following criteria:

- Practicality and ease of use
- User choice for higher assurance option
- User acceptance
- Multifactor scalability
- Unlinking ability of different enrollments performed using the same biometric

Combining Biometrics and Cryptography

A number of researchers have studied the interaction between biometrics and cryptography, two potentially complementary security technologies. Biometrics is about measuring unique personal features, such as an individual's voice, fingerprint, retina, or iris. It has the potential to identify individuals with a high degree of assurance, thus providing for a foundation of trust. Cryptography, on the other hand, concerns itself with the projection of trust, which is taking trust from where it exists to where it is needed.

A strong combination of biometrics and cryptography might have the potential to link a user with a digital signature that has been created with a high level of assurance. For example, it will become harder to use a stolen token to generate a signature, or for a user to falsely repudiate a signature by claiming the token was stolen when it was not.

As an example, the acquisition of a repeatable string from an iris biometric opens up new opportunities for privacy. One debate concerns the possible privacy abuses of biometric databases collected to support applications such as ID cards. This prospect has started to raise a number of concerns, ranging from the possibility biometric data might be correlated with health and thus leak information, to health and religious concerns.

Ongoing research shows high-quality identification of humans is possible using biometric means, but without a central database of templates. The human subject would present materials such as a passport at an enrollment station with foundational identifying capabilities, and have an iris or some other modality scanned. The biometric data need not be retained by the issuing authority. The enrollment station could use the generated biometric key to protect a Kerberos key, shared with an authentication service, or to protect a private digital-signature key whose public verification key is linked to his distinguished name by an X.509 certificate. *Note: Kerberos, a symmetric cipher system described in RFC 1510, provides for distributed system authentication.* [4] *Additional details on X.509 are found in the PKI section of Chapter 12.*

The most difficult problem involving merging cryptography and biometrics is how to generate a repeatable string from a biometric in such a way that it can be revoked. Previous attempts have almost all had unacceptable false-reject rates. Most of them also have problems with revocation, have produced too-short keys, and have not been well tested. Efforts have shown how to generate keys robustly from iris biometric measurements, using associated error-correction data that can be changed to yield different keys. This scheme produces sufficient-length keys and can produce different keys for different applications, so an attack on someone does not give an attack on all; it supports revocation; its security case is founded on extensive research in the application area, as well as a statistical lower-bound argument; and its false-reject rate is less than half a percent. This makes it feasible for many practical uses. [5]

Smart Card and RFID Issues

A *smart card* resembles a credit card in size and shape, but is completely different inside. The inside of a smart card usually contains an embedded microprocessor. The microprocessor is under a gold contact pad on one side of the card and replaces the usual magnetic stripe on a credit card or debit card.

Magnetic stripe technology remains in wide use in the United States. However, the data on the stripe can easily be read, written, deleted, or changed with off-the-shelf equipment. Therefore, the stripe is really not the best place to store sensitive information. To protect the consumer, businesses in the United States have invested in extensive online mainframe-based computer networks for verification and processing. In Europe, this infrastructure did not develop: instead, the card carries the intelligence.

The microprocessor on the smart card is there for security. The host computer and card reader actually communicate with the microprocessor. The microprocessor enforces access to the data on the card. If the host computer read and wrote the smart card's *random access memory (RAM)*, it would be no different than a diskette.

Smarts cards may have up to 8 kilobytes of RAM, 346 kilobytes of *read only memory (ROM)*, 256 kilobytes of programmable ROM, and a 16-bit microprocessor. The smart card uses a serial interface and receives its power from external sources like a card reader. The processor uses a limited instruction set for applications such as cryptography.

Smart cards can be used with a smart-card reader attachment to a personal computer to authenticate a user. Web browsers also can use smart card technology to supplement Secure Sockets Layer (SSL) for improved security of Internet transactions. Visa's smart card FAQ shows how online purchases work using a smart card and a PC equipped with a smart-card reader. Smart-card readers can also be found in mobile phones and vending machines.

RFID Privacy and Security

Radio-frequency identification (RFID) is a compact wireless technology poised to transform the world of commerce. It's an inexpensive chip that's readable up to several meters away. As a next-generation barcode, RFID will automate inventory control, cutting costs for retailers and manufacturers. RFID is already making inroads into everyday life.

This section identifies recent technical research on the problems of privacy and security for RFID. RFID tags are small wireless devices that help identify objects and people. Thanks to dropping cost, they are likely to proliferate into the billions in the next several years, and eventually into the trillions. RFID tags track objects in supply chains, and are working their way into the pockets, belongings, and even the bodies of consumers. Various approaches are proposed by scientists for privacy protection and integrity assurance in RFID systems.

Privacy Risks for Consumers and Enterprises

Privacy advocates decry the risks of RFID because it provides silent physical tracking of consumers and inventorying of their possessions. RFID introduces new privacy and security risks into businesses and a whole new dimension to corporate espionage.

A controversy was stirred by the U.S. government's decision to begin issuing passports with embedded RFID tags. Dubbed "e-passports," they were issued to numerous diplomats and tourists. The RFID tags are supposed to speed up processing and increase security. Privacy advocates immediately denounced the decision, with challenges to the policy issued by the *EFF* and the *EPIC (Electronic Privacy Information Center)*. The privacy advocates were concerned that identity thieves could break the encryption of the RFID tags and steal sensitive passport information using hand-held tag readers. The government responded by issuing further clarifications of the security features built into the e-passports, including the addition of an RF-shield cover so that the tag could only be scanned when the passport is opened. The security problems surrounding RFID technology can be grouped in several classes: [6]

- Data ownership and data-mining techniques: All methods of data collection involve questions of privacy, data ownership, and the ethical use of data-mining techniques to discover personal characteristics of an individual. For example, a drugstore's customer loyalty card data could be used to deduce private medical information about a person. This problem predates the use of RFID technology, but the sheer volume of data provided by RFID tags adds a new urgency to these discussions.
- Data theft: In the case of current databases, a would-be data thief requires computer access and considerable hacking skills to steal data. Given that RFID tags are made to broadcast information, the possibility of data theft by easily concealable RFID scanners is very real. Chip manufacturers counter this by adding security features to the chips and data, such as secure encryption schemes.
- Data corruption: Many tags are made with the ability to be rewriteable. This feature may be locked, thus turning the tag into a write-once, read-many device or left active. For example, the RFID tags used in libraries are frequently left unlocked for the convenience of librarians to reuse the tags on different books or to track check-in and check-outs. When tags that should be locked are not, the potential does exist for pranksters or malicious users to rewrite the tags with incorrect or fraudulent data.

Smart Card Alliance

The *Smart Card Alliance,* which represents a plethora of companies and governmental entities with a vested interest in the successful marriage of RFID and technology, has come out with best practice suggestions for companies implementing RF technology in ID management systems. The questions from security and privacy advocates come at a time of tremendous activity around identification cards and payment systems that utilize RFID technology.

The U.S. State Department is currently issuing RFID-based passports and the Department of Homeland Security is considering two separate initiatives, one for RFID-chipped driver's licenses and another for RFID-chipped national ID cards for people who cross the U.S. borders frequently. At the same time, many financial institutions are testing contactless smart cards, which are RFID-enabled credit cards, cell phones, PDA's, and key fobs, that enable consumers to pay for goods by simply scanning an RFID chip across a reader.

The Smart Card Alliance goes to some lengths to distance itself from the term RFID. Instead, the group refers to the technology used in identification documents and smart card systems as *RF.* RFID uses tags or labels for tracking objects or animals, or putting RFID chips on products or pallets so retailers can track them. There is very little support for security and tags that can operate up to 25 – 35 feet. Where RF differs are its security and privacy provisions. RF has cryptographic capabilities that can be designed to operate at less than four inches.

They support electronic signatures and other security features and are being used in payment and identity documents that need security.

The security guidelines are as follows:

- Implement security techniques, such as mutual authentication, cryptography, and verification of message integrity, to protect identity information throughout the application.
- Implement security techniques, such as mutual authentication of message integrity, to protect identity information throughout the application.
- Ensure protection of all user and credential information stored in central identity system databases, allowing access to specific information only according to designated access rights. Verify identification credentials for both integrity and validity.

The *REAL ID Act of 2005* stipulates that federal agencies cannot accept a driver's license or any other state-issued card for identification purposes after May 11, 2008, unless the state meets the requirements of the Act. These include verification of supporting documentation for ID issuance, use of a common machine-readable technology in the IDs and use of physical security features designed to prevent tampering, counterfeiting or duplication for fraudulent purposes. To meet these goals, smart cards have a number of advantages over other ID and authentication technologies, such as magnetic stripe, printed bar code, optical, or RFID, including:

- Strong authentication of identity: To authenticate the cardholder to the identity document, smart card technology provides support for PINs and biometrics, as well as other features for visual identity verification.
- Strong ID credential security: Sensitive data can be encrypted both on the smart card and during communications with the external reader. Digital signatures can be used to ensure data integrity and to authenticate the card and the credentials it contains, with multiple signatures required if different authorities create the data.
- Strong ID card security: When used with other technologies, such as public key cryptography and biometrics, smart cards are almost impossible to duplicate or forge. Data stored in the chip cannot be modified without proper authorization (a password, biometric authentication, or cryptographic access key).
- Strong support for privacy: The card's unique ability to verify the authority of the information requestor and its strong card and data security make it an excellent guardian of the cardholder's personal information. Information embedded on the chip can be protected so that it cannot be surreptitiously scanned, skimmed, or obtained without the knowledge of the user.

Another important advantage is that smart cards can be used for off-line identity verification, when no access to a system or remote database is possible, by using the computer built into the card itself. This is an essential capability to improve the security of using driver's licenses as Real IDs, since there are so many situations where law enforcement officials must verify an identity and authenticate a card. The flexibility built into smart cards creates additional application possibilities, including their ability to secure Internet-based services. This means that Real IDs could be used for identity verification, e-authentication, and digital signatures in a variety of government applications, improving the efficiency of government agencies, while better protecting privacy.

Biometric Security and Privacy Management

RFID-Consortium for Security and Privacy (RFID CSP) is a partnership between academic and industrial scientists specializing in RFID security and privacy. Their mission is to make RFID safe for consumers by conducting open research and educating the next generation workforce that will develop, deploy, and maintain secure RFID infrastructures. [7]

Management has identified a number of issues relating to security and privacy of RFID and smart card technology implementations. These include:

- Database issues in privacy protection
- Privacy-enhancing technologies
- Legal and criminal issues
- Privacy leakage case studies
- Identity theft and related abuses
- Consumer and business practices and trends affecting privacy
- Information ownership, competing claims, and unresolved ambiguity
- Relationships and tradeoffs between security and privacy
- Relationships between privacy management and digital rights management
- Formal models and definitions of privacy

IS/IT management has developed a list of questions concerning security and privacy, such as:

- What type of authentication is required for this application? Do we need to know "who" the individual is or not?
- How can the collection of information be limited to only what is necessary? Can any of the applications utilize and maintain anonymity (e.g., electronic cash)?
- Has the application changed (technologically or otherwise) since the creation of the application that may warrant a rethinking of the authentication needed?
- Are there risks of placing this application onto a card with other applications?
- What safeguards are employed to limit the ability to combine and warehouse data elements collected by different applications?
- What protections can be utilized to prevent the disclosure of information across applications?

In short, designers should not be afraid to think about changing the way that old applications were used if the changes will help to protect the consumer on the new format of the smart card.

While technology can be implemented with an increased focus on protecting consumer privacy, there is still a role for policy makers. Policy makers will need to look into such issues as:

- The ability of government to use the card to track individuals
- The information handling practices of the different applications on the card

- The ability of smart card companies to warehouse and package data for sale to third parties

ISO 17799 Security Domains

Many of the government regulations involving information security are addressed in the ISO 17799 security domains. The 10 domains referenced in this document are indicative of the breath and knowledge necessary to secure information assets and information systems. The following areas are addressed: [8]

- Security policy
- Organizational security
- Asset classification
- Personnel security
- Physical and environmental security
- Communications and operations management
- Access control
- Systems development and maintenance
- Business continuity management
- Compliance

Summary

- Computer security is a branch of technology known as information security as applied to computers. The objective of computer security varies and can include protection of information from theft or corruption, or the preservation of availability, as defined in the security policy. Information security means protecting information and information systems from unauthorized access, use, disclosure, disruption, modification, or destruction. Security management concerns the ability to monitor and control access to network resources, including generating, distributing, and storing encryption keys.
- Privacy is the ability of an individual or group to seclude personal information and thereby reveal information selectively. The increasing use of biometric technology raises questions about the technology's impact on privacy in the public sector, in the workplace, and at home. Addressing these questions is necessary to ensure that biometrics do not undermine or threaten privacy. There are numerous security and privacy issues that must be addressed when employing any security or access system or product. Some of the restrictions are government imposed. Identification and verification issues must be addressed.
- Several organizations have stated that significant privacy and civil liberty concerns regarding the use of biometric devices should be addressed before any widespread deployment. Six major areas of concern include storage, vulnerability, confidence, authenticity, linking, and ubiquity.

- Fundamental barriers when using biometrics can be categorized into system issues involving accuracy, scale, security, and privacy. Researchers have studied the interaction between biometrics and cryptography, two potentially complementary security technologies. A strong combination of biometrics and cryptography might have the potential to link a user with a digital signature that has been created with a high level of assurance.
- A smart card resembles a credit card in size and shape, but contains an embedded microprocessor. Smart cards can be used with a smart-card reader attachment to a personal computer to authenticate a user. The technology used in identification documents and smart card systems is RF, versus RFID, which uses tags or labels for tracking objects or animals.

Key Terms

Attempt The submission of a single set of biometric sample to a biometric system for identification or verification.

Computer security A branch of technology known as information security as applied to computers.

Cooperative user An individual who willingly provides a biometric to the biometric system for capture.

Imposter A person who submits a biometric sample in an intentional or inadvertent attempt to pass him/herself off as another person who is a legitimate enrollee.

Indifferent user An individual who knows a biometric sample is being collected and does not attempt to help or hinder the collection of the sample.

Information assurance (IA) Information operations that protect and defend information and information systems by ensuring their confidentiality, authentication, availability, integrity, and nonrepudiation.

Information security Protecting information and information systems from unauthorized access, use, disclosure, disruption, modification, or destruction.

Privacy The ability of an individual or group to keep their lives and personal affairs out of public view or to control the flow of information about themselves.

Security management The ability to monitor and control access to network resources, including generating, distributing, and storing encryption keys.

Skimming The act of obtaining data from an unknowing end user who is not willingly submitting the sample at that time.

Review Questions

1. ________________ is a branch of technology known as information security as applied to computers.
2. ________________ means protecting information and information systems from unauthorized access, use, disclosure, disruption, modification, or destruction.

3. ________________ includes measures that protect and defend information and information systems by ensuring their availability, integrity, authentication, confidentiality, and nonrepudiation.

4. ________________ is a security measure designed to establish the validity of a transmission, message, or originator, or a means of verifying an individual's authorization to receive specific categories of information.

5. ________________ has been defined as "ensuring that information is accessible only to those authorized to have access" and is one of the cornerstones of information security.

6. ________________ is the ability to ensure and have evidence that a specific action occurred in an electronic transaction.

7. ________________ is the ability of an individual or group to seclude personal information and thereby reveal information selectively.

8. Five categories of elements involving privacy expectations or what constitutes an invasion of privacy for different contexts are ________________, ________________, ________________, ________________, and ________________.

9. ________________ concerns the ability to monitor and control access to network resources, including generating, distributing, and storing encryption keys.

10. Civil liberty organizations have voiced concerns that should be addressed regarding the use of biometric devices before any widespread deployment. These six major areas of concern include: ________________, ________________, ________________, ________________, ________________, and ________________.

11. ________________ is surreptitiously obtaining data from an unknowing end user who is performing a legitimate function.

12. ________________ is the act of obtaining data from an unknowing end user who is not willingly submitting the sample at that time.

13. An ________________ is the submission of a single set of biometric sample to a biometric system for identification or verification.

14. An ________________ is a person who submits a biometric sample in either an intentional or inadvertent attempt to claim the identity of another person to a biometric system.

15. A ________________ is an individual who is not aware that a biometric sample is being collected.

16. An ________________ is an individual who actively tries to deny the capture of a biometric data.

17. Practical biometric systems also have significant failures both in terms of ________________ and ________________.

18. A ________________ resembles a credit card in size and shape; however, the inside of this card usually contains an embedded microprocessor.

19. ________________ is a compact wireless technology that includes an inexpensive chip that is readable up to several meters away.
20. The Smart Card Alliance refers to the technology used in identification documents and smart card systems as ________________.

Discussion Exercises

1. Identify four approaches to computer security.
2. Identify 10 areas involving personal information where there are privacy ramifications.
3. Research the topic of "invasion of privacy" in the areas of finance, information, Internet, medical, and the workplace. Describe the situations and issues.
4. Research the topic of privacy and civil liberties as it applies to computer, information, and biometric issues. Provide an overview of the issues.
5. Describe the issue of user acceptability as it applies to biometric identification and verification.
6. Describe the ethical concerns of biometric identification methods.
7. Describe how identity theft can be avoided by using biometric methods.
8. Six major concerns have been identified with the biometric technology. Provide an overview of each concern and provide a "real-world" example.
9. Provide a comparison between the RFID and RF technologies. Describe how they differ.
10. Describe how a smart card can be used for access control and financial transactions.

Hands-On Projects

10-1: Security and Privacy Issues

Time required—30 minutes

Objective—Research system security and privacy issues that are part of biometric application and system implementations.

Description—Use various Web tools, Web sites, and other tools to identify biometric security and privacy issues that occur in the development and production in commercial and government organizations.

1. Describe how a smart card solution could work with both RF and RFID technologies.
2. Describe how cryptography would be implemented with a smart card solution.

3. Provide an overview of the activities of the Smart Card Alliance. Describe the current initiatives.
4. Provide an overview of the activities of the RFID CUSP. Describe the current initiatives.
5. Provide an overview of each of the ISO 17799 security domains.

10-2: Identifying Security and Privacy Issues

Time required—45 minutes

Objective—Identify system security and privacy issues that can impact the level of access control and security for OnLine Access assets and resources.

Description—The owners and managers of OnLine Access have identified the necessary integrity and accessibility issues when implementing a biometric access control solution. They must now look at the system security and privacy issues involved when implementing a biometric solution.

1. Research the security and privacy issues that are inherent in access control and security solutions.
2. Identify security and access control system security and privacy issues that have been encountered in other biometric implementations.
3. Produce a report describing how these issues have impacted the implementation of a biometric solution.
4. Describe how these issues have been resolved.

chapter 11

Implementation and Operation Issues

After completing this chapter, you should be able to do the following:

- Understand the computer and network system implementation issues in the legacy environment
- Identify evaluation techniques that can be utilized to manage and control a security and access control system
- Look at the various internal and external events that can impact a biometric security system
- Become familiar with the issues relating to the successful implementation and operation of a biometric security system
- Identify those situations and activities that can negatively impact the operations of a biometric system

The real work starts after systems and applications have been selected. A critical person in the implementation of a system is the project manager. Timelines and responsibilities must be assigned and tracked. After the implementation, the next phase includes making the system work as advertised. Close coordination and liaison with the vendors is a must in both of these phases. Because biometrics is a new technology, unexpected issues in implementation and operation will probably emerge. Work-a-rounds and impromptu fixes could be the rule and not the exception.

Introduction

To ensure that a system's implementation takes place as efficiently and with as little disruption as possible, a number of tasks are necessary. Managers often subdivide these tasks into subtasks so they can be easily scheduled and tracked at some particular level of detail.

As with any system implementation, a biometric solution has the opportunity to succeed or fail. There are many "gotchas" in implementing a security or access system that uses biometric products and services. Of particular interest is the throughput of the primary system for access requirements. Additional issues can include demographics, user location, weather, HVAC (heating, ventilating, and air conditioning), and enrollment in the system. Security and access control must be part of the original system design and not an afterthought.

As with many interesting and powerful developments of technology, there are concerns about biometrics. The biggest concern is that once a fingerprint or other biometric source has been compromised, it is compromised for life, because users can never change their fingerprints. Theoretically, a stolen biometric can haunt a victim for decades. Attempts to circumvent security biometrics have shown that it can be undermined in its present form, and it cannot yet be considered a viable form of authentication.

In any security system there will be a weak link, which is human and not technological. A common problem is personnel turnover. Former employees' computer and network access must be denied promptly to avoid possible security violations that can lead to litigation. There must be a process where the human resource department notifies IT of an employee's departure, so that network access can be removed. Delays can result in data theft or corruption.

A primary task of maintaining computer and network integrity involves troubleshooting the network environment. Past studies and results indicate that a systematic approach is the most effective. Administrators should be able to identify problems based on symptoms and initiate corrective action based on this systematic approach. If change is not managed, a great deal of time will be spent fighting fires instead of preventing them. A major part of this structured approach is network management and planning, which can be accomplished by addressing the following issues:

- Security policies
- Documentation procedures and methodology
- Hardware and software standards
- Network baseline
- Data backup
- Upgrade guidelines
- Preemptive troubleshooting

Legacy System Implementations

Implementation projects can include a number of activities. These could include ordering and installing new equipment, ordering new storage devices and media, training personnel, converting data files into new formats, drawing up an overall implementation plan, and preparing for a period of either parallel or pilot production.

Organizational managers often want to introduce a new computing system or significant upgrades so they can radically change working practices. The program then becomes the means to fulfill a noncomputing end. The "success" of the system will depend on a number of variables, such as the quality of staff training and the nature of the input that the staff had into the design process. As soon as the program is used by someone other than designers and programmers, misunderstandings can arise. These undesirable situations can be ameliorated by employing the following components in a system implementation:

- A requirements document should be part of the system contract. Extracting requirements from users might not be an easy task, but if it's not done, massive requirements creep and completely unrealistic estimates of timescale and cost are the end result. Performance does matter, however some legacy and access control projects have produced sluggish results, but are not always easy to specify.
- Prototypes are useful for extracting requirements. They should, however, be discarded once development begins. If not, this prototype may have to be updated in step with the real product's changing specification, which absorbs critical resources.
- Stakeholders should be identified early on to approve the specification and be consulted on any subsequent changes.
- The baseline is a set of minimal requirements, which needs to be established and agreed upon as early as possible. Items that don't make it into the baseline can be marked as desirable, highly desirable, etc.
- Milestones and deliverables are important, as these may be all the manager cares about. The timing of their release is important for maximum effect.

Aftercare and maintenance of the system may be overlooked, so it is essential to make it clear that once the project is over, its owners must look after it on a day-to-day basis. It's up to them to ensure that the requirements include details on how to make this transition smooth.

Computer Implementation Models

Several different computer-system project management models could be implemented for an access control system, including:

- Waterfall: Nonoverlapping phases. It's okay if a detailed analysis can be successfully completed prior to programming. This might be possible for a stable project definition and development environment.
- Spiral: Revisits all phases of development. This is useful with new technologies and minimizes the costs of rethinks, but can be hard to manage.
- Reuse-oriented: Relies on a large base of reusable software. The development may consist in adapting existing code or combining the components.
- Staged delivery: Each milestone is a working release. This is useful with new technologies. Some recent failures have been blamed on their "big bang" implementation.
- Controlled iteration: Go onto the next stage if a certain percentage is complete. This is appropriate for strongly object-oriented tasks, where progress is incremental.

Numerous programs and tools are available for the implementation of computer and information technology projects. These often include procurement and training components. One such document is the national security community policy, which governs the acquisition of information assurance (IA) and IA-enabled information technology products. NSTISSP #11 is available at the National Security Telecommunications and Information Systems Security Committee Web site.

Implementation Project

The implementation process will determine if the installation will be a success or failure. A number of common steps are required when implementing a computer-based system, which include:

1. Creation of the project: The mission statement
2. Naming the implementation team: Identification of resources
3. Needs assessment or analysis: Specification of hardware and software
4. Planning phase: Determination of goals, milestones, time frames, and verification procedures
5. Installation of hardware and software: The physical introduction of equipment
6. Implementation of software: The loading of data, templates, creation of systems, policies, and procedures
7. Training: The "how to use" phase
8. Dual or parallel systems: The verification of the biometric system
9. Production, live system: The bringing of the biometric system online
10. Critique: An ongoing process which is conducted at the end of each phase

Baseline Development

A baseline for computer and network performance must be established if network monitoring is going to be used as a preemptive troubleshooting tool. A **baseline** defines a point of reference against which to measure network performance and behavior when problems occur. This baseline is established during a period when no problems are evident on the network. A baseline is useful when identifying daily network utilization patterns, possible network bottlenecks, protocol traffic patterns, and heavy-user usage patterns and time frames. A baseline can also indicate whether a network needs to be partitioned or segmented, or whether the network access speed should be increased. The three components that must be created to establish a baseline are:

- Current topology diagrams
- Response time measurements of regular events
- Statistical characterization of the critical segments

These three components will require some effort in developing; however, the payoff will come when a problem occurs in the network. A small amount of time spent each week by a number of personnel who are assigned portions of the network can accomplish this task in a relatively short time.

Training Program

Training is required in three major areas:

- Database
- Operational
- System administration

Database training covers all the necessary steps required to prepare and enter all the appropriate information into the system. Future information—such as new records or templates—will be based on the decisions made during this portion of the implementation. Specific, one-time training is necessary for the loading of the starting or base biometric data or models.

Operational training includes the development of all operation policies and procedures to be used on a daily basis to run the system. The procedures developed during the start-up phase will be proven during the operational phase. By using all the records from the biometric test, the training process will utilize every aspect of the computer system. Various personnel will be required to thoroughly test all components and phases of the verification and identification processes. When the control data and the training data agree, the biometrics system is ready for the operational status.

System administration training consists of familiarizing all involved personnel in the functions of the computer and network operating systems. Training areas include logging on to the system, use of the menu system, and setting security measures, such as passwords. Employees designated administrators should be trained in creating new user and security groups, loading new programs, maintaining backups, and creating disaster recovery procedures.

Security Evaluation

How does an administrator know the security or access control system that is deployed will perform as promised? Who should be trusted with this determination, the service provider or an independent, impartial evaluator? Security evaluation involves an assessment of the security properties of a particular computer system or communications network. A standard set of requirements is used as a metric to determine and compare how effective different systems will respond in mitigating the effects of malicious compromises or attacks. This determination and comparison is usually performed by an independent organization, which has no vested interest in the outcome of the assessment.

If specific operating environments are considered, the term security certification is reserved. When a particular government organization or agency is convinced that a given product or system possesses sufficient security protection, it awards a **security accreditation**. The specific goals that can be associated with security evaluation include a uniform measure of security, an independent assessment, criteria validity, and evaluation assurance.

Comparing the security effectiveness of different systems is difficult if varied approaches are taken to mitigating attacks. Security evaluation that includes a standard, well-defined set of security requirements provides a means of comparing the effectiveness of different security approaches. In order to achieve this goal of uniformity, security requirements requiring subjective judgment to determine compliance must be minimized.

A security evaluation offers the opportunity for an independent organization to provide an assessment of the security properties of a system. This will remove the inherent biases that can occur if a development organization performs an evaluation on its own product. The security evaluation process can be viewed as part of quality assessment processes that are common in typical software and system engineering efforts.

When a standard metric is established for determining security effectiveness, computer systems and network vendors will wish to meet these requirements in their development efforts. These requirements must be balanced by resources and cost considerations. If the security specifications are too stringent, the vendors will ignore them.

System tests, formal methods, and other life-cycle activities can be used as evidence that a particular system is secure. Security evaluation can be used as additional evidence that a system is secure. Security evaluation may provide the most convincing assurance, because it serves as a means for reviewing, assessing, and summarizing the results of all other assurance activities.

Security Cost Justification

Since computer and network security measures can require a substantial investment, it is imperative they are made with some confidence that the investments are warranted. One national study has found that large companies and government agencies were able to quantify an average loss of $1 million from security breaches. Security programs and measures, however, can also be very expensive.

One method of determining the degree of security needed is to assess the value of the database element that may be placed at risk. This value would include the cost of collecting the data and re-creating the database. The possibility exists that the original data is no longer available, which means that the database could not be reconstructed. This scenario assumes that the database was not backed up, which is also a possibility. The organization's return on investment in network security would depend on the monetary value assigned to the data and database. This exercise is not a trivial pursuit.

Security Providers

There are a number of suppliers offering different types of tools that provide security and access control for computer and network configurations. Numerous categories of biometric, video, and physical devices are available from many suppliers and implementers. A partial list includes the following:

- Baltimore Technologies
- Cisco Systems
- Entrust
- IPlanet
- RSA Security
- VeriSign

Products offered by these companies address the issues of authentication and authorization and other security functions, such as encryption.

Security Implementation Issues

Implementing a thorough security system is not as simple as installing technologies and then forgetting about them. There are a number of challenges that must be addressed for a security and access control plan to work effectively, including the following issues:

- Ease of use: Requiring a user to jump through hoops to access a system or complete a transaction will negate the value of security.

- Cost versus risk: Consideration must be given to the dollar amount of the transaction that is being protected.
- Internal security risks: There is a possibility that an internal breach will occur.
- System defaults: System defaults are usually set at the low end of security coverage.
- Standards void: Standards that ensure that technologies, such as biometrics, work together seamlessly are in the process of being developed.

Additionally, regulations such as the Gramm-Leach-Bliley Financial Services Modernization Act or Health Insurance Portability and Accountability Act (HIPAA) make implementing security systems difficult. State regulators can further complicate security issues for financial services firms. *Note: Standards and standards organizations will be discussed in Chapter 12.*

The implementation strategy for security systems can be subdivided into four separate stages:

1. Security audit: Examines processes, infrastructure, and applications to determine if security weaknesses exist
2. Technology audit: Measures the available technologies and applications against those that are needed
3. Planning and implementation: Should consider the organization's operational processes and those employees who use them
4. Training and monitoring: Users must be thoroughly trained in and monitored for acceptance of the security provisions

Network and E-Commerce Fraud Issues

Security managers and administrators must evaluate the steps being taken to prevent online fraud, security and privacy violations, and intruder attacks. Although the cost of implementing these e-security measures can be significant, the alternative is even more costly. Actions can be taken to prevent theft, protect sensitive information, and foil destructive attacks.

Online fraud can occur in a number of ways; however, identification fraud and electronic price-tag alteration are the most predominant forms. Identification fraud occurs when e-thieves use a consumer's credit card to make online purchases or when consumers make an online purchase and then later deny the transaction. E-commerce providers often have little to no recourse and are held liable when transactions conducted at their sites are later determined to be fraudulent. This means the merchants must pay higher discount rates, fees, and penalties based on their charge-back percentages. Electronic price-tag alteration occurs when hackers manipulate the shopping cart software code and alter prices. It is estimated that one-third of all shopping cart applications have software issues.

Traditional password protection, which is used to control system and database access, is not sufficient for e-commerce security. Secure e-commerce requires that both parties to a transaction be positively mutually identified. This mutual authentication process prevents intruders from acquiring valuable information or goods under false pretenses. It is also essential that e-commerce transaction information be secured against theft. Encrypting the information before it is sent and then decrypting it when it is received can protect transaction communications.

This security technique is designed to foil intruders who may intercept the transaction somewhere in the transmission path.

E-commerce transaction information, which is stored on computer systems and servers, must be carefully protected. This is usually accomplished by defensive hardware and software, which can include router, firewall, and intrusion prevention and intrusion detection mechanisms. The goal is to keep systems safe from external and internal theft attempts. Confidential data, such as credit card numbers and account information, should be encrypted. Encryption can make the information unusable in the event that e-thieves gain access to the system.

Biometric Systems Implementations

This section deals with system issues associated with implementing biometric systems. Enrollment is a significant part of a practical biometric system (see Chapter 3 for a discussion of enrollment issues, such as positive and negative enrollment, enrollment for screening, quality control of enrollment data, enrollment integrity, etc.). Many system-level issues arise when the database size is very large. In certain biometric applications, such as a national identification card or a driver's license, the database could contain tens of millions of identities and hundreds of millions of biometric samples.

As with many interesting and powerful developments of technology, there are implementation concerns with biometrics. The biggest concern is the fact that once a fingerprint or other biometric source has been compromised it is compromised for life, because users can never change their fingerprints.

Security and Privacy Issues

As biometric systems are deployed within security systems, or as part of verification and identification programs, implementation issues relating to security and privacy must be considered. The role of a biometric system is to recognize (or not recognize) an individual through specific physiological or behavioral traits. The use of the word "recognize" is significant—it is defined as "the same as that known." In other words, a biometric system does not establish the identity of an individual in any way, it merely recognizes that that person is who he says he is (in a verification or a positive identification system), or that he was not previously known to the system (in a negative identification system, for example, to avoid double enrollment in a welfare system). This tie between the actual identity of an individual and the use of biometrics is subtle and provokes much debate, particularly as it relates to privacy and other societal issues. With respect to privacy, the nature of a visa issuance and border crossing processes poses potential privacy risks. Privacy protection will best be achieved through privacy-sensitive procedures rather than through a choice of technology.

Privacy considerations are paramount in the design and implementation of effective biometric ID systems. Two of the major advocacy resources on the Web that track privacy issues are:

- **Electronic Privacy Information Center (EPIC)**, www.epic.org [1]
- **Privacy International (PI)**, www.privacyinternational.org [2]

System Deployment Issues

The key steps and issues that should be examined before the successful implementation of a biometrics project include:

- **Requirements definition:** Operational requirements surveys, application impact studies, statements of work, source selection, etc.
- **Planning considerations:** Program management, evaluation of alternative solutions, management of the permissions database, etc.
- **Life-cycle cost analysis:** Initial hardware/software costs, enrollment costs, maintenance, problem recovery, etc.
- **Social issues and concerns:** Including privacy

A new concept is anonymous biometric authentication. This approach is being developed to protect an individual's private information and to maintain the anonymity of the authentication system. [3]

Major Operation Issues

Each biometric modality and, in some cases, each product, differs greatly in approach and installation, so that direct comparison is difficult during the research and selection process. Moreover, each mode involves some trade-offs. For instance, iris identification is accurate, but can be slow and requires more cooperation from users than some other types of biometrics. There are a number of other major issues to consider in selecting the best biometric modality:

- **Ease of enrollment:** New individuals must be enrolled quickly and simply, not just to save time, but to maintain staff goodwill. Biometrics depends on goodwill just as any other type of security does. Operators are asking people to expose their eyes, allowing themselves to be fingerprinted, or permit other essentially intrusive procedures. Expect resistance for religious or political reasons, but also simply because bodies are private, and people aren't comfortable exposing body parts.
- **Recognition speed:** Speed of identification can play a similar role. For example, fingerprint identification is relatively slow and most suitable for low-volume applications, not for hundreds of workers waiting impatiently to check into the facility each morning.
- **Device size:** Size of the sensor device is most important in small areas, such as adjacent to doors.
- **Error rates:** Error rates are not a big problem with small populations, but a high error rate with a large population is a recipe for disaster, because user patience tends to decrease as error rates increase.
- **Environment:** The environment can affect the choice of modes in subtle ways. For example, if protecting a lab where the staff wears gloves, fingerprint readers probably aren't a good choice. Voice recognition, or multimodal systems, might be more viable.
- **Cost:** Especially for low-volume operations, cost is a key consideration. Biometrics saves the burden and expense of a card-based system, not to mention eliminating the headache of lost or stolen cards. People don't often forget their hands.

- **Multiple-factor authentication:** What if other constraints push the unit to biometric solutions that are comparatively less secure? Multimodal solutions using two or more different biometrics are becoming more common. Multimodality can also be more flexible, with certain kinds of access requiring only one mode and others requiring more.

A management issue concerning the infrastructure that must be considered is the ability to make repairs and add new components without negatively impacting the operation of the biometric system. **Hot swapping** is the ability to add and remove devices to a computer while the computer is running and have the operating system automatically recognize the change. **Plug-and-play** is an industry-wide standard for add-on hardware that indicates it will configure itself, eliminating the need to set jumpers and making installation of the product quick and easy.

Hurdles to Clear

Administrators must not let a biometric solution lull them into a false sense of security. They should not abandon legacy security solutions involving firewalls, encryption, passwords, and other precautions because of a biometrics implementation.

Administrators may want to avoid large, centralized databases of biometric information. Self-contained, individual fingerprint readers, for example, can verify the identity and keep the biometric data out of the centralized database. Users also feel more comfortable knowing their fingerprints are not in some massive repository. But losing a reader can be expensive and annoying. Additional hurdles include:

- **Biometric technologies have limitations.** Some portion of the population will always be physiologically unable to use certain modes. It's not just that one-armed man, either, as approximately 4 percent of the population can't use fingerprint technology because of dry skin.
- **Psychological and political issues are no less important.** Experience has shown citizens are unwilling to entrust their fingerprints to others. Others are squeamish about exposing their eyes to scanners, no matter how harmless they are. Even the chance of infection from a fingerprint scanner is objectionable to some people.
- **Biometric systems can also be costly and complicated to deploy.** That makes it all the more important to work carefully with vendors. Administrators should focus on the overall solution, not just the product, or even the specific technology. It is essential to ask for vendor support, such as ideas and possibilities, not just for products.
- **Additional requirements impact scalability.** Depending on the intent of the biometric implementation, the number of people using it will probably grow, sometimes rapidly. For example, biometric-controlled access may be mandatory first for one group working on a network, then for another, and another until all users must be enrolled. The biometric solution should be scalable to handle increases in users and locations.

Although standards for biometrics are just emerging, administrators should ensure that solutions are based on existing standards and not dependent on a vendor's proprietary technology. Using standards-based components permits a wider range of possible solutions and vendors for each component. Additionally, standards-based technology allows upgrades more easily when newer, better, faster solutions emerge. The field of biometrics is immature and new modes and implementations come along each year. Fingerprints are already being replaced by other modes. Managers should attempt to select a vendor with a reputation for keeping up with evolving standards. Biometrics, security, and other technology standards will be described in Chapter 12.

Weighing the Options

When comparing solutions, managers will likely need to conduct research to get the information they need. This may include a statement that indicates the speed at which a prospective biometric solution can handle the people who are waiting for access. The vendor may quote the verification time for the reader, which is the elapsed time from the user presentation at the device until identity verification. This is certainly part of the total time needed, but the total time it takes for a person to use the device must also be included. This small detail can impact the ongoing operation of the system.

Depending on environmental conditions at the access control location, administrators may also need to look closely at each solution's durability. Does the local environment include abrasive sand, electrostatic shock, high or low temperatures, direct sun or radiation, chemicals, rain or snow, wind-driven grit, or other difficult circumstances? If so, administrators must make sure the mode and its implementation match the need.

Biometric solutions must also integrate with existing legacy computer and network systems. Products that are interoperable will have a longer useful life and greater flexibility. Solutions should be chosen that are independent of operating system and hardware. The ability to acquire hardware from one vendor and software from another can be crucial for creating best-of-breed solutions; however, multivendor systems can be a headache. If there is a need for a new, special application, a software development kit can simplify things. The application may also require remote enrollment or management capabilities for facilities in multiple locations.

Finally, be aware that the biometrics business is pretty chaotic these days. Companies merge, acquire one another, or sometimes go out of business entirely. This has its advantages, as a single company may offer many technologies; however, there are also potential downsides. For example, long-term product support may be unpredictable and unstable. Working collaboratively with knowledgeable and imaginative systems integrators is vital in a technology so complex.

Biometrics is one technology where government agencies have an advantage over businesses. The government is by far the biggest customer for biometric security, so government agencies see the newest and best ideas first. Government agencies have a moral responsibility to pioneer and shape biometric options to create a biometric solution that's perfect for an organization. [4]

Performance and Reliability

When finger-imaging technology was first applied to reduce multiple participation fraud in government assistance programs, there were many concerns about the performance and reliability of the technology in a social-service application, as well as about the potential stigma that a finger-image requirement would place on potential clients. The experience of departments that have incorporated finger imaging into the process of applying for welfare assistance suggests many of these fears were unfounded. Finger imaging has been readily integrated into the human services programs of the affected states. However, despite the positive reaction to finger imaging from the state officials, there is still uncertainty regarding the extent to which this technology can reduce multiple participation fraud.

The states planned for implementation of their biometric identification systems in response to a wide variety of factors and considerations that were idiosyncratic to each environment. Some states reported that their respective legislative mandates, which prescribed specific dates by which biometric systems were required to be in place, allowed insufficient time for development and planning. The states developed and followed implementation schedules in accordance with internal priorities and considerations. The units uniformly described their implementation processes as largely uneventful, though they encountered a variety of minor implementation issues—most of which were associated with the logistical difficulties of mobilizing and managing such a complex initiative. Preparing staff for the implementation of the biometric systems, both philosophically and operationally, took different forms, priorities, and levels of effort in the states. At implementation, advance notification to clients and/or the general public about new biometric client identification procedures was considered important by all representatives.

The objective of providing advance notification was to inform and prepare clients for the additional application or recertification step. This entailed explaining the requirement and who is required to submit, to address client concerns, and to accelerate enrollment of the existing caseload. All states prepared informational mailings to clients advising them of the new requirement. Some states reported developing additional outreach media including multilingual (English and Spanish) videos, posters, and brochures for viewing and distribution in the local office. Most of the states also identified various outlets in the community through which they informed the general public in advance about the implementation of biometric client identification procedures.

The states with operating systems reported implementation of new biometric client identification procedures had a negligible impact on operations at the local office level. In general, units also reported the problems and obstacles encountered in operating their respective projects are not unlike those encountered in demonstrating any new technology or procedural modification. These departments also reported their systems and procedures were implemented without unexpected difficulty and were rapidly institutionalized. All the states confronted a range of basic physical space and logistical issues, including:

- Where to situate the new equipment
- How to appropriately alter job descriptions
- Who to reassign or hire to handle the new procedures
- How to adjust the flow of clients and paperwork most efficiently

Finger Imaging and Fraud Reduction

Assessing the ability of finger imaging to reduce fraud is difficult because the amount of fraud caused by duplicate participation in support programs is unknown, and because changes in caseload after the introduction of finger imaging cannot be interpreted unambiguously as reduction of fraud. The evaluations of finger imaging systems conducted by several states have produced the following findings:

- A small number of duplicate applications (approximately 1 duplicate for every 5,000 cases) have been detected by finger-imaging systems. Finger-imaging systems appear to detect more fraud in statewide implementations than in regional pilot systems. Additional matches have been found by interstate comparisons of finger-image data.

- Institution of a finger-imaging requirement can produce a significant, short-term reduction in caseload, because some existing clients refuse to comply with the requirement. The number of refusals depends on the implementation procedures and appears to be lower when finger imaging is incorporated into the recertification process.
- The most carefully controlled estimate of noncompliance among existing clients suggests that introduction of a finger-imaging requirement reduces participation by approximately 1.3 percent. However, this estimate reflects both reduced fraud and deterrence of eligible individuals and households. [5]

Sociological Concerns

As technology advances, and time progresses, more and more private companies and public utilities will use biometrics for safe, accurate identification and access control. However, these advances will raise many concerns throughout society, where many may not be educated on the methods. Several examples of the concerns that society has with biometrics include:

- Physical: Some believe this technology can cause physical harm to an individual or that the instruments used are unsanitary. For example, there are concerns that retinal scanners might not always be clean.
- Personal information: There are concerns whether personal information taken through biometric methods can be misused, tampered with, or sold—for example, by criminals stealing, rearranging, or copying the biometric data. Also, the data obtained using biometrics can be used in unauthorized ways without the individual's consent.

Performance Evaluations

Performance evaluation includes the parameters of how system requirements should be defined and the appropriate performance specifications to consider, and some insight regarding testing protocols and system evaluation. Three categories of performance evaluations include:

- Operational
- Technology
- Scenario

Operational evaluation: The primary goal is to determine the workflow impact seen by the addition of a biometric system.

Technology evaluation: The primary goal is to measure the performance of biometric systems, typically only the recognition algorithm component, in general tasks.

Scenario evaluation: The primary goal is to measure the performance of a biometric system operating in a specific application.

Two additional biometric performance measurements include:

Biometric performance estimation components:

- False match rate
- False non-match rate

- False-accept rate
- False-reject rate
- Receiver operating characteristic curve
- Equal error rate
- D-prime
- Failure to acquire
- Failure to enroll

Biometric identification performance matrix components:

- Reliability
- Selectivity
- Precision
- Identification false-accept rate
- Identification false-reject rate

An Introduction to Evaluating Biometric Systems

How and where biometric systems are deployed will depend on their performance. Knowing what to ask and how to interpret the answers can help evaluate the performance of these emerging technologies. [6]

Managers might conclude that biometric passwords will soon replace their alphanumeric counterparts with versions that cannot be stolen, forgotten, lost, or given to another person. But what if the performance estimates of these systems are far more impressive than their actual performance?

To measure the real-life performance of biometric systems, and to understand their strengths and weaknesses better, managers must understand the elements that comprise an ideal biometric system. In an ideal system, all members of the population possess the characteristic that the biometric identifies, like irises or fingerprints;

- Each biometric signature differs from all others in the controlled population
- The biometric signatures don't vary under the conditions in which they are collected
- The system resists countermeasures

Biometric-system evaluation quantifies how well biometric systems accommodate these properties. Typically, biometric evaluations require an independent party design the evaluation, collect the test data, administer the test, and analyze the results.

This section attempts to provide the reader with sufficient tools to understand the questions to ask when evaluating a biometric system, and to assist in determining if performance levels meet the requirements of the application. For example, if the plan is to use a biometric to reduce, as opposed to eliminate, fraud, then a low-performance biometric system may be sufficient. On the other hand, completely replacing an existing security system with a biometric-based one may require a high-performance biometric system, or the required performance may be beyond what current technologies can provide.

The industry focus is on biometric applications that provide the user with some control over data acquisition. These applications recognize subjects from mug shots, passport photos, and scanned fingerprints. Of the biometrics that meet these constraints; voice, face, and fingerprint systems have undergone the most study and testing, and therefore occupy the bulk of the discussion. While iris recognition has received much attention in the media lately, few independent evaluations of its effectiveness have been published.

Performance Statistics

In identification systems, a biometric signature of an unknown person is presented to a system. The system compares the new biometric signature with a database of biometric signatures of known individuals. On the basis of the comparison from this database, the system reports or estimates the identity of the unknown person. Systems that rely on identification includes those the police use to identify people from fingerprints and mug shots. Civilian applications include those that check for multiple applications by the same person for welfare benefits and driver's licenses.

In verification systems, a user presents a biometric signature and a claim that a particular identity belongs to the biometric signature. The algorithm either accepts or rejects the claim. Alternatively, the algorithm can return a confidence measurement of the claim's validity. Verification applications include those that authenticate identity during point-of-sale transactions or that control access to computers or secure buildings. Performance statistics for verification applications differ substantially from those for identification systems.

The main performance measure for identification systems is the system's ability to identify a biometric signature's owner. More specifically, the performance measure equals the percentage of queries in which the correct answer can be found in the top few matches. For example, law enforcement officers often use an electronic mug book to identify a suspect. The input to an electronic mug book is a mug shot of a suspect, and the output is a list of the top matches. Officers may be willing to examine only the top 20 matches.

For such an application, the important performance measure is the percentage of queries in which the correct answer resides in the top 20 matches. The performance of a verification system, on the other hand, is traditionally characterized by two error statistics: false-reject rate and false alarm rate. These error rates come in pairs; for each false-reject rate there is a corresponding false alarm. A false reject occurs when a system rejects a valid identity; a false alarm occurs when a system incorrectly accepts an identity. In a perfect biometric system, both error rates would be zero. Unfortunately, biometric systems aren't perfect, so administrators must determine the trade-offs.

Evaluation Protocols

An evaluation protocol determines how administrators test a system, select the data, and measure the performance. Successful evaluations are administered by independent groups and tested on biometric signatures not previously seen by a system. If administrators don't test with previously unseen biometric signatures, they are only testing the ability to tune a system to a particular data set. For an evaluation to be accepted by the biometric community, the details of the evaluation procedure must be published along with the evaluation protocol, testing procedures, performance results, and representative examples of the data set. Also, the information on the evaluation and data should be sufficiently detailed so that users, developers, and vendors can repeat the evaluation.

The evaluation itself should not be too hard or too easy. If the evaluation is too easy, performance scores will be near 100 percent, which makes distinguishing between systems nearly impossible. If the evaluation is too hard, the test will be beyond the ability of existing biometric techniques. In both cases, the results will fail to produce an accurate assessment of existing capabilities. An evaluation is just right when it spreads the performance scores over a range that lets managers distinguish among existing approaches and technologies. From the spread in the results, the best performers can be determined along with the strengths and weaknesses of the technology. [7]

Technology Evaluation

The most general type of evaluation tests the technology itself. Analysts usually perform this kind of evaluation on laboratory or prototype algorithms to measure the state of the art, to determine technological progress, and to identify the most promising approaches. The best technology evaluations are open competitions conducted by independent groups. In these evaluations, test participants familiarize themselves with a database of biometric signatures in advance of the test. They then test algorithms on a sequestered portion of the database. This practice allows systems to be tested on data the participants haven't seen previously. The use of test sets allows the exact same test to be given to all participants. Evaluations typically move from the general to the specific. The first step is to decide which scenarios or applications need to be evaluated. Once the evaluators determine the scenarios, they decide upon the performance measures, design the evaluation protocol, and then collect the data.

Scenario and Operational

Scenario evaluations measure overall system performance for a prototype scenario that models an application domain. An example is face recognition systems that verify the identity of a person entering a secure room. The primary purpose of this evaluation type is to determine whether a biometric technology is sufficiently mature to meet performance requirements for a class of applications. Scenario evaluations test complete biometric systems under conditions that model real-world applications. Because each system has its own data acquisition sensor, each system is tested with slightly different data. One scenario evaluation objective is to test combinations of sensors and algorithms. Creating a well-designed test, which evaluates systems under the same conditions, requires collecting biometric data as closely as possible in time.

To compensate for small differences in biometric signature readings taken over a given period, administrators can use multiple queries per person. However, the problem goes far beyond loss of the computer; compromised information security may incur far greater business cost. Furthermore, laptops frequently provide access to a corporate network via software connections complete with stored passwords on the laptop. The solid-state fingerprint sensor, which is small, inexpensive, and low power, solves these problems. With appropriate software, this device authenticates the four entries to laptop contents: login, screen-saver, boot-up, and file decryption.

An operational evaluation is similar to a scenario evaluation. While a scenario test evaluates a class of applications, an operational test measures performance for a specific algorithm for a specific application. For example, an operational test would measure the performance of system X on verifying the identity of people as they enter secure building Y. The primary goal of an operational evaluation is to determine if a biometric system meets the requirements of a specific application.

Although managers can choose from several general strategies for evaluating biometric systems, each type of biometric has its own unique properties. This uniqueness means that each biometric must be addressed individually when interpreting test results and selecting an appropriate biometric for a particular application. For most commercial off-the-shelf biometric systems, analysts must evaluate the system under operational conditions for each application. But doing so can be expensive and time-consuming. Before embarking on such evaluations, analysts should perform preliminary tests to determine which, if any, system has the potential to meet performance requirements. Fingerprint, face, and voice recognition systems are evaluated in the following sections.

Fingerprint Recognition

The kind of evaluation described here for fingerprint systems can be completed in most locations and similar methods can be developed for other biometrics. Commonly, fingerprint biometric technology replaces password-based security. Most systems use a single fingerprint that the account holder actively provides to the system. To logon, an individual types in the user name and places a finger on the scanner. The system then verifies the person's identity.

Upon scanning, the system generates a quality score for each fingerprint. If a scan doesn't meet a certain preset quality, the system returns an error. The most variable results are associated with the system's failure to acquire images of adequate quality. Such failures result in high image quality error rates that can be directly correlated to the false-reject rates. These errors depend on both time and the test subject. The test subject with the lowest image-quality error rate has the lowest false-reject rate. The test subject with the highest image quality error rate has the highest false-reject error rate. Testing systems for false-alarm errors in the one-in-a-thousand range is relatively easy. A small number of users can perform enough tests in a relatively short time; average test time is one to two hours to check this fingerprint system. For higher security levels, the number of users and the test time must be increased.

Evaluations in general, and technology evaluations in particular, have been instrumental in advancing biometric technology. By continuously raising the performance bar, evaluations encourage progress. Although improving biometric technologies can improve performance, inherent limitations remain that are nearly impossible to work around, except perhaps by combining multiple biometric techniques. These limitations are unique to each kind of biometric technology. The biometric community has not yet established upper limits for face and voice biometrics. How many distinguishable faces or voices are there? What is the probability that the faces of two individuals look the same? One limitation to face uniqueness is the identical twin rate of one in 10,000. Although identical twins might have slight facial differences, a face biometric system will not recognize those differences. Even if identical twins are handled as a special case, family resemblance can still create complications.

Face Recognition

In the 1990s, automatic-face-recognition technology moved from the laboratory to the commercial world largely because of the rapid development of the technology. Now, many applications use face recognition. These applications include everything from controlling access to secure areas to verifying the identity on a passport.

The most recent major evaluation of facial recognition technology was conducted through the **Feret** test. In this evaluation, research groups were given a set of facial images to develop, and were then asked to implement these enhancements. These groups were tested on a sequestered set of images, which required the participants' systems to process numerous images. The Feret evaluation measured performance for both identification and verification, and provided performance statistics for different image categories. The first category consisted of images taken on the same day under the same incandescent lighting. This category represented a scenario with the potential for achieving the best possible performance with face recognition algorithms. Each of the categories became progressively more difficult, with the final category consisting of images taken at least a year and a half apart.

The majority of face recognition algorithms appear to be sensitive to variations in illumination, such as those caused by the change in sunlight intensities throughout the day. In the majority of algorithms evaluated under Feret, changing the illumination resulted in a significant performance drop. For some algorithms, this drop was equivalent to comparing images taken over the course of a year and a half. Changing facial position can also have an effect on performance. A 15-degree difference in position between the query image and the database image will adversely affect performance. At a difference of 45 degrees, recognition becomes ineffective. Many face verification applications make it mandatory to acquire images with the same camera. However, some applications, particularly those used in law enforcement, allow image acquisition with many types of cameras. This variation has the potential to affect algorithm performance as severely as changing illumination. But, unlike the effects of changing illumination, the effects on performance of using multiple camera types have not been quantified.

Voice Recognition

Despite the inherent technological challenges, the most popular applications of voice recognition technology will likely provide access to secure data over telephone lines. Voice recognition has already been used to replace number entry on certain telecom systems. This kind of voice recognition is related to speech recognition. While *speech recognition* technology interprets what the speaker says, *speaker recognition* technology verifies the speaker's identity. Speaker recognition systems fall into two basic types: text-dependent and text-independent. In text-dependent recognition, the speaker says a predetermined phrase. This technique inherently enhances recognition performance, but requires a cooperative user. In text-independent recognition, the speaker need not say a predetermined phrase and need not cooperate or even be aware of the recognition system. Speaker recognition suffers from several limitations.

Different people can have similar voices, and anyone's voice can vary over time because of changes in health, emotional state, and age. Furthermore, variation in handsets or in the quality of a telephone connection can greatly complicate recognition. Current NIST speaker-recognition evaluations measure verification performance for conversational speech over telephone lines. In a recent NIST evaluation, the data used consisted of speech segments for several hundred speakers. Recognition systems were tested by attempting to verify speaker identities from the speech segments. To measure performance under different conditions, several samples were recorded on many lines. Not surprisingly, differences among telephone handsets can severely affect performance. Handset microphones come in two types, either carbon-button or electret (a dielectric in an induced state of electric polarization). Performance is better when the training and testing handsets are of the same type. Since voice by itself does not currently provide sufficient accuracy, voice can be combined with another biometric, like face or fingerprint recognition.

These or similar concerns apply to the majority of biometrics currently being investigated. The final decision about putting biometric systems to work depends almost entirely on the application's purpose. Do the advantages and benefits outweigh the disadvantages and costs? The performance level of a biometric system designed to detect fraud in insurance claims, for example, is not as critical as the performance level of a biometric system that entirely replaces an existing security system used by an airline. In the near future, managers will likely have more effective ways of determining the difference between the advertised and actual performance of biometric systems. Meanwhile, managers must avoid accepting the hype about each new biometric method until it can be tested thoroughly. In cases where access to test results are not available, managers must ask the vendors pointed questions about the performance of their products. As with any emerging technology, it's prudent to err on the side of caution.

Statistics and Graphics Reports

Security managers and administrators must understand the various statistical and graphical reports and measurements that are part on managing and controlling biometric systems. This section provides a brief overview of these measurements.

Degrees of freedom is a statistical measure of the uniqueness of biometric data. Technically, it is the number of statistically independent features (parameters) contained in biometric data.

Cumulative match characteristic (CMC) is a method of showing measured accuracy performance of a biometric system operating in the closed-set identification task. Templates are compared and ranked based on their similarity. The CMC shows how often the individual's template appears in the ranks (1, 5, 10, 100, etc.), based on the match rate. A CMC whorl image compares the rank (1, 5, 10, 100, etc.) versus identification rate.

Equal error rate (EER) is a statistic used to show biometric performance, typically when operating in the verification task. The EER is the location on a receiver operating characteristics (ROC) or detection error trade-off (DET) curve where the false-accept and false-reject rates are equal. In general, the lower the EER value, the higher the accuracy of the biometric system. Note, however, that most operational systems are not set to operate at this rate, so the measure's true usefulness is limited to comparing biometric system performance. The EER is sometimes referred to as the **crossover error rate (CER)**.

Receiver operating characteristics (ROC) is a method of showing measured accuracy performance of a biometric system. A verification ROC compares false-accept rate versus verification rate. An open-set identification (watchlist) ROC compares false-alarm rates versus detection and identification rate.

Detection error trade-off (DET) curve is a graphical plot of measured error rates. DET curves typically plot matching error rates (false non-match rate vs. false match rate) or decision error rates (false-reject rate vs. false-accept rate).

False match rate is a statistic used to measure biometric performance. It is similar to the false acceptance rate (FAR).

False non-match rate is a statistic used to measure biometric performance. Similar to the false rejection rate (FRR), except that the FRR includes the failure-to-acquire error rate and the false non-match rate does not.

False Rejection Rate (FRR) is a statistic used to measure biometric performance when operating in the verification task. It shows the percentage of times the system produces a false reject. A false reject occurs when an individual is not matched to his or her own existing biometric template. Example: John claims to be John, but the system incorrectly denies the claim.

False Acceptance Rate (FAR) is a statistic used to measure biometric performance when operating in the verification task. It shows the percentage of times a system produces a false accept, which occurs when an individual is incorrectly matched to another individual's existing biometric. Example: Frank claims to be John and the system verifies the claim.

False alarm rate is a statistic used to measure biometric performance when operating in the open-set identification (sometimes referred to as watchlist) task. This is the percentage of times an alarm is incorrectly sounded on an individual who is not in the biometric system's database (the system alarms on Frank when Frank isn't in the database), or an alarm is sounded but the wrong person is identified (the system alarms on John when John is in the database, but the system thinks John is Steve).

A **type I error** or FRR is an error that occurs in a statistical test when a true claim is (incorrectly) rejected. For example, John claims to be John, but the system incorrectly denies the claim.

A **type II error** or FAR is an error that occurs in a statistical test when a false claim is (incorrectly) not rejected. For example: Frank claims to be John and the system verifies the claim.

True accept rate is a statistic used to measure biometric performance when operating in the verification task. The percentage of times a system (correctly) verifies a true claim of identity. For example, Frank claims to be Frank and the system verifies the claim.

True reject rate is a statistic used to measure biometric performance when operating in the verification task. The percentage of times a system (correctly) rejects a false claim of identity. For example, Frank claims to be John and the system rejects the claim.

Managing Biometric Data

Biometrics Management and Security for the Financial Services Industry Specification X9.84 defines the minimum-security requirements for effective management of biometrics data for the financial services industry and the security for the collection, distribution, and processing of biometrics data. **Data vaulting** is the process of sending data offsite, where it can be protected from hardware failures, theft, and other threats. Several companies now offer Web backup services that compress, encrypt, and periodically transmit a customer's data to a remote vault. In most cases, the vaults have auxiliary power supplies, powerful computers, and manned security. This capability is also referred to as a remote backup service. **Vulnerability** is the potential for the function of a biometric system to be compromised by intent (fraudulent activity), design flaw (including usage error), accident, hardware failure, or external environmental condition.

Accuracy is a catch-all phrase for describing how well a biometric system performs. The actual statistic for performance will vary by task (verification, open-set identification (watch list), and closed-set identification). Additional terms and processes that are applied when managing biometric data include:

- *Benchmarking* is the process of comparing measured performance against a standard, openly available, reference.
- *Binning* is the process of parsing (examining) or classifying data in order to accelerate and/or improve biometric matching.
- A *decision* is the resultant action taken (either automated or manual) based on the comparison of a similarity score (or similar measure) and the system's threshold.
- *Threshold* is a user setting for biometric systems operating in the verification or open-set identification (watchlist) tasks. The acceptance or rejection of biometric data is dependent on the match score falling above or below the threshold. The threshold is adjustable so that the biometric system can be more or less strict, depending on the requirements of any given biometric application.
- A *challenge* is the demand for disclosure of one or more attributes related to a subject made by service authority. Challenge/response is a family of protocols in which one party (e.g., a reader) presents a question ("challenge") and another party (e.g., a credential) must provide a valid answer ("response") in order to be authenticated.
- *Challenge response* is a method used to confirm the presence of a person by eliciting direct responses from the individual. Responses can be either voluntary or involuntary. In a voluntary response, the end user will consciously react to something that the system presents. In an involuntary response, the end user's body automatically responds to a stimulus. A challenge response can be used to protect the system against attacks.

Response Time and Throughput

Response time is the time used by a biometric system to return a decision on identification or verification of a biometric sample. **Throughput rate** is the number of biometric transactions that a biometric system processes within a stated time interval.

E-commerce and biometric systems require a fast response to requests, whether by someone processing an ATM transaction or someone requiring some access control authentication. This is generally termed response time and is impacted by the network throughput. In communication networks, **throughput** is the amount of digital data per time unit that is delivered over a physical or logical link, or that is passing through a certain network node. For example, it may be the amount of data that is delivered to a certain network terminal or host computer, sensor, reader, or between two specific computers. The throughput is usually measured in bits per second (bits/s or bps), occasionally in data packets per second or data packets per timeslot. The *system throughput* or *aggregate throughput* is the sum of the data rates that are delivered to all terminals in a network. The *maximum throughput* of a node or communication link is synonymous to its capacity.

In a computer or biometric-enabled network, the throughput achieved from one computer to another may be lower than the maximum throughput, or the network access channel capacity, for several reasons. For example, factors that affect network throughput include the following situations:

- The channel capacity may be shared by other users. Increased file transfer sizes and limited channel bandwidth can cause a situation called "bottlenecking." Flow control in the TCP protocol affects the throughput if the bandwidth-delay product is larger than the TCP buffer size. In that case, the sending computer must wait for acknowledgment of the data packets before it can send more packets. This delay will impact response time for the user.
- Packet loss may be due to network congestion. Packets may be dropped in switches and routers when the packet queues are full due to traffic congestion. Poor response for the user is the result.
- Packet loss may be due to bit errors. TCP congestion avoidance controls the data rate. A so-called "slow start" occurs in the beginning of a file and after packet drops that are caused by router congestion or bit errors in wireless links. This causes retransmissions and slow response times for the users.
- User prioritization can be impacted by scheduling algorithms in routers and switches. If fair queuing is not provided, users sending large packet will get higher bandwidth. Some users may be prioritized in a weighted fair queuing (WFQ) algorithm if differentiated or guaranteed quality of service (QoS) is provided. Users can be given a low priority, therefore poor response time.
- Ethernet "back-off" waiting time after collisions is a common occurrence in CSMA-CD networks. A Local Area Network (LAN) that is improperly configured or overloaded can cause this undesirable situation resulting in poor user response.

International Implementations

A single world-wide standard biometric technology is not a practical or necessary goal. Cultural differences among nations make it impossible to reach an international consensus on a single form of biometrics that will both provide sufficient technical performance and still respect the customs and biases of the citizenry. While the United States may choose to employ fingerprint technologies because of their performance advantages or their ability to interact with federal law enforcement databases, other countries believe their citizens will only accept facial recognition or iris technologies to secure their travel documents. The U.S. design must include a solution that is flexible enough to accommodate other countries' choice of biometrics and agree upon a process by which other countries assume the responsibility to secure the integrity of their travel document issuance processes.

Kiosk Issues

Organizations may be neither sized nor staffed to accommodate the users who need to provide biometric samples. One solution that has been suggested was to design and deploy ATM-like self-service kiosks to support the remote capture of both text data and biometric samples from applicants. These kiosks would be equipped with high-quality sensors capable of capturing finger, face, and iris biometrics, and would be configured with surveillance cameras that could detect any attempt at fraud during enrollment. Similar to an ATM's operation, these kiosks would be designed to allow an individual to securely deposit her documents and any fee payments. These kiosks could be installed in secure locations throughout localities, cities, and even countries. At these facilities, personnel could be trained to assist an applicant who was having trouble with the application or biometric capture processes.

Physical Security Controls

Physical security controls pertain to the physical infrastructure, physical device security, and physical access. The physical infrastructure includes the facilities that house the computers, databases, and networks used to access these same assets. The physical network infrastructure encompasses both the selection of the appropriate media type and the path of the physical cabling. It is essential that no intruder is able to eavesdrop on the data traversing the network and that all critical systems have a high degree of availability. Physical security issues can impact most of the computer and network assets of an organization. Figure 11-1 provides a graphic emphasizing these vulnerability issues.

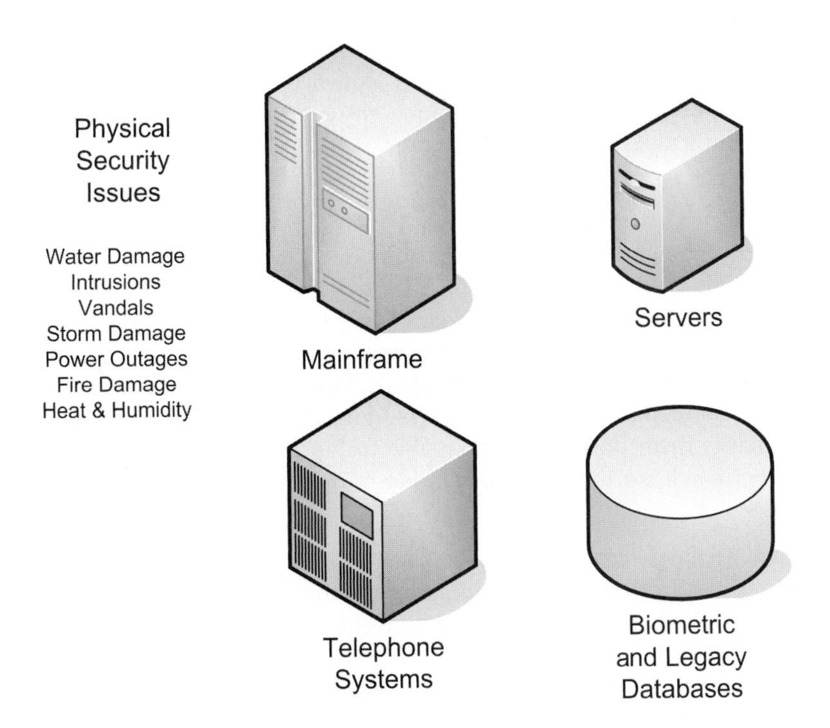

Figure 11-1 Computer and Network Vulnerabilities

Course Technology/Cengage Learning

Network Access

Telecommunications access adds another dimension to security, since data can be accessed from terminals outside the computer room. A number of questions arise concerning network access:

- How is the terminal user identified?
- Is this user authorized to access the computer?
- What operations can this user execute?
- Are the communication lines being monitored or sniffed?

Access control techniques include having a unique identification code for each user and a password to protect against these situations.

Users must be required to logon to the network; this information is automatically recorded in a central database so a record can be maintained of all network activities. When someone tries to logon unsuccessfully a number of times, the user's ability to log on should be disabled. This logon failure should invoke a log entry and the notification of a security administrator. A supervisor, who could authorize a new password, may approve reactivation of this user.

Dial-up lines are especially vulnerable to unauthorized access. The most common security techniques used with dial-up circuits are callback and handshake. With callback, the user dials the computer and provides identification, the computer then breaks the connection and dials the user back at a predetermined number that is obtained from the database. The disadvantage is that this does not work well with traveling employees who would not be at a predetermined location. The handshake technique requires a terminal with special hardware circuitry. The computer sends a special control sequence to the terminal, and the terminal identifies itself to the computer. This technique ensures that only authorized terminals access the computer, but it does not regulate the user of the terminal.

Media Access

From a security point of view, the type of cable chosen for various parts of the network can depend upon the sensitivity of the information that traverses it. The most common cable types used in networking infrastructures are coaxial cable (coax), twisted-pair copper, and optical fiber. Unlike copper or coax, optical fiber does not radiate any energy, which provides a high degree of security against eavesdropping. Fiber is also much more difficult to tap into than copper or coax. The heat-fused cladding of multimode graded index and single-mode cable makes them practically impossible to tap. Even if a tap occurs, it can be detected, because the match on the fiber-optic cable must be perfect for light transmission without disruptions.

Wire taps can sometimes be detected by using tools to measure attenuation (reduction of signal strength) on the cable. Typically, a time domain reflectometer (TDR) is used with copper or an optical TDR (OTDR) is used with fiber optics to look for abnormalities. These devices are generally used to measure signal attenuation and installed cable length, but an experienced technician can use it to locate wiretaps. It should be noted that an expert intruder can insert a tap so it is not easily detectable by either device. It is good practice to take an initial baseline signal level of the physical cable infrastructure and periodically verify the integrity of the entire physical cable plant.

The physical path of the media, which is part of the network topography, is an issue that must be addressed to ensure the availability of the network and its devices. It is imperative to have a structured cabling system that minimizes the risk of downtime. The cable infrastructure must be well secured to prevent unauthorized access. There are ITU recommendations that provide specifications and recommendations for the construction, installation, and protection of cable plants.

In the case of access into a controlled environment, such as a computer room, network wiring closet, or demarcation point, a restrictive set of access rules should apply. Organizational site guidelines must be developed and implemented to restrict access for cabling to these areas by controlling access to cable trays and overhead cable support systems. Approved methods and access points must be established prior to cable routing and installation. The security system to the area can vary from simple lockable cable feeder openings to dedicated individual fiber and cable runs, each with alarm-equipped access covers and video surveillance.

The common areas of exposure when addressing cable security include:

- Cable trays in the crawl space under raised floors
- Wall-mounted wiring access
- Overhead cable and wire trays

Wired LAN connectivity occurs where cabling terminates in a centrally located wiring closet. There may be a patch panel in the wiring closet that provides for the interconnection between the Wide Area Network (WAN) circuits and the LAN components. There may also be wiring between a terminal server or communications server and the patch panel. The wiring closet is a major point of vulnerability and must be afforded a high level of security.

Physical Device Security

Physical device security includes identifying the location of devices, limiting physical access, and the installation of environmental safeguards. All network infrastructure equipment should be physically located in restricted access areas. This means creating a secure space for wiring closets that contain switches, firewalls, modems, DSUs, splitters, and routers. It is essential that this equipment closet be well secured; however, remember that intruders can enter through a ceiling opening. This network closet is sometimes called a demarcation point, where all the telephone facilities enter the building. This is an excellent place for an intruder to place a monitor for eavesdropping on all of the organization's network traffic.

Any OSI Layer 1 (Physical) LAN device is a potential candidate for illegal access. This could include a repeater, hub, or multistation access unit (MAU). The potential increases if shielded twisted pair (STP) or unshielded twisted pair (UTP) cabling is utilized. Another place for compromise is the LAN patch panel, where most CAT 5/6 cables terminate.

Infrastructure equipment includes more than networking equipment, which must be protected. Other sensitive components of the infrastructure include numerous servers, gateways, and administration terminals. Print servers, terminal servers, file servers, and domain-name servers must all be afforded a high level of security and should not be available to the general staff. Consoles and administrative terminals must be secured from the general staff and all other unauthorized personnel, since these devices provide access into the secured infrastructure of the organization.

Often overlooked is the output from the office printer or fax machine. These devices are usually centrally located for the convenience of the staff. This is a good place to look for confidential information, so do not overlook its importance as a candidate for protection. It might be necessary to utilize a shredder and not allow any whole pages out of the facility. Do not forget that there are "dumpster divers" waiting for someone to dispose of confidential material that can be used for some illegal activity.

Physical Access Security

As mentioned earlier, wiring closets and rooms that house infrastructure devices and networking facilities must be secured from unauthorized people. These areas should be in the interior of an organization's operation. It is necessary to ensure that exterior access is also protected, which usually addresses control of nonemployees and employees alike. A guard station is usually the method used to control exterior access for many organizations. This means that a process has been developed that includes employee identification and authorization, which determines the access level of the individual.

Part of physical security policies and procedures must address contract personnel, maintenance personnel, and others who do not possess unrestricted access. These people may be required to be in controlled area, but must be escorted by an authorized employee or sign in before they enter the controlled area. Identification cards may be required if these are very sensitive areas. The larger the organization, the more important these measures become, since many employees may not even know who works in the organization.

To ensure enforceable physical security, it is essential to make sure that the employee's work areas mesh well with access restrictions. This includes the flow of personnel within an organization's premises. It is easy for an employee to leave a door chocked instead of locking and unlocking it many times a day. Installing an alarm that will sound when a door is left open can eliminate this type of situation. A logical and workable security plan that doesn't consider the daily workflow will result in employee aggravation.

Many organizations host both visitors and personnel from other departments. These people must be given access on a temporary basis, but such access should be to a controlled area away from sensitive operations. These visitors should also be required to sign in and receive a temporary identification.

Personnel Security

It is essential that personnel in sensitive positions be screened to ensure their integrity. This process must be part of corporate policy, and not an afterthought. Personnel security involves one or more of the following techniques:

- Security and background screening of prospective employees before they are hired
- An active security awareness program that emphasizes security issues and ramifications
- Training of employees in security issues and their responsibilities
- Identification of employees, contractors, and maintenance workers
- Error prevention techniques to detect accidental mistakes and malicious activities

Customer Security

Physical computer and networking assets may be protected inside a building or may be in a public area. These devices can include ATMs and cash dispersing machines. Crime rates against these devices and the customers who use them are increasing. Robberies occur shortly after customers withdraw money from an ATM, or ATM customers are assaulted and forced to withdraw funds. Physical violence is often part of these attacks. A legal question has arisen as to how far a bank must go to protect ATM customers. The location of these devices and the time of day that the transactions occur have an enormous impact on the security realized. Many banks provide video monitoring and lighting systems, but these initiatives are no solace to the customer who is attacked. Banks do not accept responsibility for these crimes and say consumers bear part of the risk when they use these machines late at night. These situations exemplify the type of societal problems that can and will occur as communications systems become more widely used by the general public.

Disaster Recovery Planning

Disaster recovery is the process, policies and procedures of restoring operations that are critical to the resumption of business—including regaining access to data, communications, workspace, and other business processes—after a natural or human-induced disaster. To increase the opportunity for a successful recovery of valuable records, a well-established and thoroughly tested disaster recovery plan must be developed. This task requires the cooperation of a well-organized committee led by an experienced chairperson. A **disaster recovery plan (DRP)** should also include plans for coping with the unexpected or sudden loss of communications and/or key personnel; the focus of which is data protection. Disaster recovery planning is part of a larger process known as business continuity planning.

Business continuity planning (BCP) is the process that an organization implements to prepare for future incidents that could jeopardize the organization's core mission and long-term health. Every business and organization can experience a serious incident which will prevent it from continuing normal operations. This situation can occur any day, and at any time. The many potential causes are varied, and include flood, explosion, computer malfunction, accident, destructive act, building fire, a regional incident like an earthquake, or a national incident such as a pandemic illness. The list is almost endless. The information below is designed to help managers plan for these scenarios. These basic disaster recovery planning efforts will help reduce both the risk and impact if and when the worst occurs.

Business Impact Analysis

The first stage in the business continuity process is normally to consider the potential impacts of each type of disaster. This is a critically important step: How can managers properly plan for a problem if they have little idea of the likely impacts on the organization of the different scenarios?

At its most basic level, **business impact analysis (BIA)** is a method of systematically assessing the possible impacts resulting from various events or scenarios. Usually, impacts resulting from other incident types, such as loss of confidentiality or data integrity, are examined at the same time. Although this need not be the case, there are certainly some advantages to undertaking a comprehensive and wider-focused exercise.

The BIA is intended to help managers understand the scale of loss which could occur. It will cover not just direct financial loss, but many other issues, such as loss of stakeholder confidence and reputation damage.

Risk Analysis

Having determined the various impacts, it is now important to consider the risks which could lead to these situations. This critical activity determines which of the identified scenarios are most likely to occur, and therefore, which should attract most attention during the BCP process. **Risk analysis** is a technique to identify and assess factors that may jeopardize the success of a project or achieving a goal. This technique also helps define preventive measures to reduce the probability of these factors from occurring, and identify countermeasures to successfully deal with these constraints when they develop to avert possible negative effects on the competitiveness of the company.

Service Level Agreements

Most organizations are dependent upon the services or products provided by others. This can be an Achilles' heel as far as continuity is concerned. However, the situation can be mitigated via the use of a binding service-level agreement. A **service level agreement (SLA)** is a formally negotiated agreement between two parties, such as a contract between customers and their service provider, or between service providers. It records the common understanding about services, priorities, responsibilities, guarantee, and collectively, the *level of service*. For example, it may specify the levels of availability, serviceability, performance, operation, or other attributes of the service (such as billing and even penalties in the case of violation of the SLA). As with the disaster recovery plan, the SLA should be carefully crafted to suit a particular situation.

Network Administration

The term **network management** has traditionally been used to specify real-time network surveillance and control; network traffic management; functions necessary to plan, implement, operate, and maintain the network; and systems management. Network management supports the users' needs for activities that enable managers to plan, organize, supervise, control, and account for the use of interconnection services. It also provides for the ability to respond to changing requirements, such as ensuring that facilities are available for predictable communications behavior and providing for information protection, which includes the authentication of sources and destinations of transmitted data.

The task of network management involves setting up and running a network, monitoring network activities, controlling the network to provide acceptable performance, and assuring high availability and fast response time to the network users. Network equipment manufacturers have developed an impressive array of network management products over the past several years. An effective network management system requires trained personnel to interpret the results. An effective network management system will include most of the following elements:

- An inventory of computers, database, network, and other equipment
- A trouble report receiving and logging process
- A trouble history file
- A trouble diagnostic, testing, and isolation facility and procedures
- A hierarchy of trouble clearance and escalation procedures

- An activity log for retaining records of all major changes
- An alarm reporting and processing facility

Not every network management system will have all of these elements, but most systems will contain the above functions to some degree. The more complex the network, the more likely the functions will be automated on a mechanized system. Figure 11-2 describes the standard process flow for trouble-shooting systems issues.

Scientific Testing of Biometric ID Systems

Testing of biometric devices requires repeat visits with multiple human subjects. Further, the generally low error rates mean that many human subjects are required for statistical confidence. Consequently, biometric testing is extremely expensive, and generally affordable only by government agencies. Few biometric technologies have undergone rigorous, developer/vendor-independent testing to establish robustness, distinctiveness, accessibility, acceptability, and availability in "real-world" (nonlaboratory) applications.

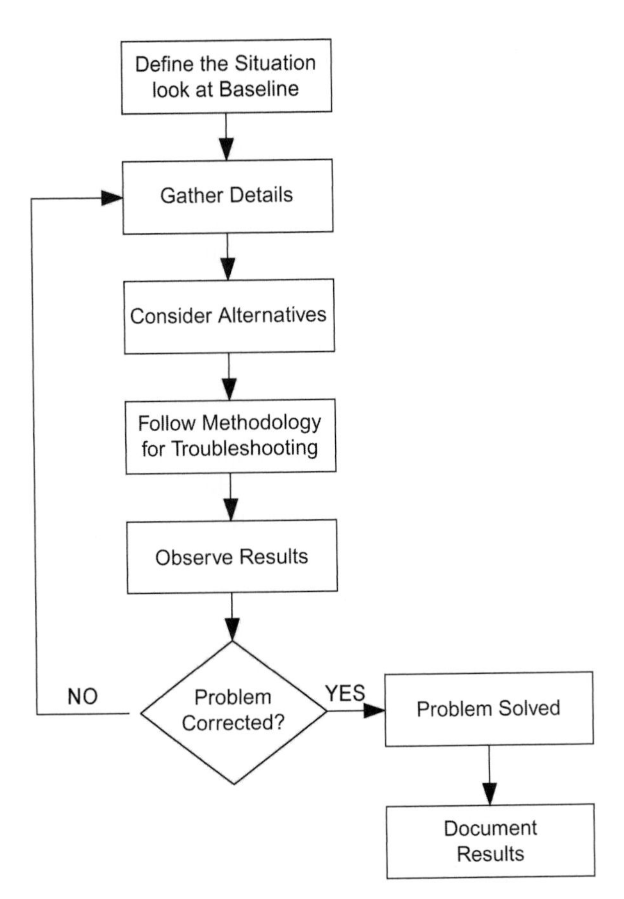

Figure 11-2 Troubleshooting Process Flow

Course Technology/Cengage Learning

Biometrics.gov is the central source of information on biometrics-related activities of the federal government. Two sister sites provide a repository of biometrics-related public information (www.biometricscatalog.org) and opportunities for discussion (www.biometrics.org). These Web sites, working together, were developed to encourage greater collaboration and sharing of information on biometric activities among government departments and agencies; commercial entities; state, regional, and international organizations; and the general public.

Summary

- To ensure that a system's implementation takes place as efficiently and with as little disruption as possible, specific tasks must include ordering and installing new equipment, preparing for data storage, training personnel, creating an implementation plan, and preparing for production. There are numerous programs and tools available for implementation of computer and information technology projects. These often include procurement and training components.
- A primary task of maintaining computer and network integrity involves troubleshooting the network environment. Past studies and results indicate that a systematic approach is the most effective. Administrators should be able to identify problems from symptoms and initiate corrective action based on this systematic approach. Security managers and administrators must evaluate the steps being taken to prevent online fraud, security and privacy violations, and intruder attacks. Although the cost of implementing these e-security measures can be significant, the alternative is even more costly.
- An evaluation protocol determines how administrators test a system, select the data, and measure the performance. Successful evaluations are administered by independent groups and tested on biometric signatures not previously seen by a system. The most general type of evaluation tests the technology itself. Analysts usually perform this kind of evaluation on laboratory or prototype algorithms to measure the state of the art, to determine technological progress, and to identify the most promising approaches.
- Privacy considerations are paramount in the design and implementation of effective biometric ID systems. As biometric systems are deployed within security systems, or as part of verification and identification programs, implementation issues relating to security and privacy must be considered.
- Physical security controls pertain to the physical infrastructure, physical device security, and physical access. The physical infrastructure includes the facilities that house the computers, databases, and networks used to access these same assets.

Key Terms

Accuracy A catch-all phrase for describing how well a biometric system performs. The actual statistic for performance will vary by task, including verification, open-set identification (watchlist), and closed-set identification.

Business impact analysis (BIA) A method of systematically assessing the possible impacts resulting from various events or scenarios.

Cumulative match characteristic (CMC) A method of showing measured accuracy performance of a biometric system operating in the closed-set identification task.

Data vaulting The process of sending data offsite, where it can be protected from hardware failures, theft, and other threats.

Degrees of freedom A statistical measure of the uniqueness of biometric data.

Detection error trade-off (DET) curve A graphical plot of measured error rates.

Equal error rate (EER) A statistic used to show biometric performance, typically when operating in the verification task.

False match rate A statistic used to measure biometric performance. It is similar to the false acceptance rate (FAR).

False non-match rate A statistic used to measure biometric performance. Similar to the false reject rate (FRR), except the FRR includes the failure-to-acquire error rate and the false non-match rate does not.

False Rejection Rate (FRR) A statistic used to measure biometric performance when operating in the verification task. It shows the percentage of times the system produces a false reject. A false reject occurs when an individual is not matched to his/her own existing biometric template.

False Acceptance Rate (FAR) A statistic used to measure biometric performance when operating in the verification task. It shows the percentage of times a system produces a false accept, which occurs when an individual is incorrectly matched to another individual's existing biometric.

False Alarm Rate A statistic used to measure biometric performance when operating in the open-set identification (sometimes referred to as watchlist) task. This is the percentage of times an alarm is incorrectly sounded on an individual who is not in the biometric system's database or an alarm is sounded but the wrong person is identified.

Hot swapping The ability to add and remove devices to a computer while the computer is running and have the operating system automatically recognize the change.

Network management Traditionally used to specify real-time network surveillance and control; network traffic management; functions necessary to plan, implement, operate, and maintain the network; and systems management.

Operational evaluation One of the three types of performance evaluations. The primary goal is to determine the workflow impact seen by the addition of a biometric system.

Physical device security Includes identifying the location of devices, limiting physical access, and the installation of environmental safeguards.

Plug-and-play An industry-wide standard for add-on hardware that indicates it will configure itself, eliminating the need to set jumpers and making installation of the product quick and easy.

Receiver operating characteristics (ROC) A method of showing measured accuracy performance of a biometric system.

Response time The time used by a biometric system to return a decision on identification or verification of a biometric sample.

Risk analysis A technique to identify and assess factors that may jeopardize the success of a project or achieving a goal.

Scenario evaluation One of the three types of performance evaluation. The primary goal is to measure performance of a biometric system operating in a specific application.

Security accreditation The specific goals that can be associated with security evaluation include a uniform measure of security, an independent assessment, criteria validity, and evaluation assurance.

Service level agreement (SLA) A formally negotiated agreement between two parties.

Technology evaluation One of the three types of performance evaluation. The primary goal is to measure performance of biometric systems, typically only the recognition algorithm component, in general tasks.

Throughput The amount of digital data per time unit that is delivered over a physical or logical link, or that is passing through a certain network node.

Throughput rate The number of biometric transactions that a biometric system processes within a stated time interval.

True accept rate A statistic used to measure biometric performance when operating in the verification task, that is, the percentage of times a system (correctly) verifies a true claim of identity.

True reject rate A statistic used to measure biometric performance when operating in the verification task, that is, the percentage of times a system (correctly) rejects a false claim of identity.

Type I error or FRR An error that occurs in a statistical test when a true claim is (incorrectly) rejected.

Type II error or FAR An error that occurs in a statistical test when a false claim is (incorrectly) not rejected.

Vulnerability The potential for the function of a biometric system to be compromised by intent (fraudulent activity); design flaw (including usage error); accident; hardware failure; or external environmental condition.

Review Questions

1. ________________ is the ability to add and remove devices to a computer while the computer is running and have the operating system automatically recognize the change.

2. ________________ is an industry-wide standard for add-on hardware that indicates it will configure itself, eliminating the need to set jumpers and making installation of the product quick and easy.

3. The primary goal of an ________________ is to determine the workflow impact seen by the addition of a biometric system.

4. The primary goal of a ________________ is to measure performance of biometric systems, typically only the recognition algorithm component, in general tasks.

5. The primary goal of a ________________ is to measure performance of a biometric system operating in a specific application.

6. The notion of a particular government organization or agency making a decision that a given product or system possesses sufficient security protection is called a ________________.

7. The ________________ tests were technology evaluations of emerging approaches to face recognition.

8. _______________ is the process of sending data offsite, where it can be protected from hardware failures, theft, and other threats.
9. _______________ is the potential for the function of a biometric system to be compromised by fraudulent activity; design flaw; accident; hardware failure; or external environmental condition.
10. _______________ is the time used by a biometric system to return a decision on identification or verification of a biometric sample.
11. _______________ is the number of biometric transactions that a biometric system processes within a stated time interval.
12. The _______________ or _______________ is the sum of the data rates that are delivered to all terminals in a network.
13. _______________ is a method used to confirm the presence of a person by eliciting direct responses from the individual.
14. _______________ includes identifying the location of devices, limiting physical access, and the installation of environmental safeguards.
15. _______________ is the process an organization implements to prepare for future incidents that could jeopardize the organization's core mission and its long-term health.
16. _______________ is a catch-all phrase for describing how well a biometric system performs.
17. _______________ is a method of systematically assessing the possible impacts resulting from various events or scenarios.
18. _______________ is a technique to identify and assess factors that may jeopardize the success of a project or achieving a goal.
19. A _______________ is a formally negotiated agreement between two parties, such as a contract that exists between customers and their service provider, or between service providers.
20. The term _______________ has traditionally been used to specify real-time network surveillance and control; network traffic management; functions necessary to plan, implement, operate, and maintain the network; and systems management.

Discussion Exercises

1. Describe the relationship between the equal error rate, receiver operating characteristics, and the detection error trade-off curve.
2. Describe the relationship between a false match rate, false non-match rate, false acceptance rate, and the failure to acquire error rate.
3. Describe the difference between the false rejection rate and the false acceptance rate.
4. Describe the difference between the true accept rate and the true reject rate.

5. Provide an overview of the X9.84 specification.
6. Research the topic of using a kiosk for a self-service biometric capture of data and samples.
7. Research the topic of tools that can be utilized for computer and network security functions.
8. Identify those devices, components, and media that can be compromised by sniffer monitors and other surveillance tools.
9. Provide an overview of the components and functions of business continuity and disaster recovery plans.
10. Provide a list of those vendors that offer network management hardware and software systems. Identify the products offered and provide an overview of each.

Hands-On Projects

11-1: Implementation and Operation Issues

Time required—30 minutes

Objective—Research implementation and operation issues that are part of biometric application and system implementations.

Description—Use various Web tools, Web sites, and other tools to identify biometric implementation and operation issues that occur in production and operation systems in commercial and government organizations.

1. Identify an organization that has implemented a biometric security solution. Provide an overview of the system, including the modalities utilized. Identify issues with the implementation.
2. Identify project management tools that could be utilized for biometric-system implementations. Describe features and functions available with the project management solutions.
3. Identify disaster recovery solutions that could be implemented with biometric systems.
4. Search for samples of service level agreements that could be applied to biometric-system implementations. Relate the samples to the biometric application.

11-2: Identifying Implementation and Operation Issues

Time required—45 minutes

Objective—Identify system implementation and operation issues that can impact the level of access control and security for OnLine Access assets and resources.

Description—The owners and managers of OnLine Access have identified the necessary security and privacy issues when implementing a biometric access control solution. They must now look at the system implementation and operation issues that can be encountered when implementing a biometric solution.

1. Research the implementation and operation issues that are inherent in access control and security solutions.
2. Identify security and access control system implementation and operation issues that have been encountered in other biometric implementations.
3. Produce a report describing how these issues have impacted the implementation of a biometric solution.
4. Describe how these issues have been resolved.

chapter 12

Standards and Legal Environment

After completing this chapter, you should be able to do the following:

- Become familiar with numerous standards and practices utilized in the computer and networking industry
- Identify standards and standards organizations that impact the biometric industry
- Become familiar with the various biometric standards and specifications involving access control processes
- Look at the legal and legislative issues associated with the biometric industry
- See how engineering and technical initiatives are progressing in the biometric technologies

Biometric standards are in a state of flux, and this situation could continue in the future. New developments occur frequently, and new products are being developed and offered by a multitude of vendors. The legal issues of privacy and individual rights must also be considered.

Introduction

A number of standards exist for security products and services. Overviews of the various relevant security standards are provided in this chapter. Security policies and procedures are required in today's e-commerce and technology centers. This chapter also includes sections that discuss security specifications and security evaluation.

With today's heterogeneous networking environments, it is necessary to know whether security protocols and software work together. It is useless to have only part of the network secure—the entire network must be secure from both internal and external threats. A successful computer network security system requires a marriage of technology and process, which is part of a security requirement assessment. Biometric standards and specifications play an important role in these endeavors.

There are also numerous legal issues including security and privacy that can develop from the use of biometric devices. There are many engineering standards, initiatives, and issues involved with biometric devices. Professional organizations can provide information and support to security administrators and managers.

Standards and Practices

Standards are agreed-upon protocols as determined by officially sanctioned standards-making organizations or user-group consensus. A **protocol** is a set of rules that govern communication between hardware and/or software components. Standards include specifications produced by accredited associations, such as ANSI, ISO, SIA, ETSI, or NIST. In the United States, the use of standards is typically optional, and multiple standards can be developed on the same subject. In some countries, the use of existing standards may be required by law and the development of multiple standards on the same subject may be restricted. A *base standard* is a set of fundamental and generalized procedures. They provide an infrastructure that can be used by a variety of applications, each of which can make its own selection from the options offered by them.

Practices are the techniques, methodologies, procedures, and processes used in organizations to accomplish a particular job. **Best practices** are methods and procedures found to be most effective; however, best practices are not rules, laws, or standards which people are required to follow.

Fair information practices are the basis for privacy best practices, both online and offline. The practices originated in the Privacy Act of 1974, which is the legislation that protects personal information collected and maintained by the U.S. government. The fair information practices include notice, choice, access, onward transfer, security, data integrity, and remedy. The Privacy Act of 1974 states in part that "No agency shall disclose any record which is contained in a system of records by any means of communication to any person, or to another agency, except pursuant to a written request by, or with the prior written consent of, the individual to whom the record pertains."

Federal Information Processing Standard

The *Federal Information Processing Standard (FIPS)* is a standard for adoption and use by federal departments and agencies that has been developed within the Information Technology Laboratory and published by the National Institute of Standards and Technology (NIST), a part of the U.S. Department of Commerce. A FIPS publication covers some topics in information technology to achieve a minimum level of quality or interoperability.

Federal Information Processing Standards Publication 201 (FIPS 201) is a U.S. federal government standard that specifies personal identity verification (PIV) requirements for federal employees and contractors. In response to Homeland Security Presidential Directive 12 (HSPD-12), the NIST Computer Security Division initiated a new program for improving the identification and authentication of federal employees and contractors for access to federal facilities and information systems.

FIPS 201 was developed to satisfy the technical requirements of HSPD 12. It was approved by the secretary of commerce and issued on February 25, 2005. FIPS 201 and NIST SP 800-78 (Cryptographic Algorithms and Key Sizes for PIV) are required for U.S. federal agencies, but do not apply to U.S. national security systems. The Smart Card Interagency Advisory Board has indicated that to comply with FIPS 201 PIV II, U.S. government agencies should use smart card technology.

In information technology (IT), *federated identity* has two general meanings:

- The virtual reunion, or assembled identity, of a person's user information (or principal), stored across multiple distinct identity management systems. Data is joined together by use of the common token, usually the user name.
- The process of a user's authentication across multiple IT systems or even organizations set forth in FIPS 201

Standards Organizations

Many standards-making organizations are involved to varying degrees in the fields of computer and network security. The most significant organizations are described in this section.

American National Standards Institute

The American National Standards Institute (ANSI) is a private, nonprofit organization that administers and coordinates the U.S. voluntary standardization and conformity assessment system. The mission of ANSI is to enhance both the global competitiveness of U.S. business and the U.S. quality of life by promoting and facilitating voluntary consensus standards and conformity assessment systems, and safeguarding their integrity. [1]

International Organization for Standardization

The International Organization for Standardization (ISO) is a nongovernmental network of the national standards institutes from 151 countries. The ISO acts as a bridging organization, in which a consensus can be reached on solutions that meet both the requirements of business and the broader needs of society (the needs of stakeholder groups like consumers and users, for example). For more information, visit http://www.iso.org.

National Institute of Standards and Technology

The *NIST*, founded in 1901, is a nonregulatory federal agency within the U.S. Department of Commerce. NIST's mission is to promote U.S. innovation and industrial competitiveness by advancing measurement science, standards, and technology in ways that enhance economic security and productivity, facilitate trade, and improve the quality of life. NIST's measurement and standards work promotes the well-being of the nation and helps improve, among many others things, the nation's homeland security. For more information, visit http://www.nist.gov/.

International Electro/technical Commission

The *International Electro/technical Commission (IEC)* is the world's leading organization that prepares and publishes International Standards for all electrical, electronic, and related technologies—collectively known as "electro-technology." IEC Standards cover a vast range of technologies from power generation, transmission, and distribution to home appliances and office equipment, semiconductors, fiber optics, batteries, nanotechnologies, solar energy, and marine energy converters (to mention just a few). Wherever there are electricity and electronics, the IEC will provide support safety and performance, the environment, electrical energy efficiency, and renewable energies. The IEC also manages conformity assessment schemes that certify that equipment, systems, or components conform to its International Standards.

Semiconductor Industry Association

The Semiconductor Industry Association (SIA) provides every chip company (large, small, integrated, or fabless) with a significant voice. Collectively, they continue to make tremendous progress in trade, technology, public policy, occupational safety and health, environmental concerns, industry statistics, and government procurement.

European Telecommunications Standards Institute

The *European Telecommunications Standards Institute (ETSI)* is an independent, not-for-profit, standardization organization of the telecommunications industry (equipment makers and network operators) in Europe, with worldwide projection. ETSI has been successful in standardizing the GSM cell phone system and the TETRA professional mobile radio system.

National Security Agency

The *National Security Agency/Central Security Service (NSA/CSS)* is the U.S. government's cryptologic intelligence agency, administered under the U.S. Department of Defense. Created in 1952, it is responsible for the collection and analysis of foreign communications and foreign signals intelligence, which involves a significant amount of cryptanalysis. This agency is also responsible for protecting U.S. government communications and information systems from similar agencies. This task involves a significant amount of cryptography.

Smart Card Standards and Interoperability

A smart card, chip card, or integrated circuit card is defined as any pocket-sized card with embedded integrated circuits which can process information. This section addresses smart card issues with emphasis on smart card policy, standards, interoperability, and education and tracks the latest in federal smart card developments. The *Smart Card Handbook* was developed to help agencies develop and deploy smart card systems. This document:

- Provides a multimedia tutorial on smart cards
- Maintains a database of government smart card projects
- Maintains a list of participants in the Smart Card Project Managers Group
- Provides a library of PowerPoint presentations and documents related to smart cards
- Lists a number of business cases and related case studies

Government Smart Card Interagency Advisory Board

The *Interagency Advisory Board (IAB)*, made up of smart card leaders from major U.S. federal agencies, unanimously voted to approve a key specification for the **personal identity verification** (**PIV**) project, which is the implementation of smart card and biometric-enabled ID credentials for all government employees. It was mandated by President George W. Bush in a document titled, "Homeland Security Presidential Directive HSPD-12." Ten agencies were represented from the IAB, enough to reach a quorum for the action. The agencies were DHS, DOD, GSA, Interior, Justice, NASA, State, Transportation, Treasury, and Veterans Affairs.

The approval of the document enables the progression of FIPS 201 as well as the meeting of requirements set forth in HSPD12. Both of these documents impact the government-wide credential. The outcome will be the implementation of a standardized federal identity credential for all employees and contractors.

The *Smart Card Alliance* was established to accelerate the widespread adoption, usage, and application of smart card technology. The Alliance brings together leading users and technologists from both the public and private sectors in an open forum to address industry issues and create new opportunities for all. Alliance membership includes over 150 U.S.-based and international organizations of industry suppliers, integrators, and end user groups.

Alliance activities include events, educational programs, industry reports and white papers, member-driven industry and technology councils, and industry outreach. The Alliance's member-driven Industry and Technology Councils focus on topics in specific industries or market segments to accelerate smart card adoption and industry growth. Its Educational Institute offers focused classroom-style smart card training that is designed to help industry professionals understand the essential technical and business issues in evaluating and planning a profitable strategy for smart cards in their own markets.

The Alliance publishes a wide variety of reports, case studies, and white papers, covering both business and technology issues in all industry segments that use smart cards. These reports are available to the public at no charge as part of the Alliance educational outreach program.

Standards

The **Open System Interconnection** (**OSI**) model defines a networking framework for implementing protocols in seven layers. Control is passed from one layer to the next, starting at the application layer in one station, proceeding to the bottom layer, over the channel to the

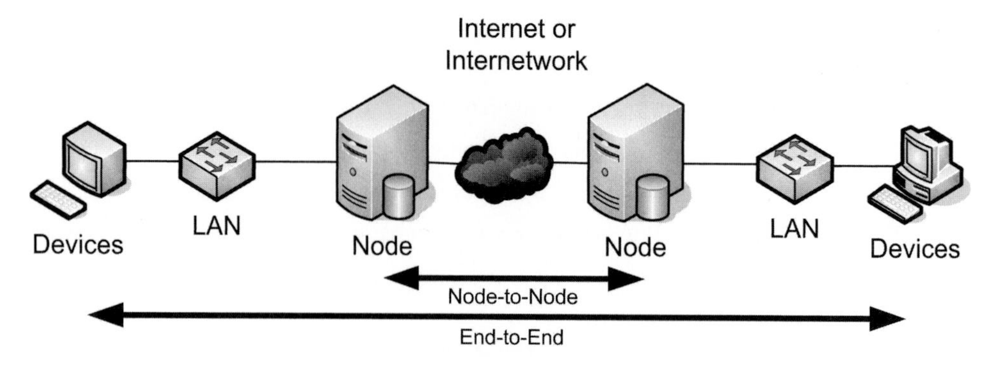

Figure 12-1 End-to-End versus Node-to-Node Security

Course Technology/Cengage Learning

next station and back up the hierarchy. The OSI basic reference model is a layered, abstract description for communications and computer network protocol design. It was developed as part of the OSI initiative and is sometimes known as the OSI seven-layer model. From top to bottom, the OSI model consists of:

- Application: Layer 7
- Presentation: Layer 6
- Session: Layer 5
- Transport: Layer 4
- Network: Layer 3
- Data Link: Layer 2
- Physical: Layer 1

A layer is a collection of related functions that provides services to the layer above it and receives service from the layer below it. For example, a layer that provides error-free communications across a network provides the path needed by applications above it, while it calls the next lower layer to send and receive packets that make up the contents of the path.

Security is implemented at the lower levels of the OSI model: the data link and network layers. Figure 12-1 depicts an end-to-end versus a node-to-node security structure. Deploying security services at these OSI layers makes most of the security services transparent to the user.

Security Standards

Numerous security standards have been developed for the computer and networking industry and environment. Of particular importance are the ISO 17799 and ISO/IEC 27001 standards. This section provides an overview of relevant standards.

ISO 17799

ISO 17799 provides information concerning compliance and implementing audit data and procedures, maintaining security awareness, and conducting post-mortem analysis for this standard. It is divided into 10 major sections:

- System access control
- System development and maintenance
- Physical and environmental security
- Compliance
- Personnel security
- Security organization
- Computer and network management
- Asset classification and control
- Security policy
- Business continuity planning

ISO/IEC 27001

ISO/IEC 27001 is the formal standard against which organizations may seek independent certification of their *information security management systems (ISMS)*. This includes frameworks to design, implement, manage, maintain, and enforce information security processes and controls systematically and consistently throughout the organizations.

The standard covers all types of organizations including commercial enterprises, government agencies, and nonprofit organizations. It specifies the requirements for establishing, implementing, operating, monitoring, reviewing, maintaining, and improving a documented ISMS within the context of the organization's overall risk-management processes. It specifies requirements for the implementation of security controls customized to the needs of individual organizations or parts thereof.

ISO/IEC 27001 provides an ISMS model for adequate and proportionate security controls to protect information assets and give confidence to interested parties. This standard is intended to be suitable for several different types of use, including:

- Use within organizations to formulate security requirements and objectives
- Use within organizations as a way to ensure security risks are managed in a cost-effectively way
- Use within organizations to ensure compliance with laws and regulations
- Use within an organization as a process framework for the implementation and management of controls to ensure the specific security objectives of an organization are met
- The definition of new information security management processes
- Identification and clarification of existing information security management processes
- Use by the management of organizations to determine the status of information security management activities

- Use by the internal and external auditors of organizations to demonstrate the information security policies, directives, and standards adopted by an organization and determine the degree of compliance with those policies, directives, and standards
- Use by organizations to provide relevant information about information security policies, directives, standards, and procedures to trading partners and other organizations that they interact with for operational or commercial reasons
- Implementation of a business enabling information security
- Use by organizations to provide relevant information about information security to customers

Private Communication Technology

Private Communication Technology (PCT) is a protocol that provides secure encrypted communications between two applications over a Transmission Control Protocol/Internet Protocol (TCP/IP) network. It works over the Internet, an intranet, or an extranet. Microsoft developed PCT in response to security weaknesses in Secure Socket Layer (SSL). SSL was restricted to 40-bit keys, so Microsoft separated the authentication and encryption functions in PCT to bypass this restriction. PCT allows applications to use 128-bit key encryption for authentication within the United States, and 40-bit key for export use. PCT is included with Microsoft Internet Explorer.

Private Communications Technology 1.0 was a protocol developed by Microsoft in the mid-1990s. PCT was designed to address security flaws in version 2.0 of Netscape's SSL protocol and to force Netscape to hand change control of the then proprietary SSL protocol to an open standards body. PCT has since been superseded by SSLv3 and Transport Layer Security. For a while it was still supported by Internet Explorer, but the latest versions have removed it. It is still found in the Windows operating system libraries, although in Windows Server 2003 it is disabled by default.

Transport Layer Security (TLS) and its predecessor, **Secure Sockets Layer** (SSL), are cryptographic protocols that provide secure communications on the Internet for such things as Web browsing, e-mail, Internet faxing, instant messaging, and other data transfers. There are slight differences between SSL and TLS, but the protocol remains substantially the same. The TLS protocol allows applications to communicate across a network in a way designed to prevent eavesdropping, tampering, and message forgery. TLS provides endpoint authentication and communications privacy over the Internet using cryptography. Typically, only the server is authenticated (i.e., its identity is ensured) while the client remains unauthenticated. This means the end user, whether an individual or an application, such as a Web browser, can be sure of the communicating entity or individual. The next level of security, in which both ends of the conversation or transmission are sure with whom they are communicating, is known as mutual authentication. Mutual authentication requires public key infrastructure (PKI) deployment to clients unless TLS-PSK or TLS-SRP are used, which provide strong mutual authentication without needing to deploy a PKI.

Secure Sockets Layer

As organizations use the Internet for more than information dissemination, they will need to rely on trusted security mechanisms. An increasingly popular general-purpose solution is to implement security as a protocol that resides between the underlying transport protocol and the application. The foremost example of this approach is the SSL and the follow-on Internet standard TLS. SSL is built into many Web browsers, including Netscape Communicator, Microsoft Internet Explorer, and other software products.

SSL is a Web protocol that sets up a secure session between Web client and server. All data transmitted over the communication channel is encrypted. S-HTTP is a similar protocol that encrypts only at the Hypertext Transfer Protocol (HTTP) level, whereas SSL encrypts all data passed between client and server at the IP socket level. SSL was originally developed by Netscape; it was then submitted to the IETF for standardization. The IETF is the group responsible for creating and maintaining the Internet protocols.

Both SSL and S-HTTP provide security benefits for e-commerce, including protection from eavesdropping and tampering. With SSL, browsers and servers authenticate each other, and then encrypt data transmitted during a session. This procedure verifies to the client that the Web server is authentic before it submits confidential information. It also allows Web servers to verify that users are authentic before granting them access to restricted and sensitive information. Digital certificates are required in this scheme.

Both Web browsers and Web servers must be SSL-enabled. When a client contacts a server, the server forwards a certificate signed by a Certificate Authority (CA). The client then uses the CA's public key to open the certificate and extracts the Web site's public key.

SSL consists of the SSL Handshake Protocol, that provides authentication services and negotiates an encryption method, and the SSL Record Protocol, which performs the task of packaging data, so it can be encrypted.

Although SSL protects information as it is passed over the Internet, it does not protect private information—such as credit card numbers—once they are stored on a merchant's server. If the server is not secure and the data is not encrypted, an unauthorized party can access that information. Hardware devices called peripheral component interconnect (PCI) cards can be installed on Web servers to secure data for an entire SSL transaction from the client to the Web server. The PCI card processes the SSL transactions, freeing the Web server to perform other tasks.

Pretty Good Privacy

Pretty Good Privacy (PGP) is an encryption and digital-signature utility for adding privacy to electronic mail and documents. PGP is an alternative to RSA's S/MIME. While S/MIME uses RSA public-key algorithms, PGP uses Diffie-Hellman public-key algorithms. They both yield similar results. A similar, but older and less robust, privacy protocol is Privacy-Enhanced Mail (PEM).

PGP was designed on the principal that e-mail, like conversations, should be private. Both sender and recipient need assurances that messages are from an authentic source. They are also looking for assurance that messages have not been altered or corrupted and that the sender cannot repudiate or disown the message. PGP can assure both privacy and nonrepudiation. It also provides a tool to encrypt information on magnetic media.

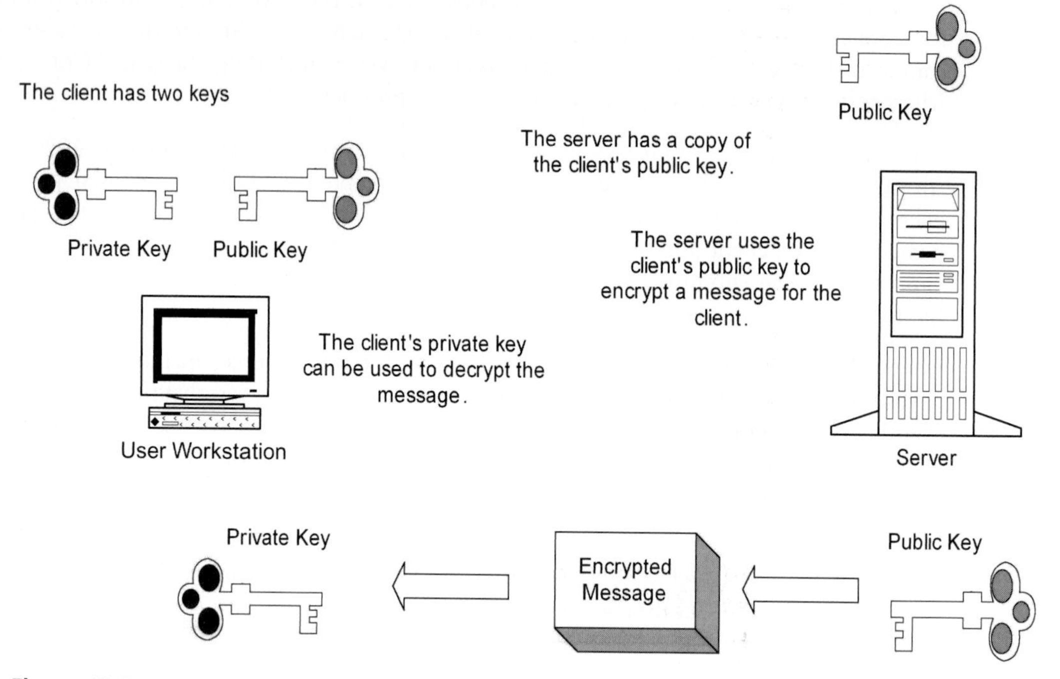

Figure 12-2 Public Key Environment

Course Technology/Cengage Learning

PGP is designed to integrate into popular e-mail programs and operate on major operating systems such as Windows and Macintosh. It uses a graphical user interface (GUI) to simplify the encryption process. The actual operation of PGP consists of:

- Authentication
- Confidentiality
- Compression
- E-mail compatibility
- Segmentation

The package uses public-key encryption techniques, in which the user generates two keys: one public and one private. These keys can then be used to encrypt and digitally sign messages. PGP supports key servers, so users can place their public keys in a central location where other users can access the keys. Figure 12-2 illustrates the public key environment. There are number of reasons to utilize PGP for e-mail and file storage applications, including the following:

- Available worldwide on a number of platforms
- Based on secure algorithms
- Works in a wide range of applicability
- Outside of standards organizations and government control

Secure Hypertext Transfer Protocol

The native protocol that Web clients and servers use to communicate is *HTTP*. This protocol is ideal for open communications, but in its native form does not provide authentication or encryption features. *Secure HTTP (S-HTTP)* works in conjunction with HTTP to enable clients and servers to engage in private and secure transactions. It is especially useful for encrypting forms-based information as it passes between clients and servers. It should be noted that S-HTTP only encrypts HTTP-level messages at the application layer, whereas SSL encrypts all data being passed between client and server at the IP socket level.

S-HTTP provides considerable flexibility in terms of what cryptographic algorithms and modes of operation can be used. Also, as the need for authentication among the Internet and Web grows, users need to be authenticated before sending encrypted files to each other. With S-HTTP, messages may be protected using digital signatures, authentication, and encryption. During the initial contact, the sender and receiver establish preferences for encrypting and processing secure messages.

A number of encryption algorithms and security techniques can be used, including DES and RC2 encryption or RSA public-key signing. Users can also choose to use a particular type of certificate, or no certificate. In situation where certificates are not available, it is possible for a sender and receiver to use a session key that has been exchanged in advance. A challenge/response mechanism is also available. The IETF Web Transaction Security Working Groups is responsible for developing S-HTTP.

Secure Electronic Transaction

Secure electronic transaction (**SET**) is an open specification for handling credit card transactions over a network, with emphasis on the Web and Internet. Secure transactions are critical for e-commerce on the Internet. Merchants must automatically and safely collect and process payments from Internet clients; therefore, a secure protocol is required to support the activities of the credit card companies. Also impacted by e-commerce requirements are consumers, vendors, and software developers. GTE, IBM, MasterCard, Microsoft, Netscape, and Visa developed SET.

SET is designed to secure credit card transactions by authenticating cardholders, merchants, and banks by preserving the confidentiality of payment data. SET includes the following features:

- Requires digital signatures to verify that the customer, the merchant, and the bank are all legitimate.
- Uses multiparty messages that allow information be encrypted directly to banks.
- Prevents credit card numbers from getting in the wrong hands.
- Requires integration into the credit card processing system.

SET includes a negotiation layer that negotiates the type of payment method, protocols, and transports. This task is the responsibility of the Joint Electronic Payment Initiative (JEPI). Payment methods could include credit cards, debit cards, electronic cash, and checks. Payment protocols, such as SET, define the message format and sequence required for completion of the payment transaction. Transports include such protocols as SSL and HTTP. Figure 12-3 depicts the SET components and their relationships.

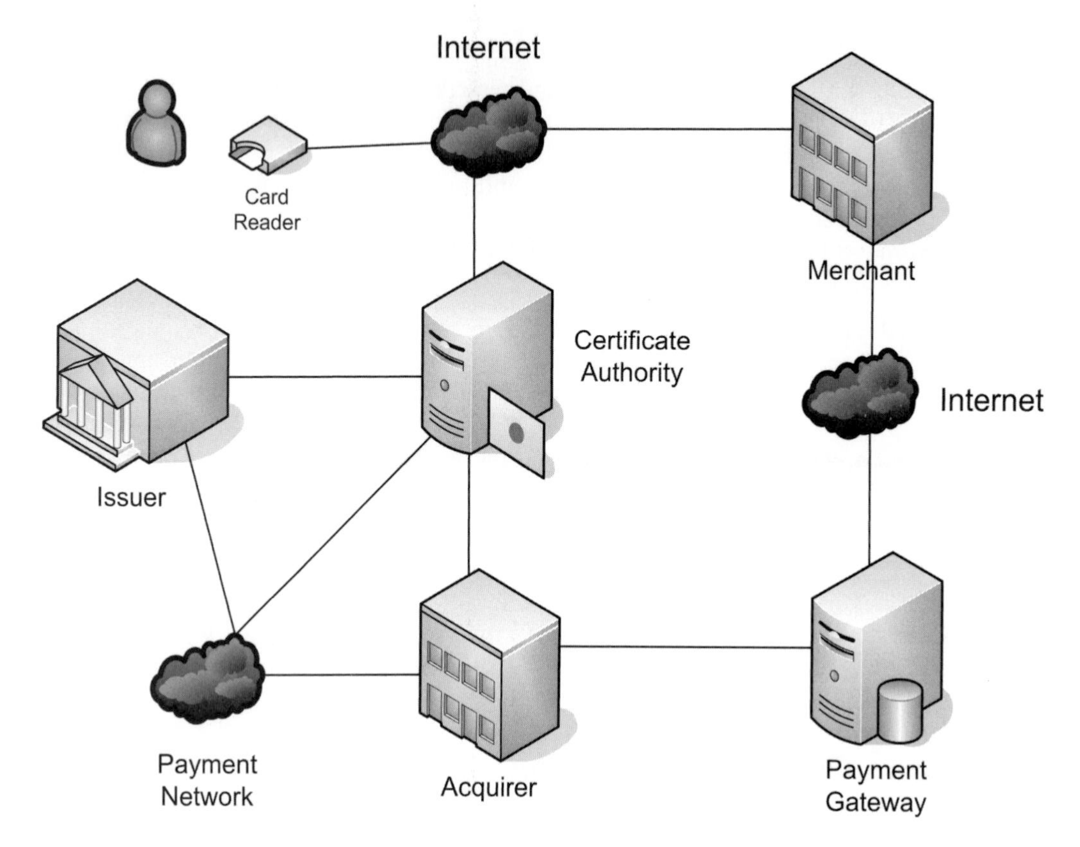

Figure 12-3 SET Components and Relationships

Course Technology/Cengage Learning

An important part of SET's success will be its overall acceptance by cardholders, credit card issuers, payment processors (acquirers), and CAs. *Note CAs provide digital signatures that are critical for verifying the authenticity of cardholders and others involved in the transactions.* Microsoft and Netscape have included SET support in their browsers. Figure 12-4 depicts the sequence of events associated with a SET transaction.

Trusted Computer System Evaluation Criteria/Orange Book

The U.S. National Security Agency has outlined the requirements for secure products in a document titled *Trusted Computer System Evaluation Criteria (TCSEC)*. TCSEC is commonly called the *Orange Book*. This standard defines access control methods for computer systems that computer vendors can follow to comply with Department of Defense (DOD) security standards. TCSEC is a collection of criteria used to grade or rate the security offered by computer systems.

The Orange Book

The Orange Book was the first guideline for evaluating security products for operating systems. Security evaluation examines the security-relevant part of a system. The initial efforts were concentrated in the national security sector. The purpose of the Orange Book was to provide:

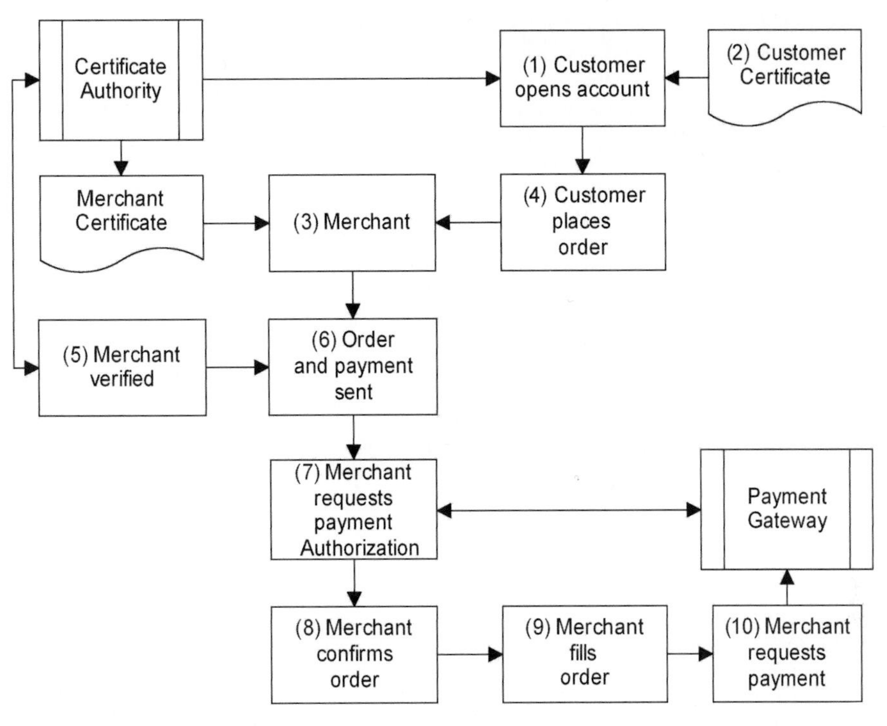

Figure 12-4 SET Transaction Flow

Course Technology/Cengage Learning

- A basis for specifying security requirements when acquiring a computer system
- A yardstick for users to assess the degree of trust that can be placed in a computer security system
- Guidance for manufacturers of computer security systems

The evaluation classes of the Orange Book are designed to address typical patterns of security requirements. These security feature and assurance requirement categories include the following classes:

- Accountability: Audit logs of security relevant events maintained
- Assurance: Operational: Security architecture and life cycle Design methodology, testing, and configuration management
- Continuous protection: Security mechanisms tamperproof
- Documentation: Guidance to install and use security features; test and design instructions
- Identification of subjects: Individual subjects identified and authenticated
- Marking of objects: Labels specify the sensitivity of objects
- Security policy: Mandatory and discretionary access control policies expressed in terms of subjects and objects

The security classes of the Orange Book are defined incrementally, which means that all requirements of one class are automatically included in those of all higher classes.

Criteria Categories	Criteria
Security policy	Device labels Discretionary access control Exportation of labeled information Labeling human-readable output Mandatory access control Object reuse Subject sensitivity labels
Accountability	Audit Identification and authentication Trusted path
Assurance	Configuration management Covert channel analysis Design specification and verification System architecture System integrity Security testing Trusted distribution Trusted recovery
Documentation	Design documentation Security features user's guide Test documentation Trusted facility manual

Table 12-1 Trusted Computer System Evaluation Criteria

The four security divisions of the Orange Book include minimal protection, discretionary, mandatory, and verified protection. Minimal protection (Class D) includes products that are submitted for evaluation, but do not meet any Orange Book requirement.

Table 12-1 shows the trusted computer system evaluation criteria. The criteria have been grouped into four major categories:

- security policy
- accountability
- assurance
- documentation

Trusted Network Interpretation/Red Book

Secure networking is defined in the *Red Book* or *Trusted Network Interpretation (TNI)*. The Red Book is part of the series of documents published by the National Computer Security Center (NCSC) that describe the requirements in the TCSEC. It describes TCSEC in terms of computer networks. It also attempts to address network security with the concepts and terminology introduced in the Orange Book. The Red Book may be viewed as a link between the Orange Book and new criteria, which has been proposed in later years. The Red Book differentiates between two types of networks: independent and centralized structures.

A network with independent components consists of different jurisdictions, management, and policies. Enforcing security in this distributed environment is a very difficult problem.

Network Security Service	Components
Communication integrity	Authentication Communication field integrity Nonrepudiation
Denial of service	Continuity of operation Protocol-based protection Network management
Compromise protection	Data confidentiality Traffic confidentiality

Table 12-2 Red Book Network Security Services

Centralized networks consist of a single accreditation authority, policy, and a network trusted computing base. The Red Book only considers the centralized network. In a computer network, security mechanisms may be distributed among different network components. New security problems and issues arise due to the vulnerability of the communications paths and the method of concurrent and asynchronous operation of the network components.

In networks it becomes much more obvious that different entities—such as users, service providers, network operators, and system administrators—are responsible for setting the security policy. In the Red Book, the responsible entity who states the security requirements, defining the security policy, and submitting a system for evaluation, is called the sponsor. Security policies are concerned with security and integrity, and control the establishment of authorized connections. Mandatory access control in networks includes mandatory integrity policies, such as integrity labels. An integrity label could indicate whether an object had ever been transmitted between network nodes.

Other security service designs supported by the Red Book include encryption and protocols. Each service requires a specification for functionality, strength, and assurance. Strength indicates how well a mechanism is expected to meet its goal. Assurance addresses good software engineering practices, validation, and verification. Table 12-2 provides a summarization of the network security services available in the Red Book.

Information Technology Security Evaluation Criteria

The TCSEC was exclusively an U.S. development. European countries recognized the need for a criterion and methodology for evaluations of security-enforcing products. The European efforts culminated in the *Information Technology Security Evaluation Criteria (ITSEC)*. ITSEC is a logical progression from the lessons learned in various Orange Book interpretations. A vendor or sponsor must define a target of evaluation (TOE), which is the focus of the evaluation. The TOE is considered in the context of an operational environment, which contains a set of threats, and security enforcement requirements. An evaluation can be made of a product or a system.

Common Criteria

European security evaluation criteria responded to the problems exposed by the Red Book by separating function and assurance requirements and considering the evaluation of entire security systems. The flexibility offered by the ITSEC may sometimes be an advantage, but there

Functionality	Assurance
Communication	Configuration management
Identification and authentication	Delivery and operation
Invocation of security functions	Development
Privacy	Life-cycle support
Protection of trusted security functions	Guidance documents
Resource utilization	Testing
Security audit	Vulnerability assessment
Trusted path	
User data protection	

Table 12-3 **Common Criteria Classes**

are drawbacks. The next link in the evolutionary chain of evaluation criteria is the United States' Federal Criteria for Information Technology Security. This "Combined Federal Criteria" was issued only once, in initial draft form. The U.S. team joined forces with the Canadian team and the ITEC and produced the *Common Criteria* for the entire world. The Common Criteria merged ideas from its various predecessors and abandoned the total flexibility of ITSEC. It followed the Federal Criteria in using protection profiles and predefined security classes. The definition of these classes contains information about security objectives, rationale, threats, and threat environment. The Common Criteria defined topics of interest to security as shown in Table 12-3.

Common Data Security Architecture

Common Data Security Architecture (CDSA) is a security reference standard that provides a way to develop applications, which take advantage of software security mechanisms. The *Open Group* has accepted CDSA for evaluation, and IBM, Intel, and Netscape are refining it. It addresses the security issues of Internet and intranet applications. It also provides an interoperable standard and an expansion platform for future security elements.

The Open Group is an international consortium of vendors, government agencies, and educational institutions that develops standards for open systems. It was formed in 1996 as the holding company for *Open Software Foundation (OSF)* and X/Open Company Ltd. These two organizations work together to deliver technology innovations and widespread of open systems specifications. An **open system** is a computer and communications system capable of communicating with other like systems, transparently. The objective is that each system implements common international standards and data communication protocols.

Payment Card Industry Payment Standard

The **Payment Card Industry Data Security Standard** (PCI DSS) was developed by the major credit card companies as a guideline to help organizations that process card payments, prevent credit card fraud, hacking, and various other security issues. A company processing, storing, or transmitting credit card numbers must be PCI DSS compliant or it risks losing the ability to process credit card payments. Merchants and service providers must validate compliance with an audit by a PCI DSS Qualified Security Assessor (QSA) Company.

In October 2007, Visa International announced new payment applications security mandates, designed to help companies comply with PCI mandates, that must be implemented by 2010.

These mandates call for new merchants that want to be authorized for payment card transactions to be using only Payment Application Best Practice (PABP)-validated applications. These new mandates will help companies achieve PABP compliance, an implementation of PCI DSS in vendor software.

Biometric Standards

Standards are a critical element that has been absent in the world of biometrics. There are virtually no standards in place for automated biometrics, including minutiae analysis, the method used by human experts to analyze fingerprints. While there are ANSI-NIST fingerprint minutiae standards, they don't seem to be of sufficient information density to be usable for all automated biometrics; thus, vendors typically use proprietary minutiae algorithms.

However, it appears that the first step in forming some kind of environment of standards has been initiated by leading computer and software vendors. During the CardTech/SecureTech Conference, it was announced that the newly formed *BioAPI Consortium* would leverage work done by several companies to create a common application programming interface (API) for existing and emerging biometric technologies. The **BioAPI specification** defines the application programming interface and service provider interface for a standard biometric technology interface.

CardTech addresses every aspect transaction technology that is linked to the use of identity technology. *SecurTech* addresses every aspect of logical and physical access security. The BioAPI Consortium was founded to develop a biometric API that brings platform and device independence to application programmers and biometric service providers. The Consortium is a group of over 120 companies and organizations that have a common interest in promoting the growth of the biometrics market. The objective is a specification and reference implementation for a standardized API that is compatible with a wide range of biometric application programs and a broad spectrum of biometric technologies.

The resulting new API standard frees system designers and integrators from developing different programs for each vendor's biometric hardware. API standards are very important; they enable vendors to adopt products more easily. Without a proven, reliable API, it's very risky for developers to design new products.

In 1998, the *International Computer Security Association (ICSA)* also announced the first six biometric products to meet its new certification standards. The ICSA tested more than 100 biometric products before certifying six that performed as advertised with real humans in real-world environments.

Standards are a key facet in making biometrics a widespread technology and in reducing the differences between products. This decreases the risk of using automated biometrics. By reducing the risk of development, standards help grow the market, which benefits vendors. Standards also promote an aura of stability and maturity attractive to investors. A series of international standards have been issued by the ISO/IEC. A brief description of these follows:

- ISO/IEC 7810: The series of international standards describing the characteristics of identification cards, including physical characteristics, sizes, thickness, dimensions, construction, materials, and other requirements.

- ISO/IEC 7812: The governing international standard for magnetic stripe identification cards, such as door entry cards, automated teller machine (ATM) cards, and credit cards.
- ISO/IEC 7816: The international standard for integrated circuit cards with contacts, as well as the command set for all smart cards.
- ISO/IEC 14443: The international standard, "Identification Cards — Contactless Integrated Circuit(s) Cards — Proximity Cards," for contactless smart chips and cards that operate (i.e., can be read from or written to) at a distance of less than 10 centimeters (4 inches). This standard operates at 13.56 MHz.
- ISO/IEC 15693: The international standard, "Identification Cards — Contactless Integrated Circuit(s) Cards — Vicinity Cards," for cards operating at the 13.56 MHz frequency which can be read from a greater distance as compared to proximity cards.
- ISO/IEC 24727: A set of programming interfaces for interactions between integrated circuit cards and external applications to include generic services for multisector use.

Proponents of biometric standards include a number of corporations and organizations. AFIS has been a leader in establishing fingerprint standards and is responsible for two standards: the interchange of finger image data and finger image compression. The Biometric Consortium has led U.S. government's coordinated biometric efforts since 1993. Its main push has been the development of the *Human Authentication (HA-API)*. The first HA-API specification was announced in late 1997. The project itself is essentially divided into two parts: the creation of a generic biometric API and a proof of concept implementation, integrating the API within a commercial network authentication system. The *Advanced Identification Services API (AIS-API)* was first announced in November 1997. This product supports the capture, storage, query, and retrieval of biometric data.

The *U.S. National Biometric Test Center (NBCT)* was established by the U.S. Department of Defense's Biometric Consortium in the second quarter of 1997. The main goal of the NBTC is to further the U.S. government's efforts in standardizing biometric testing procedures and focusing on the "real-world" performance standards. The NBTC is particularly focused on developing standard testing methodologies so that biometrics can be compared.

Public Key Infrastructure

In cryptography, a *PKI* is an arrangement that binds public keys with respective user identities by means of a CA. The user identity must be unique for each CA. The binding is established through the registration and issuance process, which, depending on the level of assurance the binding has, may be carried out by software at a CA, or under human supervision. The PKI role that assures this binding is called the Registration Authority (RA). For each user, the user identity, the public key, their binding, validity conditions, and other attributes are made unforgeable in public key certificates issued by the CA.

The Public Key Cryptography Standard #11 (PKCS #11) defines the interface for cryptography operations with hardware tokens. Message Digest 5 (MD5) is one of the most popular hashing algorithms, developed by MIT Professor Ronald L. Rivest, which produces a 128-bit hash from any input.

X.509 Certificate Policy for the U.S. Department of Defense is the international standard for security and authentication services supporting security frameworks for electronic information distribution. The term "X.509 certificate" has become the de facto name for public key certificates in use today. This specification defines the main data structure (i.e., the "certificate") used for performing these services and addressing the handling of keys.

The Online Certificate Status Protocol (OCSP) is an Internet protocol used for obtaining the revocation status of an X.509 digital certificate. It is described in RFC 2560 and is on the Internet standards track. It was created as an alternative to certificate revocation lists (CRL), specifically addressing certain problems associated with using CRLs in a PKI. Messages communicated via OCSP are encoded in ASN.1 and are usually communicated over HTTP.

Legal Issues and Acts

The primary objective of HSPD-12 is the development and deployment of a federal government-wide common and reliable identification verification system that will be interoperable among all government agencies and serve as the basis for reciprocity among those agencies.

The Financial Services Modernization Act of 1999, also known as the Gramm-Leach-Bliley Act, was enacted to facilitate affiliation among banks, securities firms, and insurance companies. The Act includes provisions to protect consumers' personal financial information held by financial institutions.

The Health Insurance Portability and Accountability Act of 1996 (HIPAA) was passed to protect health insurance coverage for workers and their families and to encourage the development of a health information system by establishing standards and requirements for the secure electronic transmission of certain health information. HIPAA mandates that the design and implementation of the electronic systems guarantee the privacy and security of patient information gathered as part of providing health care.

Employer Legal Responsibilities

It is common practice in many organizations to escort former employees to the door and collect all corporate items such as smart cards and credit cards. Having security measures in place means that someone must administer those measures. This task is usually the responsibility of the IT department, which means that these employees must be trustworthy, competent, and conscientious. Former employees' network access must be denied promptly to avoid possible security violations that can lead to litigation. There must be a process where the human resources department notifies IT of an employee's departure, so that network access can be removed. Delays can result in data theft or corruption. *Note: An organization's legal counsel should be contacted for current local practices, procedures, and legal advice.*

Current Biometric Initiatives

A number of legal initiatives have been developed or are in development following the terrorist attacks of 2001. Additional initiatives are also directed toward the immigration issue. Both have implications for U.S. border security. This section provides a brief introduction of these initiatives.

Terrorism, Immigration, and Border Security

Sec. 403(c) of the *USA-PATRIOT Act* specifically requires the federal government to "develop and certify a technology standard that can be used to verify the identity of persons" applying for or seeking entry into the United States on a U.S. visa "for the purposes of conducting background checks, confirming identity, and ensuring that a person has not received a visa under a different name."

The *Enhanced Border Security and Visa Entry Reform Act of 2002*, Sec. 303(b)(1), requires that only "machine-readable, tamper-resistant visas and other travel and entry documents that use biometric identifiers" shall be issued to aliens by October 26, 2004. The Immigration and Naturalization Service (INS) and the State Department currently are evaluating biometrics for use in U.S. border control pursuant to EBSVERA.

Even prior to September 11, however, large-scale civilian biometric identification systems were being pushed. Both the *Personal Responsibility and Work Opportunity Act of 1995 (PRWOA)*, a welfare reform law, and the *Immigration Control and Financial Responsibility Act of 1996 (ICFRA)*, an immigration reform law, called for the use of "technology" for identification purposes.

The PRWOA requires the states to implement an electronic benefits transfer program "using the most recent technology available which may include personal identification numbers (PINs), photographic identification and other measures to protect against fraud and abuse." This law covers, for example, the Food Stamps program.

The ICFRA requires the U.S. president to "develop and recommend a plan for the establishment of a data system or alternative system to verify eligibility for employment in the United States, and immigration status in the United States for purposes of eligibility for benefits under public assistance programs or government benefits." This system "must be capable of reliably determining with respect to an individual whether the individual is claiming the identity of another person."

The *Illegal Immigration Reform and Immigrant Responsibility Act of 1996 (IIRAIRA)* requires the INS to include on alien border crossing cards "a biometric identifier (such as the fingerprint or handprint of the alien) that is machine readable." The State Department collects fingerprints and photographs of aliens for these cards.

The *REAL ID Act of 2005* provided legislation intended to deter terrorism by establishing national standards for state-issued driver's licenses and nondriver's identification cards in addition to other key executables.

Other Identification Initiatives

The *Truck and Bus Safety and Regulatory Reform Act of 1988 (TBSRRA)* requires "minimum uniform standards for the biometric identification of commercial drivers."

The *Sarbanes-Oxley Act of 2002* introduced changes to regulations that apply to financial practice and corporate governance for public companies. The Act introduced new rules that were intended "to protect investors by improving the accuracy and reliability of corporate disclosures made pursuant to the securities laws."

Engineering/Specifications

A **specification** is defined as the set of documentation that reflects agreements on products, practices, or operations produced by one or more organizations (or groups of cooperating entities), some for internal usage only, others for use by groups of people, groups of companies, or an entire industry. Relevant specifications that apply to the subject of computers and network security are described in this section.

Computer and Network Specifications

- *Secure hash algorithm (SHA)* is one of the most popular hashing algorithms, designed for use with the Digital Signature Standard by the NIST and the NSA. SHA-1 produces a 160-bit hash.
- *Secure Multipurpose Internet Mail Extensions (S/MIME)* is a protocol for exchanging digitally signed and/or encrypted mail.
- *Triple DES* is a block cipher formed from the Data Encryption Standard (DES) cipher by using it three times.
- *RSA* refers to public/private key encryption technology that uses an algorithm developed by Ron Rivest, Adi Shamir, and Leonard Adleman and that is owned and licensed by RSA Security.

Biometric Fingerprint Specifications

Various specifications are directed at the biometric fingerprint modality. These include:

- The *Electronic Fingerprint Transmission Specification (EFTS)* is a document that specifies requirements to which agencies must adhere to communicate electronically with the Federal Bureau of Investigation's Integrated Automated Fingerprint Identification System (IAFIS). This specification facilitates information sharing and eliminates the delays associated with fingerprint cards. See also Integrated Automated Fingerprint Identification System (IAFIS).
- The Federal Bureau of Investigation's *IAFIS,* implemented in July 1999, replaces a paper-based system for identifying and searching criminal history records. IAFIS supports a law enforcement agency's ability to digitally record fingerprints and electronically exchange information with the FBI. Manual steps have been reduced and processing speeds increased.
- The *DOD Electronic Biometrics Transmission Specification* Version 1.2 document is the latest updated version of the DOD EBTS, which describes FBI Electronic Fingerprint Transmission Specification transactions that are necessary to utilize the DOD Automated Biometric Identification System.

Biometric Element and System Specifications

A catch-all of miscellaneous biometric specifications are included in this section.

- The *DOD Biometric Collection, Transmission, and Storage Standards Technical Reference* document provides a comprehensive technical reference that lists published biometric standards and describes their applicability to the biometric functions described in the Capstone Concept of Operations (CONOPS) for DOD Biometrics in Support of Identity Superiority.
- As defined in NIST SP 800-73, transitional products that meet the "transitional" interface specification. Transitional products can be used as part of a migration strategy by federal agencies that have already initiated a large-scale deployment of smart cards as identity badges.
- The *EPC Generation 2 (EPC Gen 2)* specification was developed by EPCglobal for the second-generation RFID air-interface protocol. EPC Gen 2 was developed to support supply chain applications (e.g., tracking inventory). The current ratified standard operates in the ultra-high-frequency (UHF) range (860–960 MHz), supports operation at long distances (e.g., 25–30 feet), and has minimal support for security (e.g., static passwords to access or kill information on the RFID device).
- ICAO established international standards for *machine readable travel documents (MTRD)*. An MRTD is an international travel document (e.g., a passport or visa) containing eye-and machine-readable data. ICAO Document 9303 is the international standard for MRTDs.
- *Wavelet scalar quantization* is a compression algorithm used to reduce the size of reference templates.
- *X9.84 Biometrics Management and Security for the Financial Services* is an industry specification that defines the minimum-security requirements for effective management of biometrics data for the financial services industry and the security for the collection, distribution, and processing of biometrics data.
- *Person Data Exchange Standard (PDES)* is a specification of the U.S. government intelligence community that specifies XML tagging of person data, including biometric data.

Card-Related Specifications

Identification cards and integrated circuit cards specifications, Part 13, provides commands for application management in a multiapplication environment.

ISO/IEC 7816-13:2007 specifies commands for application management in a multiapplication environment. These commands cover the entire life cycle of applications in a multiapplication integrated circuit card, and the commands can be used before and after the card is issued to the cardholder. This specification does not cover the implementation within the card and/or the outside world.

ISO 14443 defines a proximity card used for identification that usually uses the standard credit card form factor defined by ISO 7810 ID-1; however, other form factors are also possible. The *radio frequency identification (RFID)* reader uses an embedded microcontroller (including its own microprocessor and several types of memory) and a magnetic loop antenna that operates at 13.56 MHz (RFID frequency). More recent ICAO standards for machine-readable travel documents specify a cryptographically signed file format and authentication protocol for storing biometric features, such as photos of the face, fingerprints, and/or iris.

The *personal computer/smart card (PC/SC)* specification defines how to integrate smart-card readers and smart cards with the computing environment and how to allow multiple applications to share smart card devices. *Federal Agency Smart Credential Number (FASC-N)* is the data element that is the main identifier on the PIV card and that is used by a physical access control system.

Biometric Measurements

During the verification or identification process errors can occur. There are two critical measurement factors which indicate the level of accuracy, or reliability, of any given biometric. They are the false rejection rate (FRR) and false acceptance rate (FAR).

The identifying power of a particular biometric encompasses FRR, or a type I error, and FAR, or a type II error. FRRs and FARs are complementary in determining how stringent a biometric device is in allowing user access. As a result, biometric devices commonly include features to allow for variable threshold or sensitivity settings. For example, if the FAR threshold is increased to make it more difficult for impostors to gain access, it also will become harder for authorized people to gain access. As FAR goes down, FRR rises. Alternatively, if the FAR threshold is lowered as to make it very easy for authorized users to gain access, then it will be more likely that an impostor will slip through. Hence, as FRR goes down, FAR rises.

In understanding the impact of FRR and FAR rates, consider an ATM access system: a "false acceptance" means the company may lose a few dollars, whereas a "false rejection" means a valuable customer may be lost. Another relevant example in understanding the inverse relationship of FRR and FAR rates involves a car alarm. When the car alarm is very sensitive, the probability of the bad guys stealing it is low. Yet the chance of accidentally setting off the alarm is high. Reduce the sensitivity, and the number of false alarms goes down, but the chance of someone stealing the car increases.

While the terms "false reject" and "false accept" are still commonly used in quantifying a biometrics' ability to rightfully identify an individual, the federal government has recently adopted a new standard of error rate measurement. Dr. Jim Wayman, Director of the U.S. National Biometric Test Center, has promoted the terms "false match" and "false non-match" as the new de facto terminology in determining a biometrics identifying power. Apparently, the problem with the terms "false accept" and "false reject" and "type I" and "type II" errors is that their meaning depends upon the claim of the user. For example, depending upon the biometric application, users make either a positive or a negative claim to identity. In a positive identification system, a rejection occurs if a person in not matched to a claimed record. In a negative identification system, a rejection occurs if a person is matched to a nonclaimed record. Consequently, the words "false reject/accept" have opposite meanings, depending upon who is speaking. Other biometric methods offer scalability between security and ease of access with a trade-off between the FRR and FAR. Figure 12-5 depicts this relationship.

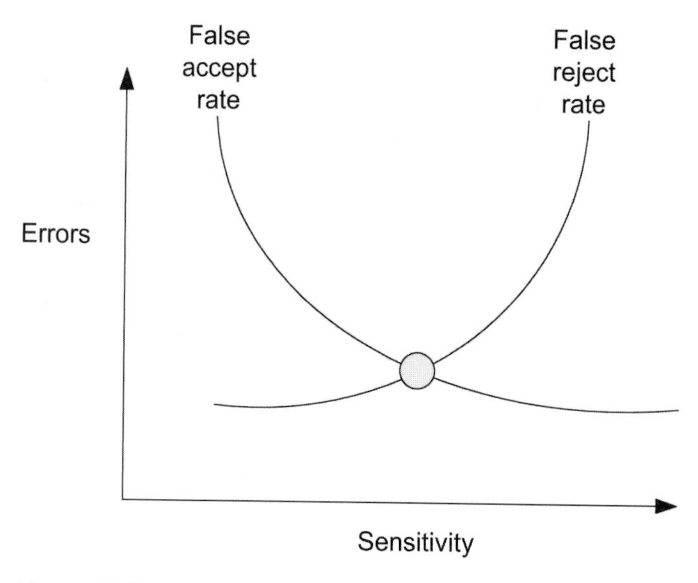

Figure 12-5 FAR versus FRR

Course Technology/Cengage Learning

What this means is the higher level of security desired, the more false rejects that will be encountered. On the other hand, if the objective is to reduce false rejects, this immediately creates a much higher FAR, leading to unwanted persons being identified in error.

Evaluation Programs

Various programs are directed at biometric evaluation programs, including:

- *FacE REcognition Technology (FERET)* is a face recognition development and evaluation program sponsored by the U.S. government from 1993 through 1997. For more information, visit http://www.frvt.org/FERET/default.htm.
- *Face Recognition Grand Challenge (FRGC)* is a face recognition development program sponsored by the U.S. government from 2003 through 2005. For more information, visit http://www.frvt.org/FRGC/.
- *Face Recognition Vendor Test (FRVT)* is a series of large-scale independent technology evaluations of face recognition systems. The evaluations have occurred in 2000, 2002, and 2005. For more information, visit http://www.frvt.org/FRVT2005/default.aspx.
- *Iris Challenge Evaluation (ICE)* is a large-scale development and independent technology evaluation activity for iris recognition systems sponsored by the U.S. government in 2005. For more information, visit http://iris.nist.gov/ICE/.
- *Fingerprint Vendor Technology Evaluation (FpVTE)* is an independently administered technology evaluation of commercial fingerprint matching algorithms. For more information, visit http://fpvte.nist.gov/.

Technology

The *U.S. Visitor and Immigrant Status Indicator (US-VISIT)* is a continuum of security measures that begins overseas, at the Department of State's visa issuing posts, and continues through arrival and departure from the United States of America. Using biometrics, such as digital, inkless finger scans and digital photographs, the identity of visitors requiring a visa is now matched at each step to ensure that the person crossing the U.S. border is the same person who received the visa. For visa-waiver travelers, the capture of biometrics first occurs at the port of entry to the United States. By checking the biometrics of a traveler against its databases, US-VISIT verifies whether the traveler has previously been determined inadmissible, is a known security risk (including having outstanding wants and warrants), or has previously overstayed the terms of a visa. These entry and exit procedures address the United States' critical need for tighter security and ongoing commitment to facilitate travel for the millions of legitimate visitors welcomed each year to conduct business, learn, see family, or tour the country.

Common Biometric Exchange File Format

The *Common Biometric Exchange File Format (CBEFF)* [2] describes a set of data elements necessary to support biometric technologies in a common way. This standard provides the ability for a system to identify, and interface with, multiple biometric systems, and to exchange data between system components. The result promotes interoperability of biometric-based application programs and systems developed by different vendors by allowing biometric data interchange. CBEFF's initial conceptual definition was achieved through a series of workshops cosponsored by the NIST and the BC. A technical development team, formed as a result of these Workshops, developed CBEFF in coordination with industrial organizations such as:

- BioAPI Consortium
- X9.F4 Working Group
- International Biometric Industry Association
- Interfaces Group of TeleTrusT
- End users

CBEFF provides forward compatibility accommodating for technology improvements and allows for new formats to be created. CBEFF implementations simplify integration of software and hardware provided by different vendors. Further development, such as a CBEFF smart card format, is proposed under the umbrella of the recently formed Biometrics Interoperability, Performance, and Assurance Working Group co-sponsored by NIST and the BC.

The expected enormous growth in the use of biometric-based systems and applications highlights the need for exchange and interoperability of biometric data. It is conceivable that many biometric-based systems and applications are expected to support multiple biometric devices and biometric data. Products with that level of support for biometric-based authentication exist today. A Common Biometric Exchange File Format promotes interoperability of biometric-based application programs and systems developed by different vendors by allowing biometric data interchange. CBEFF defines a common set of data elements necessary to support these biometric technologies. These data can be placed in a single file used to exchange biometric information between different system components or between systems.

The expected benefits of CBEFF are the ability to identify different biometric data structures, both public or proprietary, supporting multiple biometric types within a system or application, the ability to reduce the need for additional software development, and the ability to promote development cost savings.

CBEFF describes a set of required and optional fields, a "domain of use" to establish the applicability of a standard or specification that meets CBEFF requirements, and a process by which new technology or systems can create formats that meet these requirements. CBEFF allows for these standards or specifications to define a format and for these formats to define the data encoding. Adoption of CBEFF and compliance to those standards or specifications promote interoperability of biometric-based application programs and systems developed by different vendors by allowing biometric data interchange.

CBEFF's content reflects some current developments within the biometric industry including the release of BioAPI Specification version 1.0 and the development of draft ANSI standard X9.84, "Biometric Information Management and Security."

By focusing on the description of the biometric data elements, details such as data encoding, data and noncommon elements can be left up to a standard or specification that meets CBEFF requirements. By describing a process to establish new formats, the CBEFF can allow for biometrics data to be placed in new technologies and systems.

The common set of data elements described in CBEFF can be placed in a single file record or data object used to exchange biometric information between different system components (the Common Biometric Exchange File). Formatting the data, by allowing individual components to be referenced, will allow an application to easily recognize important processing information about the biometric data such as the type of biometric available, version number, vendor's name, etc.

Formatting the data will also provide pointers to the proper biometric data. These characteristics foster interoperability between different types of biometric systems, allow for the exchange of biometric-related information between different systems, and allow systems with different requirements to translate between different formats.

CBEFF accommodates any biometric technology. It includes the definition of format and content for data elements, such as:

- A biometric data header that contains such information as version number, length of data, whether the data is encrypted or not, etc., for each biometric type available to the application or system
- Biometric data (content not specified)
- Any other required biometric data or data structures
- CBEFF also describes the means for obtaining a unique value for identifying the format (owner and type) of the biometric data

The common biometric data format does not attempt to achieve compatibility among different biometric technologies, but merely identifies them and facilitates their co-existence in a system or application. It is conceivable that industrial or user groups may agree upon common standard template formats within the biometric data structures defined in CBEFF.

Professional Organizations

Although poised for substantial growth as the marketplace begins to accept biometrics, recent events have demonstrated that the fledgling industry's growth could be severely constricted by misinformation and a lack of public awareness. In particular, concerns about privacy can lead to ill-informed regulations that reasonably restrict the use of biometrics. The lack of common and clearly articulated industry positions on issues such as safety, privacy, and standards further increase odds that governments will react inappropriately to uninformed and even unfounded assertions regarding biometric technology's function and use. Two organizations, the International Biometric Industry Association and the Biometric Consortium, aim to improve this situation. *Note: The biometrics industry is in a state of flux; therefore it is prudent to obtain the most current information available through research initiatives.*

International Biometric Industry Association

A Washington, D.C.-based trade association, the *International Biometric Industry Association (IBIA)* seeks to give the young industry a seat at the table in the growing public debate on the use of biometric technology. The IBIA focuses on educating lawmakers and regulators about how biometrics can help deter identity theft and increase personal security. In addition to helping provide a lobbying voice for biometric companies, the IBIA's board of directors has taken steps to establish a strong code of ethics for its members. In addition to certifying that the consortium will adhere to standards for product performance, each member must recognize the protection of personal privacy as a fundamental obligation of the biometric industry. Besides promoting a position on member ethics, the IBIA recommends:

- Safeguards to ensure that biometric data is not misused to compromise any information
- Policies that clearly set forth how biometric data will be collected, stored, accessed, and used
- Limited conditions under which agencies of national security and law enforcement may acquire, access, store, and use biometric data
- Controls to protect the confidentiality and integrity of databases containing biometric data

The IBIA is open to biometric manufacturers, integrators, and end users. [3]

Biometric Consortium

The **Biometric Consortium** (BC) serves as a focal point for research, development, testing, evaluation, and application of biometric-based personal identification/verification technology. The BC organizes a premier biometrics conference every fall. Information about past conferences, current government and standards activity, a bulletin board service, and other biometric resources can be found throughout this Web site. BC is an open forum to share information throughout government, industry, and academia. For more information, visit http://www.biometrics.org.

In December 1995, the Facilities Protection Committee chartered the BC. With more than 500 members from government, industry, and academia, the BC serves as one of the U.S. government's focal points for research, development, testing, evaluation, and application of biometric-based systems. More than 60 different federal agencies and members from 80 other organizations participate in the BC. The BC cosponsors several biometric-related projects, including some of the activities at NIST's Information Technology Laboratory and work at the National Biometric Test Center at San Jose State University. The BC also cosponsors NIST's biometrics and smart cards laboratory, which addresses a wide range of issues related to the interoperability, evaluation, and standardization of biometric technologies and smart cards, especially for authentication applications like e-commerce and enterprise-wide network access.

The annual BS conference on the convergence of technologies for the next century highlighted and explored new applications in e-commerce, network security, wireless communications, and health services. It also addressed the convergence of biometrics and related technologies, such as smart cards and digital signatures. The BC's Web site and its open listserv are two of the consortium's richest resources. [4]

National Biometric Security Project

The *National Biometric Security Project (NBSP)* [5] is an independent, nonprofit organization created to increase national security and personal identity protection by enhancing identity assurance through biometrics. To reflect its expanded biometric application services, the NBSP Enterprise recently reestablished its Test, Research and Data Center under the new name Biometric Services International, LLC (BSI). Located in Morgantown, West Virginia, BSI is a wholly owned, nonprofit subsidiary of NBSP and is the only laboratory exclusively focused on biometrics to achieve the coveted ISO/IEC 17025:2005 accreditation for testing. BSI's application and acquisition services have been expanded to address biometric deployment considerations. These include requirements definitions; articulation of program goals and objectives, vulnerability assessments, application impact studies, life-cycle cost analyses, orientation, technical and certification training, and privacy impact assessments.

There are a number of biometric communication venues. Some of these are fee-based and some are cost free. A short list follows:

- Fee-based
  - Biometrics Discussion Group
  - Biometrics Institute Discussion Forum
  - European Biometric Forum
- Cost free
  - BioAPI Consortium Electronic Discussion Groups
  - Biometric Consortium Bulletin Board
  - Biometrics Electronic Discussion Group
  - Biometrie
  - Yahoo! Biometric eGroup

Additional Biometric Player Organizations

- The *International Civil Aviation Organization Machine Readable Travel Documents (ICAO) MRTD* establishes international standards for travel documents. An MRTD is an international travel document (e.g., a passport or visa) containing eye-and machine-readable data. ICAO Document 9303 is the international standard for MRTDs.

- The *Physical Access Interagency Interoperability Working Group (PAIIWG)* is a not-for-profit organization establishing and supporting "the **EPCglobal** Network™. The objective is a global standard for real-time, automatic identification of information in the supply chain of any company, anywhere in the world and leading the development of industry-driven standards for the Electronic Product Code™ (EPC). The purpose is to support the use of RFID in today's fast-moving and information-rich trading networks. Additional information can be found at http://www.epcglobalinc.org.
- The *International Committee for Information Technology (INCITS)* organization promotes the effective use of information and communication technology through standardization in a way that balances the interests of all stakeholders and increases the global competitiveness of the member organizations. For more information, visit http://www.INCITS.org/.

There are also numerous biometric-related groups, consultants and system integrators, and privacy organizations.

Summary

- Standards are agreed-upon protocols as determined by officially sanctioned standards-making organizations or user-group consensus. Standards include specifications produced by accredited associations, such as ANSI, ISO, SIA, ETSI, or NIST. Many standards-making organizations are involved to varying degrees in the fields of computer and network security. Practices are the techniques, methodologies, procedures, and processes used in organizations to accomplish a particular job.
- Numerous security standards have been developed for the computer and networking industry and environment. Of particular importance are the ISO 17799 and ISO/IEC 27001 standards. Equally important are the OSI standards provided by the ISO. ANSI and NIST are involved in the development of standards for the security and biometric environments.
- Numerous security protocols, practices, and standards provide the base for computer and network system implementations and operations. These include PCT, TLS, SSL, PGP, and SET. Evaluation criteria are provided in the Orange Book and Red Book.
- Significant standards activity is ongoing in the payment card industry. Issues include interoperability and the use of biometric-enabled ID credentials. The IAB is involved in the PIV project.
- Various professional organizations are active in the biometrics industry and environment. Of particular importance are the International Biometric Industry Association, Biometric Consortium, and the National Biometric Security Project. There are also numerous working groups involved in these processes.

Key Terms

Best practices Methods and procedures found to be most effective; however, best practices are not rules, laws, or standards which people are required to follow.

Biometric Consortium (BC) Serves as a focal point for the research, development, testing, evaluation, and application of biometric-based personal identification/verification technology.

Common Data Security Architecture (CDSA) A security reference standard that provides a way to develop applications, which take advantage of software security mechanisms.

ISO/IEC 27001 The formal standard against which organizations may seek independent certification of their information security management systems.

Open system A computer and communications system capable of communicating with other like systems, transparently.

Open System Interconnection (OSI) The model defines a networking framework for implementing protocols in seven layers.

Payment Card Industry Data Security Standard (PCI DSS) Developed by the major credit card companies as a guideline to help organizations that process card payments, prevent credit card fraud, hacking, and various other security issues.

Personal identity verification (PIV) Refers to the implementation of smart card and biometric-enabled ID credentials for all government employees.

Practices The techniques, methodologies, procedures, and processes used in organizations to accomplish a particular job.

Pretty Good Privacy (PGP) An encryption and digital-signature utility for adding privacy to electronic mail and documents.

Private Communication Technology (PCT) A protocol that provides secure encrypted communications between two applications over a TCP/IP network.

Protocol A set of rules that govern communication between hardware and/or software components.

Secure electronic transaction (SET) An open specification for handling credit card transactions over a network, with emphasis on the Web and Internet.

Secure Sockets Layer (SSL) Cryptographic protocols that provide secure communications on the Internet for such things as Web browsing, e-mail, Internet faxing, instant messaging, and other data transfers.

Specification Defined as the set of documentation that reflects agreements on products, practices, or operations produced by one or more organizations, some for internal usage only, others for use by groups of people, groups of companies, or an entire industry.

Standards Agreed-upon protocols as determined by officially sanctioned standards-making organizations or user-group consensus.

Transport Layer Security (TLS) See predecessor SSL.

Review Questions

1. ________________ are agreed-upon protocols as determined by officially sanctioned standards-making organizations or user-group consensus.
2. A ________________ is a set of rules that govern communication between hardware and/or software components.
3. A ________________ is a set of fundamental and generalized procedures.
4. ________________ are the techniques, methodologies, procedures, and processes used in organizations to accomplish a particular job.
5. ________________ are methods and procedures found to be most effective.
6. ________________ are the basis for privacy best practices, both online and offline.
7. ________________ is a U.S. federal government standard that specifies personal identity verification requirements for federal employees and contractors.
8. The __________ is a private, nonprofit organization that administers and coordinates the U.S. voluntary standardization and conformity assessment system.
9. The ________________ mission is to promote U.S. innovation and industrial competitiveness by advancing measurement science, standards, and technology in ways that enhance economic security, enhance productivity, facilitate trade, etc.
10. The ________________ was developed to help agencies develop and deploy smart card systems.
11. The ________________ defines the application programming interface and service provider interface for a standard biometric technology interface.
12. An ________________ is a computer and communications system capable of communicating with other like systems, transparently.
13. The ________________ was developed by the major credit card companies as a guideline to help organizations that process card payments, prevent credit card fraud, hacking, and various other security issues.
14. ________________ addresses every aspect transaction technology that is linked to the use of identity technology.
15. ________________ addresses every aspect of logical and physical access security.
16. In cryptography, a ________________ is an arrangement that binds public keys with respective user identities by means of a certificate authority.
17. The ________________ describes a set of data elements necessary to support biometric technologies in a common way.
18. The ________________ reader uses an embedded microcontroller, including its own microprocessor and several types of memory, and a magnetic loop antenna.

19. The ______________ is the data element that is the main identifier on the PIV card and that is used by a physical access control system.
20. There are two critical measurement factors which indicate the level of accuracy, or reliability, of any given biometric are ______________ and ______________.

Discussion Exercises

1. Provide an overview of the personal identity verification (PIV) project.
2. Describe the functions that occur at each layer of the OSI model as they relate to integrity and security.
3. Provide an overview of the current NIST activities that relate to biometrics.
4. Provide a current status of the PCI DSS standard. Describe the current security mandates.
5. Provide an overview of the contents of the Smart Card Handbook.
6. Describe the current activities of the BioAPI Consortium.
7. Describe the major components of the Gramm-Leach-Bliley Act.
8. Describe the major requirements of HIPAA.
9. Develop a flow chart of the process for dismissing an employee who is suspected of a theft against the organization.
10. Describe the capabilities of the RFID card.

Hands-On Projects

12-1: Standards and Legal Environment

Time required—30 minutes

Objective—Research standards and legal environment issues that are part of biometric application and system implementations.

Description—Use various Web tools, Web sites, and other tools to identify biometric standards and legal environment issues that occur in commercial and government organizations.

1. Review standards documents ISO 17799 and ISO/IEC 27001. Develop a comparison of these two security standards.
2. Develop a feature/function matrix of PCT, TLS, and SSL.
3. Describe the differences of HTTP, S-HTTP, and HTTPS.
4. Summarize the differences between the Orange Book and the Red Book.
5. Provide an overview of the PKI features and benefits.

12-2: Identifying Standards and Legal Environment Issues

Time required—45 minutes

Objective—Identify standards and legal environment issues that can impact the level of access control and security for OnLine Access assets and resources.

Description—The owners and managers of OnLine Access have identified the necessary implementation and operation issues when implementing a biometric access control solution. They must now look at the standards and legal environment issues that can be encountered when implementing a biometric solution.

1. Research the standards and legal environment issues that are inherent in access control and security solutions.
2. Identify security and access control system standards and legal environment issues that have been encountered in other biometric implementations.
3. Produce a report describing how these issues have impacted the implementation of a biometric solution.
4. Describe how these issues have been resolved.

12-3: Recommendations for the Use of a Biometric Security and Access Control System

Time required—2 hours

Objective—Produce an analysis that summarizes the issues involved when developing and implementing a security and access control system using a biometric solution. Issues are highlighted that could impact the level of security and access control for the assets and resources of the shell company OnLine Access.

Description—The owners and managers of OnLine Access have identified numerous issues that could impact the development, implementation, and operation of the organization when implementing a biometric access control solution. They must now decide if biometrics is a viable solution for security and access control of their assets and resources.

1. Summarize the results obtained from all of the research that are inherent in access control and security solutions.
2. Identify security and access control system hardware and software options that are viable in the OnLine Access environment.
3. Produce a report identifying and describing the pros and cons that must be addressed when deciding whether to implement a biometric solution.
4. Develop a management document that makes a recommendation for further action.

Appendix

Biometrics Vendors

findBIOMETRICS has developed a list of vendors that offer biometric solutions in the following application areas: (http://www.findbiometrics.com/)

- Physical access control
- Logical access control
- Justice/law enforcement
- Time and attendance
- Border control/airports
- HIPAA
- Financial/transactional

This list is subject to change as new vendors enter the market, merge, and cease to be providers. It is essential that research be conducted to ascertain the current providers, as the market is very dynamic and volatile. A review of this provider list shows that a number of vendors provide solutions in multiple biometric-application areas.

Physical Access Control

ADC Systems

Akcess BioMetrics Corp.

Asia Software

bioscrypt

bio-SECURE LLC

BioGuard

Biometrics4ALL, Inc.

BSI2000, Inc.

Cogent

Daon

Datastrip
evivetech
Eyenetwatch
Fingerprint Cards AB
Green Bit
Guardware LLC
identiMetrics
Ident Technologies GmbH
ImageWare
IriTech, Inc.
Kaba Benzing
LG Electronics
Lumidigm
Privaris, Inc.
Sagem Morpho
SENSE Holdings, Inc.
Swiss Tech India Pvt. Ltd.
ZK Software

Logical Access Control

Automa
Bioscrypt
BIO-key
Bio SwipeDrive
Ceelox, Inc.
Cherry
Cogent
Daon
Datastrip
DigitalPersona
Douglas USA
Eyenetwatch
Fingerprint Cards AB
Fujitsu

Futronic

Green Bit

identiMetrics

Ident Technologies GmbH

ImageWare

IriTech, Inc.

ISL Biometrics

LIS Solutions

Lumidigm

Pay By Touch

Privaris, Inc.

Sagem Morpho

SENSE Holdings, Inc.

Sig-Tec Corporation

Tacoma Technology Inc.

Tutis Technologies Ltd.

Zvetco Biometrics

Time and Attendance

Accu-Time Systems, Inc.

ADC Systems

Akcess BioMetrics Corporation

Anviz

Automa

Beijing IDworld Co., Ltd.

Bioscrypt

Cogent

Data-Afrique

DigitalPersona

evivetech

Eyenetwatch

Genus®

Green Bit

Ident Technologies GmbH

Identica Corporation
identiMetrics
ImageWare
LG Electronics
M.B.A. Technology
Ocean Computers, LLC.
RAI Technology, Inc.
Reanex Technologies Pte Ltd
RFLogics
Sagem Morpho
Siliconimageworks Inc.
Speedmissions
Spica International
Staff on Time
TimeRack, Inc.
Universal Biometric
ZK Software

Border Control/Airport

Asia Software
Aware, Inc.
Biometrics4ALL
BIO-key
Bioscrypt
Cogent
Comnetix Inc.
Cross Match Technologies Inc.
Daon
Datastrip
Fingerprint Cards AB
Identica Corporation
ImageWare
IriTech, Inc.
LG Electronics

Lumidigm
Neurotechnology
RFLogics
Sagem Morpho
US Biometrics

Financial/Transactional

ADC Systems
Akoura Biometrics, Inc.
BIO-key
Bioscrypt
Ceelox
Cogent
Comnetix Inc.
DigitalPersona
Ident Technologies GmbH
IdentiPHI
ImageWare
LG Electronics
Privaris
RFLogics
Sagem Morpho
TouchCredit Online Payment Solutions

Justice/Law Enforcement

AFIX Technologies, Inc.
Aware, Inc.
Beijing IDworld Co., Ltd.
Biometrics4ALL, Inc.
Bioscrypt
Cross Match Technologies Inc.
Datastrip
DigitalPersona
Cogent
Green Bit

ImageWare
LG Electronics
Lumidigm
M2SYS Technology
NEC
Sagem Morpho

HIPAA

ADC Systems
Akoura Biometrics, Inc.
BIO-key
Biometrics 2000 Corporation
Bioscrypt
Ceelox
Cherry
Cogent
Datastrip
Identica Corporation
IdentiPHI
ImageWare
Bioscrypt
LG Electronics
RFLogics
SecuriMetrics, Inc.
Sagem Morpho
Treadstone 71

Glossary

10-print card A paper form used to collect both an individual's personal and demographic information along with flat and rolled ink impression fingerprint images. It is mainly used in conjunction with an automated fingerprint identification system (AFIS).

10-print match or identification An absolute positive identification of an individual by corresponding all 10 fingerprints to those in a system of record. Usually performed by an AFIS system and verified by a human fingerprint examiner.

AAA Authentication, Authorization, and Accounting.

ABIS Automated Biometric Identification System.

Access control The process of granting access to information system resources only to authorized users, programs, processes, or other systems.

ACL Access Control List.

Acceptability A biometric feature perceived as nonintrusive by the user.

Accessibility A biometric feature easily presented to an imaging sensor.

Accuracy A catch-all phrase that describes how well a biometric system performs. The actual statistic for performance will vary by task, including verification, open-set identification (watch list), and closed-set identification.

Acquisition device The hardware used to acquire biometric samples.

Automatic Identification and Data Capture (AIDC) The term used to describe data collection by means other than manual notation or keyboard input.

Algorithm A sequence of instructions that tells a biometric system how to solve a particular problem. An algorithm will have a finite number of steps and is typically used by the biometric engine to compute whether a biometric sample and template are a match. Cryptographic algorithms are used to encrypt sensitive data files, to encrypt and decrypt messages, and to digitally sign documents.

ANSI American National Standards Institute.

Application A hardware/software system implemented to satisfy a set of requirements. In this context, an application incorporates a biometric system to satisfy a subset of requirements related to the verification or identification of an end user's identity so that the end user's identifier can be used to facilitate the end user's interaction with the system.

Application profile Conforming subsets or combinations of base standards used to provide specific functions. Application profiles identify the use of particular options available in base standards, and provide a basis for the interchange of data between applications and interoperability of systems.

Application Program Interface (API) A set of services or instructions used to standardize an application. An API is computer code used by an application developer. Any biometric system that is compatible with the API can be added or interchanged by the application developer. APIs are often described by the degree to which they are high level or low level. High level means that the interface is close to the application; low level means that the interface is close to the device.

Arch This type of fingerprint pattern occurs when the friction ridges enter from one side, make a rise in the center, and exit on the opposite side.

Asynchronous multimodality Systems that require that a user verify through more than one biometric in sequence. Asynchronous multimodal solutions include one, two, or three distinct authentication processes. A typical user interaction will consist of a verification on-finger scan; then face-if-finger is successful.

Attempt The submission of a single set of biometric sample to a biometric system for identification or verification.

Audit trail In computer/network systems: Record of events (protocols, written documents, and other evidence) which can be used to trace the activities and usage of a system. Such material is crucial when tracking down successful attacks/attackers, determining how the attacks happened, and being able to use this evidence in a court of law.

Authenticate To verify (guarantee) the identity of a person or entity. To ensure that the individual or organization is really is who it says it is.

Authentication (1) The process whereby a user proves he is who he claims to be. (See *authenticate*). **(2)** A security measure designed to establish the validity of a transmission, message, or originator, or a means of verifying an individual's authorization to receive specific categories of information.

Authentication factors Pieces of information used to verify a person's identity for security purposes. The three most commonly recognized factors are:

- Something you know, such as a password or personal identification number (PIN)
- Something you have, such as a credential, card, or token
- Something you are, such as a fingerprint or other biometric

Authorization The assignment of a privilege or privileges (i.e., access to a building or network) verifying that a known person or entity has the authority to perform a specific operation. Authorization is provided after authentication.

Autocorrelation A proprietary finger scanning technique. Two identical finger images are overlaid in the auto-correlation process to create light and dark areas known as Moiré fringes.

Automated Fingerprint Identification System (AFIS) A highly specialized biometric system that compares a single finger image with a database of finger images. AFIS is predominantly used for law enforcement, but is also being put to use in civil applications. In law enforcement, finger images are typically collected from crime scenes (these fingerprints are known as "latents") or directly from criminal suspects after they are arrested. In civilian applications, finger images may be captured by placing a finger on a scanner or by electronically scanning inked impressions on paper.

Availability The amount of time that a computing resource is available for use by a user or system.

Base standard Fundamental and generalized procedures. They provide an infrastructure that can be used by a variety of applications, each of which can make its own selection from the options offered by them.

Behavior Described as one of a recurring or characteristic pattern of observable actions or behavioral responses.

Best practices Methods and procedures found to be most effective; however, best practices are not rules, laws, or standards which people are required to follow.

Bifurcation The point in a fingerprint where a ridge divides or splits to form two ridges that continue past the point of division for a distance that is at least equal to the spacing between adjacent ridges at the point of bifurcation.

Binning The process of parsing (examining) or classifying data to accelerate and/or improve biometric matching.

BioAPI Biometrics Application Programming Interface.

BioAPI specification Defines the application programming interface and service provider interface for a standard biometric technology interface. Developed by the BioAPI Consortium.

Biometric (1) *adjective* A measurable, physical characteristic or personal behavioral trait used to recognize the identity, or verify the claimed identity, of an individual. Facial images, fingerprints, and iris scan samples are all examples of biometrics. **(2)** *noun* One of various technologies that utilize behavioral or physiological characteristics to determine or verify identity. In the singular: "Finger scan is a commonly used biometric." The plural form also is also acceptable: "Retina scan and iris scan are eye-based biometrics."

Biometric Consortium (BC) Serves as a focal point for the research, development, testing, evaluation, and application of biometric-based personal identification/verification technology.

Biometric data The extracted information taken from the biometric sample and used to either build a reference template or to compare against a previously created reference template. Data encoding a feature or features used in biometric verification.

Biometric fingerprinting Digital image capture of friction ridges and/or a template from friction ridges.

Biometric fusion The process of combining information from multiple biometric readings, either before, during, or after a decision has been made regarding identification or authentication from a single biometric.

Biometric information The stored electronic information pertaining to a biometric. This information can be given in terms of raw or compressed pixels or in terms of some characteristic (i.e., patterns).

Biometric Information Security Management System (ISMS) Includes the biometric policy, security practices, operational security procedures, organizational structure, assigned responsibilities, and resources needed to protect biometric assets.

Biometric Policy (BP) A set of rules that indicate the applicability of a biometric reference template for use by some community or class of applications that have common security requirements.

Biometric Practice Statement (BPS) Describes the security practices that an organization follows during the biometric reference template life cycle, including the business, legal, regulatory, and technical considerations.

Biometric reference data Data stored on a card for the purpose of comparison with the biometric verification data.

Biometric sample The identifiable, unprocessed image or recording of a physiological or behavioral characteristic, acquired during submission, and used to generate biometric templates. Also referred to as biometric data.

Biometric system An automated system capable of capturing a biometric sample from a user, extracting biometric data from that sample, comparing the biometric data with that contained in one or more reference templates, deciding how well they match, and indicating whether or not an authentication of identity has been achieved. Also the integrated biometric hardware and software used to conduct biometric identification or verification.

Biometric template The formatted digital record used to store an individual's biometric attributes. This record typically is a translation of the individual's biometric attributes and is created using a specific algorithm.

Biometric verification The process of verifying, using a one-to-one comparison, the biometric verification data against biometric reference data.

Breeder document A document used as an original source of identity to apply for (or breed) other forms of identity credentials.

BSS Biometric Storage System.

Business Continuity Planning (BCP) A term used involving concerns with failure of IT equipment, or the ability to employ it effectively.

Business Impact Analysis (BIA) A method of systematically assessing the possible impacts resulting from various events or scenarios.

Capacitive sensor Consists of a row/column configuration of tiny metal electrodes. Every column is linked to a pair of sample-and-hold circuits. The fingerprint image is recorded in sequence row by row as each metal electrode acts as one capacitor plate and the contacting finger acts as the second plate. The values of the array are determined by the contour (ridges and valleys) of the fingerprint. The sensor quickly captures several images of the fingerprint and selects the highest quality image.

Capture The process of taking a biometric sample from the user. In the capture, a sample of the user's biometric is acquired using the required sensor (i.e., camera, microphone, fingerprint scanner).

Card (1) A type of physical form factor designed to carry electronic information and/or human readable data.
(2) Under FIPS 201, a dual interface smart card-based ID badge for both physical and logical access that contains within it an integrated circuit chip.

Card holder An individual to whom an ID card is issued or assigned.

Card Management System (CMS) A smart card/token and digital credential management solution that is used to issue, manage, personalize, and support cryptographic smart cards and PKI certificates for identity-based applications throughout an organization.

Card reader Any device that reads encoded information from a card, token, or other identity device and communicates to a host such as a control panel/processor or database for further action.

Card serial number An identifier which is guaranteed to be unique among all identifiers used for a specific purpose. (See *unique identifier*).

Certificate A digital representation of information that at least 1) identifies the certification authority issuing it; 2) names or identifies its subscriber; 3) contains the subscriber's public key; 4) identifies its operational period, and 5) is digitally signed by the certification authority issuing it. (See *digital certificate*).

Certificate Authority (CA) A trusted third party that is responsible for issuing and revoking digital certificates within the public key infrastructure.

Certificate Revocation List (CRL) A list of certificates that have been revoked before their expiration by a certificate authority.

Certification The process of verifying the correctness of a statement or claim and issuing a certificate as to its correctness.

Chain of trust An attribute of a secure ID system that encompasses all of the system's components and processes and assures that the system as a whole is worthy of trust. A chain of trust should guarantee the authenticity of the people, issuing organizations, devices, equipment, networks, and other components of a secure ID system. The chain of trust must also ensure that information within the system is verified, authenticated, protected, and used appropriately.

Challenge The demand for disclosure of one or more attributes related to a subject made by service authority.

Challenge/response A family of protocols in which one party (i.e., a reader) presents a question ("challenge") and another party (i.e., a credential) must provide a valid answer ("response") in order to be authenticated.

Characteristic A distinguishing feature or attribute. Biometric technologies are becoming the foundation of an extensive array of highly secure identification and personal verification solutions.

Checksum A computed value that depends on the contents of a message. The checksum is transmitted with the message. The receiving party can then recompute the checksum to verify that the message was not corrupted during transmission.

Chip Electronic component that performs logic, processing, and/or memory functions.

CIA Confidentiality, Integrity, Accessibility.

CITeR Center for Identification Technology Research.

Cloning The process of creating an identical copy of something.

Closed-set identification A biometric task where an unidentified individual is known to be in the database and the system attempts to determine his/her identity.

Common Biometric Exchange Formats Framework (CBEFF) specification Describes a set of data elements necessary to support biometric technologies in a common way. This data can be placed in a single file used to exchange biometric information between different system components or between systems. The result promotes interoperability of biometric-based application programs and systems developed by different vendors by allowing biometric data interchange.

Common Data Security Architecture (CDSA) A security reference standard that provides a way to develop applications that take advantage of software security mechanisms.

Comparison The process of comparing biometric data with a previously stored reference template or templates. (See *biometric data*).

Computer security A branch of technology known as information security as applied to computers.

Confidence level The degree of likelihood that an identifier refers to a specific individual.

Confidentiality One of the cornerstones of information security, and defined by the International Organization for Standardization (ISO) as "ensuring that information is accessible only to those authorized to have access."

Contact smart card A smart card that connects to the reading device through direct physical contact between the smart card chip and the smart card reader. (See *ISO/IEC 7816*).

Contactless smart card A smart card that communicates with a reader through a radio frequency interface.

Control panel The access control system component that connects to all door readers, door locks, and the access control server. The control panel validates the reader and accepts data. Depending on the overall system design, the control panel may then send the data to the access control server or may have enough local intelligence to determine the user's rights and make the final access authorization.

Control point Any device which is controlled by a physical access system (for example, doors, turnstiles, gates, lights, cameras, elevators). There may be multiple control points for a single access requirement.

Cooperative Refers to the behavior of the "threat" or would-be deceptive user.

Cooperative user An individual who willingly provides a biometric to the biometric system for capture.

Credential (1) Evidence attesting to one's rights, privileges, or evidence of authority. **(2)** In FIPS 201, the PIV card and data elements associated with an individual that authoritatively binds an identity (and, optionally, additional attributes) to that individual. A smart card can store multiple digital credentials.

Crossover Error Rate (CER) A comparison metric for different biometric devices and technologies; the error rate at which FAR equals FRR. The lower the CER, the more accurate and reliable the biometric device.

Cumulative Match Characteristic (CMC) A method of showing measured accuracy performance of a biometric system operating in the closed-set identification task.

Data Items representing facts, text, graphics, bit-mapped images, sound, and analog or digital live-video segments.

Database (DB) A collection of information organized in such a way that a computer program can quickly select desired pieces of data.

Database Management System (DBMS) A complex set of software programs that controls the organization, storage, management, and retrieval of data in a database.

Data integrity The condition in which data is identically maintained during any operation, such as transfer, storage, and retrieval.

Data mining The process of searching for unknown information or relationships in large databases.

Data vaulting The process of sending data off site, where it can be protected from hardware failures, theft, and other threats. Several companies now offer Web backup services that compress, encrypt, and periodically transmit a customer's data to a remote vault. In most cases, the vaults have auxiliary power supplies, powerful computers, and manned security.

Data warehouse A database that consolidates an organization's data.

Data warehousing A software strategy in which data are extracted from large transactional databases and other sources and stored in smaller databases, which makes analysis of the data easier.

Decision The result of the comparison between the score and the threshold. The decisions a biometric system can make include match, non-match, and inconclusive, although varying degrees of strong matches and non-matches are possible.

Degrees of freedom A statistical measure of the uniqueness of biometric data.

Delta point Part of a fingerprint pattern that looks similar to the Greek letter delta.

Demarcation The physical point where a regulated network service is provided to the user.

Detection error trade-off (DET) curve A graphical plot of measured error rates.

DHS Department of Homeland Security.

Difference score A value returned by a biometric algorithm indicating the degree of difference between a biometric sample and a reference.

Digital certificate (or public key certificate) Digital documents (i.e., information such as the name of the person or an organization and their address) attesting to the binding of a public key to an individual or other entity. Digital certificates allow verification of the claim that a specific public key does in fact belong to a specific individual.

Digital signature The number derived by performing cryptographic operations on the text to be signed. This operation, or hash function (also called hash algorithm), is performed on the binary code of the text. The result is known as the message digest, and always has a fixed length. A signature algorithm is applied to the message digest, resulting in the digital signature.

Disaster recovery The process, policies, and procedures of restoring operations critical to the resumption of operations, including regaining access to data (records, hardware, software, etc.), communications (incoming, outgoing, toll-free, fax, etc.), workspace, and other business processes after a natural or human-induced disaster.

Disaster Recovery Plan (DRP) Includes plans for coping with the unexpected or sudden loss of communications and/or key personnel, focusing on data protection.

Door reader The device on each door that communicates with a card or credential and sends data from the card to the controller for decision on access rights.

Door strike The electronic lock on each door that is connected to the controller.

DOS Denial Of Service.

Dots Per Inch (DPI) The number of pixels used to define an image. A measure of image quality is resolution in dots per inch or DPI.

DSV Dynamic Signature Verification.

Dual interface card A smart card that has a single smart card chip with two interfaces (a contact and a contactless interface) using shared memory and chip resources.

Eavesdropping The interception of communications between a reader and a credential during transmission by unintended recipients. Messages can be protected against eavesdropping by employing a security service usually implemented by encryption.

EFTS Electronic Fingerprint Transmission Specification.

Electronic access control Uses computers to solve the limitations of mechanical locks and keys.

Electronic cash A scheme that makes purchasing easier over the Web.

Electronic Data Interchange (EDI) A process whereby standardized forms of e-commerce documents are transferred between diverse and remotely located computer systems.

Encryption The scrambling of data so that it becomes difficult to unscramble or decipher. Scrambled data is called cipher text, as opposed to unscrambled data, which is called plaintext. Unscrambling cipher text is called decryption. Data encryption is done by the use of an algorithm and a key. The key is used by the algorithm to scramble and unscramble the data. There are two main types of encryption—asymmetric (or public key) and symmetric (or secret key).

Enrollee A person who has a biometric reference template stored in a biometric package.

Enrollment The process of collecting biometric samples from a user and the subsequent preparation, encryption, and storage of biometric reference templates representing that person's identity.

Enterprise Single Sign-On (E-SSO) A system designed to minimize the number of times that a user must type their ID and password to sign into multiple applications. The E-SSO solution automatically logs users in and acts as a password filler where automatic login is not possible. Each client is typically given a token that handles the authentication; in other E-SSO solutions each client has E-SSO software stored on their computer to handle the authentication. An E-SSO authentication server is also typically implemented into the enterprise network.

ePassport A travel document that contains an integrated circuit chip based on international standard ISO/IEC 14443 and that can securely store and communicate the ePassport holder's personal information to authorized reading devices.

Equal Error Rate (EER) A statistic used to show biometric performance, typically when operating in the verification task.

ETSI European Telecommunications Standards Institute.

Excite field The RF field or electromagnetic field constantly transmitted by a contactless door reader. When a contactless card is within range of the excite field, the internal antenna on the card converts the field energy into electricity that powers the chip. The chip then uses the antenna to transmit data to the reader.

Extraction The process by which the biometric sample captured in the previous block is transformed into an electronic representation. During enrollment this electronic representation is known as the biometric template. During the authentication process, it is known as the live sample.

Extranet A private, Internet-like network that a company operates to conduct business with its employees, customers, vendors, and suppliers.

FaceIt® A facial recognition software engine that allows computers to rapidly and accurately detect and recognize faces.

Failure To Acquire (FTA) Results from the failure of a biometric system to capture and/or extract usable information from a biometric sample.

Failure to acquire rate The frequency of a failure to acquire.

Failure To Enroll (FTE) Occurs when there is a failure of a biometric system to form a proper enrollment reference for an end user.

Failure to enroll rate The probability that a biometric system will have a failure to enroll.

False acceptance When a biometric system incorrectly identifies an individual or incorrectly authenticates an imposter against a claimed identity.

False Acceptance Rate (FAR) The probability that a biometric system will incorrectly identify an individual or will fail to reject an imposter.

False alarm rate A statistic used to measure biometric performance when operating in the open-set identification (sometimes referred to as watch list) task.

False Non-Match Rate (FNMR) Alternative term for false rejection rate. Used in the context of negative claim of identity.

False rejection When a biometric system fails to identify an enrollee or fails to verify the legitimate claimed identity of an enrollee.

False Rejection Rate (FRR) The probability that a biometric system will fail to identify an enrollee, or verify the legitimate claimed identity of an enrollee.

Faraday cage An enclosure formed by conducting material, or by a mesh of such material, that blocks out external static electrical fields. Any electric field will cause the charges to rearrange so as to completely cancel the field's (RF signal) effects in the cage's interior.

FASC-N Federal Agency Smart Credential Number.

Feature(s) Distinctive mathematical characteristic(s) derived from a biometric sample; used to generate a reference.

Feature extraction The automated process of locating and encoding distinctive characteristics from a biometric sample in order to generate a template.

Federal Agency Smart Credential Number (FASC-N) The data element that is the main identifier on the PIV card and that is used by a physical access control system.

Federated identity In information technology (IT), federated identity has two general meanings: **(1)** The virtual reunion, or assembled identity, of a person's user information (or principal), stored across multiple distinct identity management systems. Data is joined together by use of the common token, usually the user name; **(2)** The process of a user's authentication across multiple IT systems or even organizations. (See *FIPS 201*.)

FERET FacE REcognition Technology.

Filtering Involves partitioning the database based on information not contained in the biometric itself.

Fingerprint The image left by the minute ridges and valleys found on the hand of every person. In the fingers and thumbs, these ridges form patterns of loops, whorls, and arches.

Fingerprint scanning Acquisition and recognition of a person's fingerprint characteristics for identifying purposes. This process allows the recognition of a person through quantifiable physiological characteristics that detail the unique identity of an individual.

Fingerprint sensor Part of a biometric device used to capture a fingerprint image for subsequent processing.

FIPS Federal Information Processing Standard.

FIPS 201 Federal Information Processing Standard 201.

Firewall Permits the flow of e-mail traffic, product and service information, job postings, and so forth, while restricting unauthorized access to private information residing on the intranet.

Form factor The physical device that contains the smart card chip. Smart chip–based devices can come in a variety of form factors, including plastic cards, key fobs, wristbands, wristwatches, PDAs, and mobile phones.

FpVTE FingerPrint Vendor Technology Evaluation.

FRGC Face Recognition Grand Challenge.

Friction ridge The ridges on the skin of the fingers, toes, palms, and soles of the feet that make contact with an incident surface under normal touch. On the fingers, the unique patterns formed by friction ridges make up fingerprints.

Fused biometrics Single or multiple sensors are used to collect different biometrical information, such as a face image and a fingerprint image.

Fusion The union of different biometric modalities resulting in an integrated system.

Gallery The biometric system's database, or set of known individuals, for a specific implementation or evaluation experiment.

Guideline Typically, a collection of system specific or procedural specific "suggestions" for best practice.

HA-API Human Authentication Application Program Interface.

Habituated Describes the level of experience and expertise the intended user possesses in using the system.

Hacking The act of gaining illegal or unauthorized access to a computer system or network.

Hamming distance The number of noncorresponding digits in a string of binary digits; used to measure dissimilarity.

Hardware The physical part of a computer, including the digital circuitry, as distinguished from the computer software, which executes within the hardware.

Hash algorithm A software algorithm that computes a value (hash) from a particular data unit in a manner that enables detection of intentional/unauthorized or unintentional/accidental data modification by the recipient of the data.

Hash function A reproducible method of turning some kind of data into a relatively small number that may serve as a digital "fingerprint" of the data.

Head-end system The physical access control server, software, and database(s) used in a physical access control system.

HIPAA Health Insurance Portability and Accountability Act.

Hot swapping The ability to add and remove devices to a computer while the computer is running and have the operating system automatically recognize the change.

HTTP HyperText Transfer Protocol.

Hybrid card A smart card that contains two smart card chips—both contact and contactless chips—that are not interconnected.

IAB Interagency Advisory Board.

IAFIS Integrated Automated Fingerprint Identification System.

ICAO International Civil Aviation Organization.

ICC Integrated Circuit Card. ICC typically refers to a plastic (or other material) card containing an integrated circuit which is compatible to ISO/IEC 7816.

ICE Iris Challenge Evaluation.

ICSA International Computer Security Association.

Identification (1) The one-to-many process of comparing a submitted biometric sample against all biometric reference templates on file to determine whether it matches any of the templates and, if so, the identity of the enrollee whose template was matched. The biometric system using the one-to-many approach seeks to find an identity in a database rather than authenticate a claimed identity. **(2)** Procedures and mechanisms which allow an entity external to a system resource to notify that system of its identity.

Identification card A card identifying its holder and issuer which may carry data required as input for the intended use of the card and for transactions based thereon.

Identification mode The biometric system that identifies a person from the entire enrolled population by searching a database for a match based solely on the biometric.

Identifier A unique data string used as a key in the biometric system to name a person's identity and its associated attributes. An example of an identifier would be a passport number.

Identity The common-sense notion of personal identity, including a person's name, personality, physical body, and history (which are attributes such as nationality, educational achievements, employer, security clearances, and financial and credit history, etc.). In a biometric system, identity is typically established when the person is registered in the system through the use of so-called "breeder documents," such as a birth certificate, passport, etc.

Identity (ID) management A broad administrative area that deals with identifying individuals in a system, such as a country, network, or enterprise.

Identity Management System (IDMS) Composed of one or more computer systems or applications that manage the identity registration, verification, validation, and issuance process, as well as the provisioning and deprovisioning of identity credentials.

Identity theft The appropriation of another's personal information to commit fraud, steal the person's assets, or pretend to be the person.

IDS Intrusion Detection System.

IEC International Electro/technical Commission.

Imposter A person who submits a biometric sample in an intentional or inadvertent attempt to pass himself off as another person who is a legitimate enrollee.

Indifferent user An individual who knows that a biometric sample is being collected and does not attempt to help or hinder the collection of the sample.

Information Data that has been processed to add or create meaning and knowledge for the person receiving it.

Information assurance (IA) Information operations that protect and defend information and information systems by ensuring their confidentiality, authentication, availability, integrity, and nonrepudiation.

Information security Protecting information and information systems from unauthorized access, use, disclosure, disruption, modification, or destruction.

Integrity Implies protection against unauthorized modification or destruction of information. This defines a state in which information has remained unaltered from the point it was produced by a source, during transmission, storage, and eventual receipt by the destination.

Interoperability (1) The ability of two or more systems or components to exchange information and to use the information that has been exchanged. **(2)** For the purposes of FIPS 201, the ability for any government facility or information system, regardless of the PIV issuer, to verify a cardholder's identity using the credentials on the PIV card.

Intranet An organization's internal computer network.

Intrusion detection The art of detection and responding to computer and network misuse.

Intrusion Prevention System (IPS) A preemptive approach to network security used to identify potential threats and respond to them swiftly.

ISMS Information Security Management System.

ISO International Standards Organization.

ISO 19092 Provides a technology-specific extension to the ISO/IEC 17799/27001 Code of practice for information security management.

ISO/IEC 27001 The formal standard against which organizations may seek independent certification of their information security management systems.

ITSEC Information Technology Security Evaluation Criteria.

Key In encryption and digital signatures, a value used in combination with a cryptographic algorithm to encrypt or decrypt data.

Latent print The transferred impression of friction ridge detail that is not readily visible; a generic term used for questioned friction ridge detail.

Least privilege A principle that may be used to guide the design and administration of a computer system or enterprise network.

Live capture The process of capturing a biometric sample by an interaction between an end user and a biometric system.

Live scan A scan of a fingerprint or palm print taken directly from a subject's hand.

Liveness detection A technique used to ensure the biometric sample submitted is from an end user.

Loop The fingerprint pattern that occurs when the friction ridges enter from either side, curve sharply, and pass out near the same side they entered.

MAC (1) Mandatory Access Control. **(2)** Message Authentication Code.

Machine Readable Travel Documents (MRTD) ICAO establishes international standards for travel documents. An MRTD is an international travel document (i.e., a passport or visa) containing eye- and machine-readable data. ICAO Document 9303 is the international standard for MRTDs.

Mandatory access control An access control technique that assigns a security level to all resources (e.g., information, parts of a building), assigns a clearance level to all potential users requiring access, and ensures that only users with the appropriate clearance level can access a requested resource.

Man-in-the-middle attack An attack on an authentication protocol in which the attacker is positioned between the individual seeking authentication and the system verifying the authentication. In this attack, the attacker attempts to intercept and alter data traveling between the parties.

Match A decision that a biometric sample and a stored template originates from the same human source, based on their high level of similarity.

Matching The comparison of biometric templates to determine their degree of similarity or correlation. A match attempt results in a score that, in most systems, is compared against a threshold. If the score exceeds the threshold, the result is a match; if the score falls below the threshold, the result is a non-match.

MD5 Message Digest 5.

Memory card Typically a smart card or any pocket-sized card with an embedded integrated circuit or circuits containing nonvolatile memory storage components and perhaps some specific security logic.

Message authentication code (MAC) A short piece of information used to support authentication of a message. A MAC algorithm accepts as input a secret key and an arbitrary-length message to be authenticated, and outputs a MAC (sometimes known as a tag or checksum). The MAC value protects both a message's integrity as well as its authenticity by allowing verifiers (who also possess the secret key) to detect any changes to the message content.

Metadata Information that describes data in a database.

Microcontroller A highly integrated computer chip that contains all of the components comprising a controller.

Microprocessor card Typically a smart card or any pocket-sized card with an embedded integrated circuit or circuits containing memory and microprocessor components.

Mimic The presentation of a live biometric measure in an attempt to fraudulently impersonate someone other than the submitter.

Minutiae point The point where a friction ridge begins, terminates, or splits into two or more ridges. Minutiae are friction ridge characteristics that are used to individualize a fingerprint image.

Modality A type or class of biometric system.

Model A detailed description or scaled representation of one component of a larger system that can be created, operated, and analyzed to predict actual operational characteristics of the final produced component.

Multiapplication card A smart card that runs multiple applications using a single card (for example, physical access, logical access, data storage, and electronic purse).

Multibiometrics A multimodal biometric system, which integrates face recognition, fingerprint verification, and speaker verification in making a personal identification.

Multifactor authentication The use of multiple techniques to authenticate an individual's identity. This usually involves combining two or more of the following: something the individual has (i.e., a card or token); something the individual knows (i.e., a password or personal identification number); and something the individual is (i.e., a fingerprint or other biometric measurement).

Multifactor reader A smart card reader that includes a PIN pad, biometric reader, or both to allow multifactor authentication.

Multimodal biometrics Fusing two or more biometric modalities, like face images and voice timber.

Multimodal biometric system A biometric system in which two or more of the modality components, such as biometric characteristic, sensor type, or feature extraction algorithm, occur in multiple components.

Multisensory Refers to combining data from two or more biometric sensors, such as synchronized reflectance-based and temperature-based face images.

Multitechnology card An ID card possessing two or more ID technologies that are independent and that don't interact or interfere with one another. An example is a card that contains a smart card chip and a magnetic stripe.

Multitechnology reader A card reader/writer that can accommodate more than one card technology in the same reader (i.e., both ISO/IEC 14443 and ISO/IEC 15693 contactless smart card technologies or both 13.56 MHz and 125 kHz contactless technologies).

Mutual authentication For applications requiring secure access, the process that is used for the smart card–based device to verify that the reader is authentic and to prove its own authenticity to the reader before starting a secure transaction.

NBCT National Biometric Test Center.

NBSP National Biometric Security Project.

NCIC National Crime Information Center.

Network Access Control (NAC) A series of technologies that may include software, hardware, or servers that monitor devices accessing a network.

Neural network A type of algorithm that learns from past experience to make decisions.

NIST National Institute of Standards and Technology.

Noncooperative Refers to the behavior of the "threat" or would-be deceptive user.

Nonhabituated Describes the level of experience and expertise the intended user possesses in using the system.

Nonrepudiation The ability to ensure and have evidence that a specific action occurred in an electronic transaction. The message originator cannot deny sending a message or that a party in a transaction cannot deny the authenticity of her signature.

NOS Network Operating System.

NSA National Security Administration.

OCSP Online Certificate Status Protocol.

Off-card Refers to data that is not stored on the ID card or to a computation that is not performed by the integrated circuit on the ID card.

OLAP Online Analytical Processing.

On-card Refers to data that is stored on the ID card or to a computation that is performed by the integrated circuit chip on the ID card.

One-Time Password (OTP) Passwords that are used once and then discarded. Each time the user authenticates to a system, a different password is used, after which that password is no longer valid.

One-to-many (1-N) A synonym for identification.

One-to-one (1-1) A synonym for verification.

Open ID A decentralized digital identity system, in which any user's online identity is given by, for example, a URL (such as for a blog or a home page) and can be verified by any server running the protocol. Users are able to clearly control what pieces of information can be shared such as their name, address, or phone number.

Open-set identification A biometric task that more closely follows operational biometric system conditions to 1) determine if someone is in a database and 2) find the record of the individual in the database. (See *watch list*).

Open system A computer and communications system capable of communicating with other like systems, transparently.

Open System Interconnection (OSI) The model defines a networking framework for implementing protocols in seven layers.

Operational evaluation One of the three types of performance evaluations. The primary goal is to determine the workflow impact seen by the addition of a biometric system.

Optical sensor Optics-based fingerprint systems translate the illuminated images of fingerprints into digital code for further software processing such as enrollment and authentication.

OSF Open Software Foundation.

PACS Physical Access Control System.

PAIIWG Physical Access Interagency Interoperability Working Group.

Password A form of secret authentication data that is used to control access to a resource. The password is kept secret from those not allowed access, and those wishing to gain access are tested on whether or not they know the password and are granted or denied access accordingly. Typically referred to as "something you know" for single factor authentication. A password is a unique string of characters that a user types in as an identification code. The system compares the code against a stored list of authorized passwords and users; if the code is legitimate, the system allows the user access at whatever security level has been approved for the owner of that password.

Pattern classification A technique aimed at reducing the computational overhead of pattern matching.

Pattern matching Involves comparing the live sample to the reference sample in the database.

Pattern recognition Aims to classify data (patterns) based on either *a priori* knowledge or on statistical information extracted from the patterns.

Payment Card Industry Data Security Standard (PCI DSS) Developed by the major credit card companies as a guideline to help organizations that process card payments prevent credit card fraud, hacking, and various other security issues.

PCT Private Communication Technology.

PDES Person Data Exchange Standard.

Personal Computer/Smart Card (PC/SC) Defines how to integrate smart card readers and smart cards with the computing environment and how to allow multiple applications to share smart card devices.

Personal Identification Number (PIN) (1) A number used in conjunction with an access control system as a secondary credential by the user to ensure the holder of the access control card is the authorized user. **(2)** A secret that an individual memorizes and uses to authenticate his or her identity or to unlock certain information stored on an ID card (i.e., the biometric information). PINs are generally only decimal digits.

Personally Identifiable Information (PII) Any piece of information which can potentially be used to uniquely identify, locate, or contact a person or to steal the identity of a person.

Personal Identity Verification (PIV) Refers to the implementation of smart card and biometric-enabled ID credentials for all government employees.

Personal Identity Verification (PIV) card The physical artifact (i.e., identity card, smart card) issued to an individual that contains printed and stored identity credentials (i.e., photograph, cryptographic keys, digitized fingerprint representation) so that the claimed identity of the cardholder can be verified against the stored credentials by another person (human-readable and verifiable) or an automated process (computer-readable and verifiable).

Pharming A cyber attack that directs people to a fraudulent Web site by poisoning the domain name system server.

Phishing A cyber attack that directs people to a fraudulent Web site to collect personal information for identity theft.

Physical Access Control System (PACS) A system composed of hardware and software components that controls access to physical facilities such as buildings, rooms, airports, and warehouses.

Physiology The branch of biology that deals with the functions and vital processes of their parts and organs.

Pixels Per Inch (PPI) The spatial resolution of a digital image. Often referred to incorrectly as dots per inch. (See *DPI*).

Platen The surface on which a finger is placed during optical finger image capture.

Plug-and-play An industry-wide standard for add-on hardware that indicates that it will configure itself, eliminating the need to set jumpers and making installation of the product quick and easy.

Policy Typically, a document that outlines specific requirements or rules that must be met.

Practices The techniques, methodologies, procedures, and processes used in organizations to accomplish a particular job.

Pretty Good Privacy (PGP) An encryption and digital-signature utility for adding privacy to electronic mail and documents.

Privacy The ability of an individual or group to keep their lives and personal affairs out of public view, or to control the flow of information about themselves.

Privacy-invasive A privacy-invasive system facilitates or enables the usage of personal data in a fashion inconsistent with generally accepted privacy principles.

Privacy-neutral A privacy-neutral system is one in which privacy is not an issue, or in which the potential privacy impact is slight. Privacy-neutral systems are difficult to misuse from a privacy perspective, but do not have the capability to protect personal privacy.

Privacy-protective A privacy-protective system is one used to protect or limit access to personal information, or which provide a means for an individual to establish a trusted identity.

Privacy-sympathetic A privacy-sympathetic system is one that limits access to and usage of personal data and in which decisions regarding design issues such as storage and transmission of biometric data are informed, if not driven, by privacy concerns.

Private Communication Technology (PCT) A protocol that provides secure encrypted communications between two applications over a TCP/IP network.

Private key The secret part of an asymmetric key pair that is used to create digital signatures and, depending upon the algorithm, to decrypt messages, files or other information encrypted (for confidentiality) with the corresponding public key.

Privilege An authorization or right granted by an application authority for an individual or group to perform an action.

Protocol A set of rules that govern communication between hardware and/or software components.

Proximity cards A generic name for contactless integrated circuit devices typically used for security access or payment systems. It can refer to 125 kHz RFID devices or 13.56 MHz contactless smart cards. (See *ISO/IEC 14443*).

Public key The public part of an asymmetric key pair that is used to verify signatures created with its corresponding private key. Depending on the algorithm, public keys are also used to encrypt messages, files, or other information that can then be decrypted with the corresponding private key. The user releases this key to the public who can use it to encrypt messages to be sent to the user and to verify the user's digital signature. Compare with private key.

Public key certificate A digital document that is issued and digitally signed by the private key of a certificate authority (CA) and that binds an attribute of a subject to a public key.

Public (asymmetric) key cryptography A type of cryptography that uses a pair of mathematically-related cryptographic keys. The public key can be made available to anyone and can encrypt information or verify a digital signature. The private key is kept secret by its holder and can decrypt information or generate a digital signature.

Public Key Infrastructure (PKI) The architecture, organization, techniques, practices, and procedures that collectively support the implementation and operation of a certificate-based, public-key cryptographic system. There are four basic components to the PKI: the certificate authority (CA), which is responsible for issuing and verifying digital certificates; the registration authority (RA), which provides verification to the CA prior to issuance of digital certificates; one or multiple directories to hold certificates (with public keys); and a system for managing the certificates.

Radio Frequency (RF) Any frequency within the electromagnetic spectrum associated with radio wave propagation. Many wireless communications technologies are based on RF, including radio, television, mobile phones, wireless networks, and contactless payment cards and devices.

Radio Frequency Identification (RFID) Technology that is used to transmit information about objects wirelessly using radio waves. RFID technology is composed of two main pieces: the device that contains the data and the reader that captures such data. The device has a silicon chip and an antenna and the reader also has an antenna. The device is activated when put within range of the reader. The term RFID has been most commonly associated with tags used in supply chain applications in the manufacturing and retail industries.

Reader Any device that communicates information or assists in communications from a card, token, or other identity document and transmits the information to a host system, such as a control panel/processor or database for further action.

Receiver Operating Characteristics (ROC) A method of showing measured accuracy performance of a biometric system.

Recognition The generic term used in the description of biometric systems (e.g., face recognition or iris recognition) relating to their fundamental function.

Record Consists of the template and other information about the end user (e.g. name, access permissions).

Response A message returned by the integrated circuit chip to the terminal after the processing of a command message received by the chip.

Response time The time used by a biometric system to return a decision on identification or verification of a biometric sample.

RFID tag (labels) Simple, low-cost, and disposable electronic devices that are used to identify animals, track goods logistically, and replace printed bar codes at retailers. RFID tags include an integrated circuit that typically stores a static number (an ID) and an antenna that enables the chip to transmit the stored number to a reader.

Ridge ending Occurs at the point on a fingerprint or palm print where a friction ridge begins or ends without splitting into two or more continuing ridges.

Risk analysis A technique to identify and assess factors that may jeopardize the success of a project or achieving a goal.

Robustness A characterization of the strength of a security function, mechanism, service, or solution and the assurance or confidence that it is implemented and functioning correctly.

Role The actions and activities assigned to or required or expected of a person or group.

Role-Based Access Control (RBAC) Provides access to resources based on a user's assigned role.

S/MIME Secure Multipurpose Internet Mail Extensions.

Scenario evaluation One of the three types of performance evaluation. The primary goal is to measure performance of a biometric system operating in a specific application.

Schema Data is stored in a database according to a predefined format and an established set of rules.

Schengen visa A system that uses biometrics to ensure that an applicant has not previously applied for a visa at another embassy or under an assumed identity.

Score A number indicating the degree of similarity or correlation of a biometric match. Traditional authentication methods—passwords, PINs, keys, and tokens—are binary, offering only a strict yes/no response. This is not the case with most biometric systems. Nearly all biometric systems are based on matching algorithms that generate a score subsequent to a match attempt. This score represents the degree of correlation between the verification template and the enrollment template.

Screening Involves a reasonable examination of persons, cargo, vehicles, or baggage for the protection of the vessel, its passengers and crew.

Secret key A key used with symmetric cryptographic techniques by a set of specified entities.

Secure Electronic Transaction (SET) An open specification for handling credit card transactions over a network, with an emphasis on the Web and Internet.

Secure Sockets Layer (SSL) Cryptographic protocols that provide secure communications on the Internet for such things as Web browsing, e-mail, Internet faxing, instant messaging, and other data transfers.

Security management Concerns the ability to monitor and control access to network resources, including generating, distributing, and storing encryption keys.

Sensor Hardware found on a biometric device that converts biometric input into a digital signal and conveys this information to the processing device.

Sensor aging The gradual degradation in performance of a sensor over time.

Service Level Agreement (SLA) A formally negotiated agreement between two parties.

SHA Secure Hash Algorithm.

S-HTTP Secure Hypertext Transfer Protocol.

SIA Semiconductor Industry Association.

Similarity score A value returned by a biometric algorithm that indicates the degree of similarity or correlation between a biometric sample and a reference.

Single error rates Error rates state the likelihood of an error (false match, false non-match, or failure to enroll) for a single comparison of two biometric templates or for a single enrollment attempt. This can be thought of as a "single" error rate.

Skimming The practice of obtaining information from a data storage device without the owner's knowledge. Skimming is typically associated with magnetic stripe-based credit cards.

Smart card A device that includes an embedded integrated circuit that can be either a secure microcontroller or equivalent intelligence with internal memory or a memory chip. The card connects to a reader with direct physical contact or with a remote contactless radio frequency interface. With an embedded microcontroller, smart cards have the unique ability to store large amounts of data, carry out their own on-card functions (i.e., encryption and mutual authentication), and interact intelligently with a smart card reader.

Sniffer A LAN protocol analyzer that supports a variety of hardware to include Ethernet and Token Ring LAN topologies.

Sniffing The act of auditing or watching computer network traffic. Hackers may use sniffing programs to capture data that is being communicated on a network (i.e., usernames and passwords).

Software Includes the programs or other "instructions" that a computer needs to perform specific tasks.

Specification Defined as the set of documentation that reflects agreements on products, practices, or operations produced by one or more organizations (or groups of cooperating entities), some for internal usage only, others for use by groups of people, groups of companies, or an entire industry.

Spoofing The ability to fool a biometric sensor into recognizing an illegitimate user as a legitimate user (verification) or into missing an identification of someone who is in the database.

Standard Typically, a collection of system-specific or procedural-specific requirements that must be met by everyone. Specifications produced by accredited associations, such as ANSI, ISO, SIA, ETSI, or NIST. In the United States, the use of standards is typically optional and multiple standards can be developed on the same subject.

Structured Query Language (SQL) A standard language for making interactive queries from and updating a database.

Subject A person, system, or object with associated attributes.

Submission The process whereby a user provides behavioral or physiological data in the form of biometric samples to a biometric system. A submission may require looking in the direction of a camera or placing a finger on a platen. Depending on the biometric system, a user may have to remove eyeglasses, remain still for a number of seconds, or recite a pass phrase in order to provide a biometric sample.

Symmetric cryptographic technique A cryptographic technique using the same secret key for both the originator's and the recipient's operation. Without the secret key, it is computationally infeasible to compute either operation.

Symmetric keys Keys that are used for symmetric (secret) key cryptography. In a symmetric cryptographic system, the same secret key is used to perform both the cryptographic operation and its inverse (for example, to encrypt and decrypt, or to create a message authentication code and to verify the code).

Synchronous multimodality The use of multiple biometric technologies in a single authentication process. For example, biometric systems exists which use face and voice simultaneously, reducing the likelihood of fraud and reducing the time needed to verify.

TCSEC Trusted Computer System Evaluation Criteria.

Technology evaluation One of the three types of performance evaluation. The primary goal is to measure performance of biometric systems, typically only the recognition algorithm component, in general tasks.

Template Data that represents the biometric measurement of an enrollee, used by a biometric system for comparison against subsequently submitted biometric samples. Biometric data after it has been processed from its original representation (using a biometric feature extraction algorithm) into a form that can be used for automated matching purposes (using a biometric matching algorithm).

Three-factor authentication Consists of some method of determining "something you are," "something you have," and "something you know."

Threshold A predefined number, often controlled by a biometric system administrator, which establishes the degree of correlation necessary for a comparison to be deemed a match.

Throughput The amount of digital data per time unit that is delivered over a physical or logical link or that is passing through a certain network node.

Throughput rate The number of biometric transactions that a biometric system processes within a stated time interval.

TNI Trusted Network Interpretation.

Token A physical device that carries an individual's credentials. The device is typically small, for easy transport, and usually employs a variety of physical and/or logical mechanisms to protect against modifying legitimate credentials or producing fraudulent credentials. Examples of tokens include picture ID cards (i.e., a state driver's license), smart cards, and USB devices.

Token-based authentication A security technique that authenticates users who are attempting to log in to some secure computer or network resource.

TPM Trusted Platform Module.

Transponder A wireless communications device that detects and responds to an RF signal.

Transport Layer Security (TLS) See predecessor, *Secure Sockets Layer (SSL).*

True accept rate A statistic used to measure biometric performance when operating in the verification task.

True reject rate A statistic used to measure biometric performance when operating in the verification task.

Trust The composite of availability, performance, and security, which includes the ability to execute processes with integrity, secrecy, and privacy.

Trusted Computing Group (TCG) A vendor consortium founded in 2003 to promote the use of vendor-neutral specifications.

Trust model A framework for an organizational secure systems policy that addresses resource authorization.

Trust relationship The link between two entities on a network (e.g. two servers) that enables a user with an account in one domain to have access to resources in another domain.

TWIC Transportation Worker Identification Credential.

UPS Uninterruptible Power Supply.

Validation The process of demonstrating that the system under consideration meets in all respects the specification of that system.

Valley The area of a fingerprint surrounding a friction ridge that does not make contact with an incident surface under normal touch.

Verification (1:1, matching, authentication) The process of establishing the validity of a claimed identity by comparing a verification template to an enrollment template. Verification requires that an identity be claimed, after which the individual's enrollment template is located and compared with the verification template. Verification answers the question, "Am I who I claim to be?" Some verification systems perform very limited searches against multiple enrollee records.

Verification mode The biometric system that authenticates a person's claimed identity from their previously enrolled pattern.

Virtual Private Network (VPN) A private data network that makes use of the public telecommunications infrastructure, which maintains privacy through the use of tunneling and security procedures.

VoiceVault Uses spoken words to calculate vocal measurements of an individual's vocal tract.

VPR Vascular Pattern Recognition.

Vulnerability The potential for the function of a biometric system to be compromised by intent (fraudulent activity); design flaw (including usage error); accident; hardware failure; or external environmental condition.

Wallet An electronic cash component that resides on the user's computer or on another device such as a smartcard.

Watch list A term sometimes referred to as open-set identification, describes one of the three tasks biometric systems perform.

Wavelet scalar quantization A compression algorithm used to reduce the size of reference templates.

Web Access Management (WAM) (1) A system that replaces the sign-on process on various Web applications, typically using a plug-in on a front-end Web server. **(2)** Systems that replace the sign-on process on various Web applications, typically using a plug-in on a front-end Web server. The systems authenticate a user once, and maintain that user's authentication state even as the user navigates between applications. These systems normally also define user groups and attach users to privileges on the managed systems.

Whorl The fingerprint pattern that occurs when the ridges are circular or nearly circular.

End Notes

Chapter 1

1. Technovelgy.com, "Characteristics of Successful Biometric Identification Methods," http://www.technovelgy.com/ct/Technology-Article.asp?ArtNum=11
2. Ibid.
3. Electronic Frontier Foundation, "Biometrics Resources," http://w2.eff.org/Privacy/Surveillance/biometrics/biometric-resources.html

Chapter 2

1. GlobalSecurity.org, "Emerging Biometric Technologies," http://www.globalsecurity.org/security/systems/biometrics-emerging.htm
2. Ibid.
3. Technovelgy.com, "Characteristics of Successful Biometric Identification Methods," http://www.technovelgy.com/ct/Technology-Article.asp?ArtNum=11
4. University of Cambridge, "Computer Laboratory," http://www.cl.cam.ac.uk/
5. Biometrics Catalog, "Introduction to Biometrics," http://www.biometricscatalog.org/Introduction/default.aspx
6. Ibid.
7. University of Cambridge, "Computer Laboratory," http://www.cl.cam.ac.uk/
8. L-1 Identity Solutions, "Biometrics Products," http://www.l1id.com/pages/17-biometrics
9. GlobalSecurity.org, "Emerging Biometric Technologies," http://www.globalsecurity.org/security/systems/biometrics-emerging.htm

Chapter 3

1. GlobalSecurity.org, "Common Biometric Exchange File Format (CBEFF)," http://www.globalsecurity.org/security/systems/biometrics-cbeff.htm
2. L-1 Identity Solutions, "Biometrics Products," http://www.l1id.com/pages/17-biometrics
3. The Biometrics Consortium, "Examples of Biometric Systems," http://www.biometrics.org/html/examples/examples.html
4. Ibid.
5. The Biometrics Consortium, "About the Biometrics Consortium," http://www.biometrics.org/
6. Ibid.
7. Hong, Jin-Hyuk, Yun, Eun-Kyung, and Cho, Sung-Bae, "A Review of Performance Evaluation for Biometrics Systems," *International Journal of Image and Graphics*, March 24, 2004, http://candy.yonsei.ac.kr/publications/Papers/IJIG2004.pdf
8. Ashbourn, Julian, "About the Author," http://www.jsoft.freeuk.com/author.htm
9. Diodati, Mark, "Network Security Tactics, Preparing for Integrated Physical and Logical Access Control: The Common Authenticator," SearchSecurity.com, September 18, 2007, http://searchsecurity.techtarget.com/tip/0,289483,sid14_gci1268220,00.html
10. Spence, Bill, "Biometrics Role in Physical Access Control," FindBiometrics.com, http://www.findbiometrics.com/Pages/feature%20articles/physac.html
11. The Biometrics Consortium, "Examples of Biometric Systems," http://www.biometrics.org/html/examples/examples.html
12. GlobalSecurity.com, "Homeland Security, Biometric Programs," http://www.globalsecurity.org/security/systems/biometrics-programs.htm
13. IBM, "Solutions, Customs, Ports and Border Management," http://ibm.com/industries/government/CustomsPortsBorders
14. Ibid.
15. Find Biometrics.com, "Biometric Solutions," http://www.findbiometrics.com

Chapter 4

1. Techweb.com, TechEncyclopedia, "DBMS," http://www.techweb.com/encyclopedia/defineterm.jhtml?term=DBMS
2. About.com, Databases, "Database Security Issues," http://databases.about.com/od/security/Database_Security_Issues.htm
3. Chandler, Adam, "DHS Expands Its Biometric Database," eGovernment Resource Centre, July 27, 2006, http://www.egov.vic.gov.au/index.php?env=-inlink/detail:m1832-1-1-8-s:l-3209-1-1-
4. Ibid.
5. Hruska, Joel, "FBI Planning World's Largest Biometric Database," ARS Technica, December 24, 2007, http://arstechnica.com/tech-policy/news/2007/12/fbi-planning-worlds-largest-biometric-database.ars
6. IEEE Xplore, "MCYT Baseline Corpus: A Bimodal Biometric Database," http://ieeexplore.ieee.org/Xplore/login.jsp?url=/iel5/2200/28252/01263277.pdf?arnumber=1263277
7. Cryptome.org, "DHS Biometric Storage System," http://cryptome.org/dhs040607.htm
8. NIST Scientific and Technical Databases, "Biometrics," http://www.nist.gov/srd/biomet.htm
9. Techweb.com, TechEncyclopedia, "DBMS," http://www.techweb.com/encyclopedia/defineterm.jhtml?term=DBMS

Chapter 5

1. Proctor, Paul E., *The Practical Intrusion Detection Handbook* (Upper Saddle River, NJ: Prentice Hall, 2000).
2. Globalsecurity.com, "Homeland Security, Biometric Programs," http://www.globalsecurity.org/security/systems/biometrics-programs.htm
3. Wayman, James, "Biometrics and How They Work," San Jose State University, http://www.cfp2002.org/proceedings/proceedings/jlwcfp2002.pdf
4. Globalsecurity.com, "Hand Geometry and Handwriting," http://www.globalsecurity.org/security/systems/biometrics-hand.htm
5. Ibid.
6. Ibid.
7. Globalsecurity.com, "Voice Verification," http://www.globalsecurity.org/security/systems/biometrics-voice.htm
8. Technology.com, http://www.Technology.com
9. Shah, Rahul, and Masad, Bashar, "Biometric Access Control Gets Networked," RTCMagazine.com, http://www.rtcmagazine.com/articles/view/100725
10. The Biometrics Consortium, "About the Biometrics Consortium," http://www.biometrics.org/
11. Governmentforum365.co.uk, "IBM Selected for Project Semaphore," http://www.governmentforum365.co.uk/government-news/ibm-selected-for-project-sempahore.aspx
12. U.S. Department of Homeland Security, "US-VISIT Traveler Information," http://www.dhs.gov/xtrvlsec/programs/content_multi_image_0006.shtm
13. Advanced Biometric Technology, "Biometrics Is the Key," http://www.biometrics.co.za/
14. Proctor, Paul E., *The Practical Intrusion Detection Handbook*.
15. Globalsecurity.com, "Voice Verification," http://www.globalsecurity.org/security/systems/biometrics-voice.htm
16. Ibid.
17. The Biometric Consortium, "Examples of Biometric Systems," http://www.biometrics.org/html/examples/examples.html
18. Globalsecurity.com, "Transportation Worker Identification Credential (TWIC)," http://www.globalsecurity.org/security/systems/twic.htm
19. Wayman, James, http://www.cfp2002.org/proceedings/proceedings/jlwcfp2002.pdf
20. Globalsecurity.com, "Homeland Security, Biometric Programs," http://www.globalsecurity.org/security/systems/biometrics-programs.htm
21. Wayman, James, "Biometrics Personal Identification in Networked Society – Technical Testing and Evaluation of Biometric Identification Devices," Springerlink, http://www.springerlink.com/content/p72u557k5437g1x5/
22. Namboodiri, Anoop, Uludag, Umut, and Nandakumar, Karthik, "Handbook of Multibiometrics (International Series on Biometrics)," ACM Portal, http://portal.acm.org/citation.cfm?id=1137836

Chapter 6

1. University of Cambridge, "Computer Laboratory," http://www.cl.cam.ac.uk/
2. Sphericalcube.com, "Biometrics," http://sphericalcube.com/biometrics/biometrics.php
3. "MegaMatcher 2.0 SDK Is Now Available for the Development of Large-Scale AFIS and Multi-Biometric Face-Fingerprint Identification Systems," Neurotechnology, http://www.neurotechnologija.com/press_release_megamatcher_2_0.html
4. U.S. Army, www.army.mil; "Army Public Affairs," http://www.army.mil/info/institution/publicaffairs/
5. Reachin, "Reachin Technologies," http://www.reachin.se/
6. Neurotechnology, http://www.neurotechno logija.com/press_release_megamatcher_2_0.html

Chapter 7

1. SANS, "The SANS Security Policy Projects," http://www.sans.org/resources/policies/#name
2. West Virginia University, "Research," http://www.citer.wvu.edu/research/
3. India Society for Medical Statistics (ISMS), http://www.jipmer.edu/isms/html/ISMS.htm
4. Dimitriadis, Christos, K., and Polemi, Despina, "Biometrics—Risks and Controls," ISACA, http://www.isaca.org/Template.cfm?Section=Governance&CONTENTID=21329&TEMPLATE=/ContentManagement/ContentDisplay.cfm
5. Ibid.
6. Simonds, Lauren, eSecurityplanet, "Five Tips for National Cyber Security Month," October 15, 2007, http://www.esecuritypla net.com/trends/article.php/3705106/Five-Tips-for-National-Cyber-Security-Month.htm
7. SC Magazine, http://www.scmagazine.com/
8. NIST National Institute of Standards and Technology, Information Technology Laboratory, "Federal Information Security Management Act (FISMA) Implementation Project," http://csrc.nist.gov/groups/SMA/fisma/index.html.

Chapter 8

1. L-1 Identity Solutions, "Recent News," http://www.l1id.com/
2. Dublin, Joel, "Identity-Enabled Network Devices Promise Extra Layer of Authentication," Searchsecurity.com, September 25, 2007, http://www.searchsecurity.com.au/tips/tip.asp?DocID=1272834
3. Liu, Simon, and Silverman, Mark, "A Practical Guide to Biometric Security Technology," ACM Portal,

http://portal.acm.org/citation.cfm?coll=GUIDE&dl=GUIDE&id=613088

4. Shah, Rahul, and Masad, Bahar, "Biometric Access Control Gets Networked," RTCMagazine.com, http://www.rtcmaga zine.com/home/article.php?id=100725&pg=1
5. Spence, Bill, "Biometrics Role in Physical Access Control," FindBiometrics.com, http://www.findbiometrics.com/Pages/feature%20articles/physac.html
6. Frost & Sullivan, "World Hand Geometry Biometrics Markets," Marketresearch.com, http://www.marketresearch.com/product/display.asp?productid=967248&g=1
7. Find Biometrics.com, http://www.findBIOMETRICS.com

Chapter 9

1. Burtongroup.com, "Access Control Search," http://www.burtongroup.com/Client/Search/Search.aspx?searchTerm=access+control
2. Electronic Frontier Foundation, "Biometrics: Who's Watching You?" www.eff.org/Privacy/Surveillance/biometrics/
3. Engineering and Physical Sciences Research Council, "Details of Grant, Template-Free Biometric Encryption for Data Integrity Assurance," http://gow.epsrc.ac.uk/ViewGrant.aspx?GrantRef=EP/C00793X/1; West Virginia University, "Research," http://www.citer.wvu.edu/research/
4. West Virginia University, "Center for Identification Technology Research (CITeR)," http://www.citer.wvu.edu/

Chapter 10

1. Manley, M.E., McEntee, C. A., Molet, C. A., Park, J.S., "Wireless Security Policy Development for Sensitive Organizations," Information Assurance Workshop, IEEE Xplore, http://ieeexplore.ieee.org/Xplore/login.jsp?url=http%3A%2F%2Fieeexplore.ieee.org%2Fiel5%2F10007%2F32124%2F01495946.pdf&authDecision=-203
2. SANS Institute InfoSec Reading Room, "Strengthening Authentication with Biometric Technology," http://www.sans.org/reading_room/whitepapers/authentication/strengthening_authentication_with_biometric_technology_1226
3. Electronic Frontier Foundation, "What Is EFF?" http://www.eff.org
4. Gollmann, Dieter, *Computer Security,* 2nd ed. (Boston. MA, Wiley, 2006).
5. University of Cambridge, "The Computer Laboratory," http://www.cl.cam.ac.uk/
6. Faber, Paul, "RFID Strategy — RFID Privacy and Security Issues," *Industry Week*, http://www.industryweek.com/ReadArticle.aspx?ArticleID=13371
7. RFID Consortium for Security and Privacy, http://www.rfid-cusp.org/
8. Greene, Sari Stern, *Security Policies and Procedures* (Upper Saddle River, NJ: Prentice Hall, 2006).

Chapter 11

1. Epic.org, Electronic Privacy Information Center, http://www.epic.org/
2. Privacy International (PI), http://www.privacyinternational.org/
3. Find Biometrics.com, "IBG's In-Depth Look at Biometric Implementation Issues to Secure Borders," http://www.findbiometrics.com/Pages/feature%20articles/ibg.htm
4. DeJesus, Edmund X., "The Basics of Biometrics," Government Computer News (GCN), http://www.gcn.com/print/26_22/44932-1.html
5. USDA, United States Department of Agriculture, Food, and Nutrition Source, "Food Security," http://www.fns.usda.gov/fsec/
6. Phillips, P.J., Martin, A., Wilson, C.L., and Przybocki, M., "An Introduction to Evaluating Biometric Systems," IEEE Xplore, http://ieeexplore.ieee.org/xpl/freeabs_all.jsp?arnumber=820040
7. Phillips, P.J., Martin, A., Wilson, C.L., and Przybocki, M., "An Introduction to Evaluating Biometric Systems," *Computer,* http://www.frvt.org/DLs/FERET7.pdf

Chapter 12

1. NSI, American National Standards Institute, http://www.ansi.org/
2. Globalsecurity.com, "Common Biometric Exchange File Format (CBEFF)," http://www.globalsecurity.org/security/systems/biometrics-cbeff.htm
3. International Biometrics Industry Association (IBIA), http://www.ibia.org
4. The Biometric Consortium, http://www.biometrics.org
5. National Biometric Security Project (NBSP), http://www.nationalbiometric.org/

Index

A

C

D

E

G

H

N

S